Sample Preparation-Quo Vadis: Current Status of Sample Preparation Approaches-2nd Edition

Sample Preparation-Quo Vadis: Current Status of Sample Preparation Approaches-2nd Edition

Editors

Victoria Samanidou
Irene Panderi

MDPI • Basel • Beijing • Wuhan • Barcelona • Belgrade • Manchester • Tokyo • Cluj • Tianjin

Editors
Victoria Samanidou
Laboratory of
Analytical Chemistry
University of Thessaloniki
Thessaloniki
Greece

Irene Panderi
Department of Pharmacy
National and Kapodistrian
University of Athens
Athens
Greece

Editorial Office
MDPI
St. Alban-Anlage 66
4052 Basel, Switzerland

This is a reprint of articles from the Special Issue published online in the open access journal *Molecules* (ISSN 1420-3049) (available at: www.mdpi.com/journal/molecules/special_issues/sample_preparation_2021).

For citation purposes, cite each article independently as indicated on the article page online and as indicated below:

LastName, A.A.; LastName, B.B.; LastName, C.C. Article Title. *Journal Name* **Year**, *Volume Number*, Page Range.

ISBN 978-3-0365-5722-9 (Hbk)
ISBN 978-3-0365-5721-2 (PDF)

Cover image courtesy of Victoria Samanidou

Contents

About the Editors . . . . **vii**

Preface to "Sample Preparation-Quo Vadis: Current Status of Sample Preparation Approaches-2nd Edition" . . . . **ix**

Victoria Samanidou and Irene Panderi
Editorial for Special Issue "Sample Preparation-Quo Vadis: Current Status of Sample Preparation Approaches-2nd Edition"
Reprinted from: *Molecules* **2022**, *27*, 6142, doi:10.3390/molecules27196142 . . . . **1**

José S. Câmara, Rosa Perestrelo, Cristina V. Berenguer, Carolina F. P. Andrade, Telma M. Gomes and Basit Olayanju et al.
Green Extraction Techniques as Advanced Sample Preparation Approaches in Biological, Food, and Environmental Matrices: A Review
Reprinted from: *Molecules* **2022**, *27*, 2953, doi:10.3390/molecules27092953 . . . . **3**

Aliya Nur Hasanah, Nisa Safitri, Aulia Zulfa, Neli Neli and Driyanti Rahayu
Factors Affecting Preparation of Molecularly Imprinted Polymer and Methods on Finding Template-Monomer Interaction as the Key of Selective Properties of the Materials
Reprinted from: *Molecules* **2021**, *26*, 5612, doi:10.3390/molecules26185612 . . . . **33**

Natalia Manousi, Abuzar Kabir, Kenneth G. Furton, George A. Zachariadis and Aristidis Anthemidis
Multi-Element Analysis Based on an Automated On-Line Microcolumn Separation/Preconcentration System Using a Novel Sol-Gel Thiocyanatopropyl-Functionalized Silica Sorbent Prior to ICP-AES for Environmental Water Samples
Reprinted from: *Molecules* **2021**, *26*, 4461, doi:10.3390/molecules26154461 . . . . **55**

Natasa P. Kalogiouri, Petros D. Mitsikaris, Athanasios N. Papadopoulos and Victoria F. Samanidou
Microwave-Assisted Extraction Coupled to HPLC-UV Combined with Chemometrics for the Determination of Bioactive Compounds in Pistachio Nuts and the Guarantee of Quality and Authenticity
Reprinted from: *Molecules* **2022**, *27*, 1435, doi:10.3390/molecules27041435 . . . . **69**

Mariana N. Oliveira, Oriana C. Gonçalves, Samir M. Ahmad, Jaderson K. Schneider, Laiza C. Krause and Nuno R. Neng et al.
Application of Bar Adsorptive Microextraction for the Determination of Levels of Tricyclic Antidepressants in Urine Samples
Reprinted from: *Molecules* **2021**, *26*, 3101, doi:10.3390/molecules26113101 . . . . **81**

Jianing Zhang, Fengjie Yu, Yunmin Tao, Chunping Du, Wenchao Yang and Wenbin Chen et al.
Micro Salting-Out Assisted Matrix Solid-Phase Dispersion: A Simple and Fast Sample Preparation Method for the Analysis of Bisphenol Contaminants in Bee Pollen
Reprinted from: *Molecules* **2021**, *26*, 2350, doi:10.3390/molecules26082350 . . . . **99**

Sam-ang Supharoek, Watsaka Siriangkhawut, Kate Grudpan and Kraingkrai Ponhong
A Simple and Reliable Dispersive Liquid-Liquid Microextraction with Smartphone-Based Digital Images for Determination of Carbaryl Residues in *Andrographis paniculata* Herbal Medicines Using Simple Peroxidase Extract from *Senna siamea* Lam. Bark
Reprinted from: *Molecules* **2022**, *27*, 3261, doi:10.3390/molecules27103261 . . . . **109**

Maria S. Synaridou, Vasilis Tsamis, Georgia Sidiropoulou, Constantinos K. Zacharis, Irene Panderi and Catherine K. Markopoulou
Fluorimetric Analysis of Five Amino Acids in Chocolate: Development and Validation
Reprinted from: *Molecules* **2021**, *26*, 4325, doi:10.3390/molecules26144325 . . . . **125**

Stefani Fertaki, Georgios Bagourakis, Malvina Orkoula and Christos Kontoyannis
Measuring Bismuth Oxide Particle Size and Morphology in Film-Coated Tablets
Reprinted from: *Molecules* **2022**, *27*, 2602, doi:10.3390/molecules27082602 . . . . **141**

Panagiota Papaspyridakou, Michail Lykouras, Christos Kontoyannis and Malvina Orkoula
Comparative Study of Sample Carriers for the Identification of Volatile Compounds in Biological Fluids Using Raman Spectroscopy
Reprinted from: *Molecules* **2022**, *27*, 3279, doi:10.3390/molecules27103279 . . . . **153**

About the Editors

Victoria Samanidou

Dr Victoria Samanidou is Full Professor and Director of the Laboratory of Analytical Chemistry in the Department of Chemistry of Aristotle University of Thessaloniki, Greece.

Her research interests focus on the development of sample-preparation methods using sorptive extraction prior to chromatographic analysis. She has co-authored 203 original research articles in peer reviewed journals and 60 reviews, 67 editorials/in view, and 54 chapters in scientific books (h-index 42, Scopus Author ID 7003896015, 5940 citations, Web of Science Researcher ID AAE-4121-2020). She is an editorial board member of more than 30 scientific journals and guest editor in more than 30 Special Issues. She has peer reviewed more than 701 manuscripts for 149 scientific journals.

In 2016, she was included in the top 50 power list of women in Analytical Science, as proposed by Texere Publishers.

In 2020, 2021, and 2022, she was included in the 2% top world scientists in the field of Analytical Chemistry (career -, as well as single year) published in PLOS Biology based on citations from SCOPUS.

In 2021, she was included in the "The Analytical Scientist" 2021 Power List of top 100 influential people in analytical science.

In 2016, she was elected as President of the Steering Committee of the Association of Greek Chemists—Regional Division of Central & Western Macedonia.

Irene Panderi

Irene Panderi (Eirini Panderi) is a pharmacist with a PhD in Pharmaceutical Analysis. She is a Professor and Director of the Laboratory of Pharmaceutical Analysis at the Faculty of Pharmacy, National and Kapodistrian University of Athens in Greece. She has over 20 years' experience in leading research in the fields of Pharmaceutical Analysis and Analytical Chemistry developing novel analytical methods for the analysis of drugs, metabolites and bioactive compounds in pharmaceuticals, cosmeceuticals, biological fluids and human tissues. These methods are applied to clinical testing, toxicology studies, doping control, and quality control of drugs. She is a member of the American Society of Mass Spectrometry, the Panhellenic Society of Pharmacists, the Panhellenic Society of Medicinal Chemistry, and the Hellenic Society of Mass Spectrometry and the Scientific Committee on Consumers Safety, EU (SCCS, DG-Sante). She is author and co-author of more than 95 original research articles in peer-reviewed journals, more than 30 opinions in international journals, and 8 reviews and chapters in scientific books with H-index 23 (Scopus). She has supervised 5 completed PhD thesis, 55 completed MSc theses, and several dissertation theses, and she is supervisor of 9 ongoing MSc theses and 2 ongoing PhD theses. She is a member of the editorial board several scientific journals, and she has served as a reviewer of more than 300 scientific articles in 30 international scientific journals.

Preface to "Sample Preparation-Quo Vadis: Current Status of Sample Preparation Approaches-2nd Edition"

Sample preparation is and will always be the most important step in chemical analysis. Numerous techniques, methods, methodologies, and approaches are published in the literature offering a wide range of analytical tools to the lab practitioner. Analytical scientists all over the world are trying to develop protocols for a plethora of analytes in various sample matrices. In the last decade, sample pre-treatment advances have followed green chemistry and green analytical chemistry demands, focusing on miniaturization and automation, using the least possible amount of organic solvents. The question is how far we have been till now, and what the future perspectives are. To answer this question, analytical chemists were invited to share their experience in the field and report on the recent advances in sample-preparation approaches. The outcome of our invitation was eleven excellent manuscripts, including four review articles and seven original research articles in the first edition of the Special Issue "Sample Preparation-Quo Vadis: Current Status of Sample Preparation Approaches".

The second edition is a collection of ten significant contributions to the field of sample preparation. It includes two highly interesting and comprehensive review articles and eight innovative research articles.

The Guest Editors wish to thank all authors for their fine contribution.

Victoria Samanidou and Irene Panderi

Editors

Editorial

Editorial for Special Issue "Sample Preparation-Quo Vadis: Current Status of Sample Preparation Approaches-2nd Edition"

Victoria Samanidou [1],* and Irene Panderi [2],*

1. Laboratory of Analytical Chemistry, Department of Chemistry, Aristotle University of Thessaloniki, GR 54124 Thessaloniki, Greece
2. Laboratory of Pharmaceutical Analysis, Division of Pharmaceutical Chemistry, Faculty of Pharmacy, National and Kapodistrtian University of Athens Panepistimiopolis-Zografou, GR 15771 Athens, Greece

* Correspondence: samanidu@chem.auth.gr (V.S.); irenepanderi@gmail.com (I.P.)

Citation: Samanidou, V.; Panderi, I. Editorial for Special Issue "Sample Preparation-Quo Vadis: Current Status of Sample Preparation Approaches-2nd Edition". *Molecules* **2022**, *27*, 6142. https://doi.org/10.3390/molecules27196142

Received: 1 August 2022
Accepted: 18 August 2022
Published: 20 September 2022

Publisher's Note: MDPI stays neutral with regard to jurisdictional claims in published maps and institutional affiliations.

Sample preparation is and will always be the most important step in chemical analysis. Numerous techniques, methods, methodologies and approaches are published in the literature offering a wide range of analytical tools to the lab practitioner.

Analytical scientists all over the world try to develop protocols for a plethora of analytes in various sample matrices.

In the last decade, sample pre-treatment advances followed green chemistry and green analytical chemistry demands, focusing on miniaturization and automation, using the lowest possible amounts of organic solvents.

The question is how far we have gone until now as well as what the future perspectives are.

To answer this question, analytical chemists were invited to share their experience in the field and report on the recent advances in sample preparation approaches.

Eleven excellent manuscripts were the outcome of our invitation including four review articles and seven original research articles in the first edition of the Special Issue "Sample Preparation-Quo Vadis: Current Status of Sample Preparation Approaches".

The second edition is a collection of ten significant contributions to the field of sample preparation. It includes two highly interesting and comprehensive review articles and eight innovative research articles.

José S. Câmara et al. present a comprehensive overview on the Green Extraction Techniques as Advanced Sample Preparation Approaches in Biological, Food, and Environmental Matrices (https://www.mdpi.com/1420-3049/27/9/2953 (accessed on 31 July 2022)).

Aliya Nur Hasanah et al. report on the Factors Affecting Preparation of Molecularly Imprinted Polymer and Methods on Finding Template-Monomer Interaction as the Key of Selective Properties of the Materials (https://www.mdpi.com/1420-3049/26/18/5612 (accessed on 31 July 2022)).

A feature paper on the Multi-Element Analysis Based on an Automated On-Line Microcolumn Separation/Preconcentration System Using a Novel Sol–Gel Thiocyanatopropyl-Functionalized Silica Sorbent Prior to ICP-AES for Environmental Water Samples is presented by Natalia Manousi et al. (https://www.mdpi.com/1420-3049/26/15/4461 (accessed on 31 July 2022)).

Microwave-Assisted Extraction Coupled to HPLC-UV Combined with Chemometrics for the Determination of Bioactive Compounds in Pistachio Nuts and the Guarantee of Quality and Authenticity is proposed by Natasa P. Kalogiouri et al. (https://www.mdpi.com/1420-3049/27/4/1435 (accessed on 31 July 2022)).

Mariana N. Oliveira et al. propose the Application of Bar Adsorptive Microextraction for the Determination of Levels of Tricyclic Antidepressants in Urine Samples (https://www.mdpi.com/1420-3049/26/11/3101 (accessed on 31 July 2022)).

Micro Salting-Out Assisted Matrix Solid-Phase Dispersion: A Simple and Fast Sample Preparation Method for the Analysis of Bisphenol Contaminants in Bee Pollen is presented by Jianing Zhang et al. (https://www.mdpi.com/1420-3049/26/8/2350 (accessed on 31 July 2022)).

A Simple and Reliable Dispersive Liquid-Liquid Microextraction with Smartphone-Based Digital Images for Determination of Carbaryl Residues in *Andrographis paniculata* Herbal Medicines Using Simple Peroxidase Extract from *Senna siamea* Lam. Bark is proposed by Sam-ang Supharoek et al. (https://www.mdpi.com/1420-3049/27/10/3261 (accessed on 31 July 2022)).

Maria S. Synaridou et al. present a Fluorimetric Analytical method for the determination of Five Amino Acids in Chocolate (https://www.mdpi.com/1420-3049/26/14/4325 (accessed on 31 July 2022)).

Stefani Fertaki et al. report on the Measuring Bismuth Oxide Particle Size and Morphology in Film-Coated Tablets (https://www.mdpi.com/1420-3049/27/8/2602 (accessed on 31 July 2022)).

Panagiota Papaspyridakou et al. report a Comparative Study of Sample Carriers for the Identification of Volatile Compounds in Biological Fluids Using Raman Spectroscopy (https://www.mdpi.com/1420-3049/27/10/3279 (accessed on 31 July 2022)).

The Guest Editors wish to thank all authors for their fine contribution and invite analytical chemists to submit their relative work to the third edition until the 31st of January 2023. For more information, the readers are advised to visit the site: https://susy.mdpi.com/academic-editor/special_issues/process/1103908.

Author Contributions: Contribution was equal to both authors. All authors have read and agreed to the published version of the manuscript.

Funding: This research received no external funding.

Conflicts of Interest: The authors declare no conflict of interest.

Review

Green Extraction Techniques as Advanced Sample Preparation Approaches in Biological, Food, and Environmental Matrices: A Review

José S. Câmara [1,2,*], Rosa Perestrelo [1], Cristina V. Berenguer [1], Carolina F. P. Andrade [1], Telma M. Gomes [1], Basit Olayanju [3], Abuzar Kabir [3,4], Cristina M. R. Rocha [5,6], José António Teixeira [5,6] and Jorge A. M. Pereira [1,*]

[1] CQM—Centro de Química da Madeira, Natural Products Research Group, Universidade da Madeira, Campus Universitário da Penteada, 9020-105 Funchal, Portugal; rmp@staff.uma.pt (R.P.); cristina.berenguer@staff.uma.pt (C.V.B.); carolinafatimaandrade@hotmail.com (C.F.P.A.); telma_gomes_20@hotmail.com (T.M.G.)

[2] Departamento de Química, Faculdade de Ciências Exatas e Engenharia, Universidade da Madeira, Campus da Penteada, 9020-105 Funchal, Portugal

[3] Department of Chemistry and Biochemistry, Florida International University, Miami, FL 33199, USA; olayanju.b@fiu.edu (B.O.); akabir@fiu.edu (A.K.)

[4] Department of Pharmacy, Faculty of Allied Health Science, Daffodil International University, Dhaka 1207, Bangladesh

[5] CEB—Centre of Biological Engineering, Universidade do Minho, Campus de Gualtar, 4710-057 Braga, Portugal; cmrocha@ceb.uminho.pt (C.M.R.R.); jateixeira@deb.uminho.pt (J.A.T.)

[6] LABBELS–Associate Laboratory, Universidade do Minho, Campus de Gualtar, 4710-057 Braga, Portugal

[*] Correspondence: jsc@staff.uma.pt (J.S.C.); jorge.pereira@staff.uma.pt (J.A.M.P.); Tel.: +351-291705112 (J.S.C.); +351-291705119 (J.A.M.P.)

Citation: Câmara, J.S.; Perestrelo, R.; Berenguer, C.V.; Andrade, C.F.P.; Gomes, T.M.; Olayanju, B.; Kabir, A.; M. R. Rocha, C.; Teixeira, J.A.; Pereira, J.A.M. Green Extraction Techniques as Advanced Sample Preparation Approaches in Biological, Food, and Environmental Matrices: A Review. *Molecules* **2022**, *27*, 2953. https://doi.org/10.3390/molecules27092953

Academic Editor: Constantinos K. Zacharis

Received: 14 April 2022
Accepted: 2 May 2022
Published: 6 May 2022

Publisher's Note: MDPI stays neutral with regard to jurisdictional claims in published maps and institutional affiliations.

Abstract: Green extraction techniques (GreETs) emerged in the last decade as greener and sustainable alternatives to classical sample preparation procedures aiming to improve the selectivity and sensitivity of analytical methods, simultaneously reducing the deleterious side effects of classical extraction techniques (CETs) for both the operator and the environment. The implementation of improved processes that overcome the main constraints of classical methods in terms of efficiency and ability to minimize or eliminate the use and generation of harmful substances will promote more efficient use of energy and resources in close association with the principles supporting the concept of green chemistry. The current review aims to update the state of the art of some cutting-edge GreETs developed and implemented in recent years focusing on the improvement of the main analytical features, practical aspects, and relevant applications in the biological, food, and environmental fields. Approaches to improve and accelerate the extraction efficiency and to lower solvent consumption, including sorbent-based techniques, such as solid-phase microextraction (SPME) and fabric-phase sorbent extraction (FPSE), and solvent-based techniques (μQuEChERS; micro quick, easy, cheap, effective, rugged, and safe), ultrasound-assisted extraction (UAE), and microwave-assisted extraction (MAE), in addition to supercritical fluid extraction (SFE) and pressurized solvent extraction (PSE), are highlighted.

Keywords: green extraction techniques; microextraction techniques; sample preparation; biological samples; food samples; environmental samples

---

1. Introduction

Over the last decades of the last century, technological improvements in chromatographic instruments boosted a remarkable evolution in the analytical chemistry field. Sophisticated configurations hyphenating fast and efficient chromatographic separations with powerful detection systems able to deliver unprecedented time of analysis and analytical performance become the forefront of this revolution where sample preparation was forgotten. For another decade, conventional processes, often involving large volumes of sample and organic solvents and laborious and many times cumbersome protocols prone to originate many experimental errors, continued to be used as standard procedures. Meanwhile, growing concerns with the environmental footprint and planet sustainability are promoting a green agenda affecting the most diverse human activities. The application of the green chemistry principles to analytical chemistry has been elegantly defined under the SIGNIFICANCE acronym [1]. Accordingly, the green analytical chemistry (GAC) envisages the simplest experimental layout involving minimal or no sample preparation and maximum integration of the analytical instruments used, preferentially in an automated way to limit operator intervention, energy consumption, and waste production. In this context, miniaturization of the sample extraction procedure, therefore decreasing sample and solvent requirements, as well as, wastes produced, was an obvious consequence of the GAC principles. This trend has fostered the development of a myriad of microextraction approaches, hereby considered green extraction (GreETs) approaches. These GreETs span almost all, if not all, fields of application, covering the microextraction of selected analytes from biological samples to food matrices or environmental matrices.

On this basis, this review will provide an updated overview of the most important and used green extraction approaches reported in the literature since 2016, their principles, advantages, limitations, and examples of application. Sorbent-based techniques, such as solid-phase microextraction (SPME), stir-bar sorptive extraction (SBSE), fabric-phase sorbent extraction (FPSE), and solvent-based techniques, including µQuEChERS (micro quick, easy, cheap, effective, rugged, and safe), single-drop microextraction (SDME), hollow-fiber liquid-phase microextraction (HF-LPME), and dispersive liquid–liquid microextraction (DLLME), have been considered. Additionally, the use of emerging green solvents such as ionic liquids (ILs) and deep eutectic solvents (DES) as an alternative to conventional solvents will be discussed. Finally, a brief overview of other promising green and sustainable approaches, such as pulsed electric-field-assisted extraction (PEFAE), supercritical fluid extraction (SFE), and subcritical water extraction (SWE), will also be provided.

2. Sample Preparation: A Key Step to Getting the Correct Data

There has been an unprecedented growth in measurement techniques over the last few decades. Instrumentation, such as spectroscopy, microscopy, and chromatography, as well as microdevices and sensors, have undergone phenomenal developments. In contrast, the importance of sample treatment in the analytical layout seems to have been neglected. However, in the last decade, especially driven by the need of pharmaceutical and environmental industries, an exponential growth, and a rapid evolution in this industry, was observed. Some important steps in analytical chemistry to allow accurate, efficient, and fast determinations are commonly used in the sample preparation process, including, for example, extraction (recovering analytes from samples), clean-up (removal of compounds that can interfere with analysis), and solvent evaporation/concentration (concentration of analytes using an N_2 stream), are shown in Figure 1. The procedures depend on the sample, the matrix, and the concentration level at which the analysis needs to be carried out.

Sample preparation is the source of about 30% of the experimental errors and of about 60% of the time spent on tasks in the analytical lab. For these reasons, independently of the high performance of the analytical instrument, the sampling procedure and the sample handling and pretreatment methodologies, following a carefully outlined process, are of utmost importance to acquire high-quality analytical results with high selectivity and low sensitivity limits and to ensure high accuracy and reproducibility. In addition, the selective isolation of the analytes of interest and the removal of interfering sample components are vital for eliminating the interferences and matrix effect and protecting the instrumental equipment from possible damages. However, as referred, these procedures were not always seen as key steps in the analytical process, and for that fact, the methodology followed in sample preparation did not receive the same attention as the analytical instrumentation, considered, until the last years, being the bottleneck of the whole analytical procedure. Indeed, the most widely used and commonly accepted classical extraction techniques (CETs) were liquid–liquid extraction (LLE), Soxhlet extraction, and solid-phase extraction (SPE). CETs, however, tend to be slow and labor intensive and use high amounts of hazardous organic solvents causing serious environmental concerns and present low extraction efficiency. Despite this reality, sample preparation techniques did not receive much attention until quite recently. In the last decades, to overcome the drawbacks of CETs, several novel microextraction techniques (Figure 1), which offer faster, cheaper, and "greener" pretreatment of complex samples; utilization of hazardous reagents; and less solvents with generation of less waste, maximizing the safety for operators and the environment, have been reported as alternatives to CETs.

These techniques, hereby designated as green extraction techniques (GreETs), exhibit attractive characteristics, such as simplicity, versatility, high extraction efficiency, and environmentally friendly profile, and have experienced increased development and implementation and stimulated significant progress in laboratory sample treatment. Some of them, due to their importance and growing application in the biological, food, and environmental fields, are highlighted below.

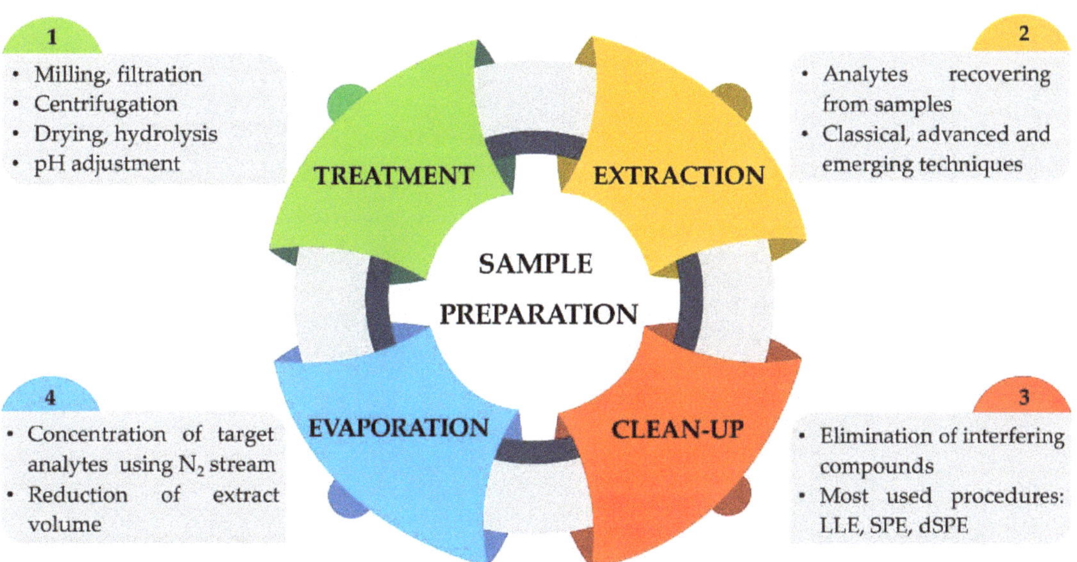

Figure 1. Different steps involved in sample preparation.

3. Green Extraction (GreETs) Techniques

The most important and used GreET meeting all green analytical chemistry (GAC) requirements, based on miniaturized SPE techniques, such as microextraction in packed syringe (MEPS), solid-phase microextraction (SPME) in direct (DI) and headspace modes (HS), stir-bar sorptive extraction (SBSE), and matrix solid-phase dispersion, in addition to liquid-phase extraction techniques, including single-drop microextraction (SDME), hollow-fiber liquid-phase microextraction (HF-LPME), dispersive liquid–liquid microextraction (DLLME), QuEChERS, solidification of floating organic drop microextraction (SFOME), and ultrasound-assisted back extraction (UABE), will be given more emphasis.

3.1. Miniaturized Sorbent-Based Techniques

SPE is one of the most used conventional extraction and preconcentration methods for the analysis of food, biological, and environmental samples [2,3]. However, this technique requires relatively large amounts of organic solvents and additional clean-up steps, which limits the automation, decreases sample throughput, and potentiates the contamination of the extracts [3]. Moreover, SPE uses large sample amounts, involving long extraction times [2,4]. Recently, new extraction methods were developed using modern techniques with less or no organic solvents to minimize environmental pollution and overcome the limitations of the conventional methods (Figure 2).

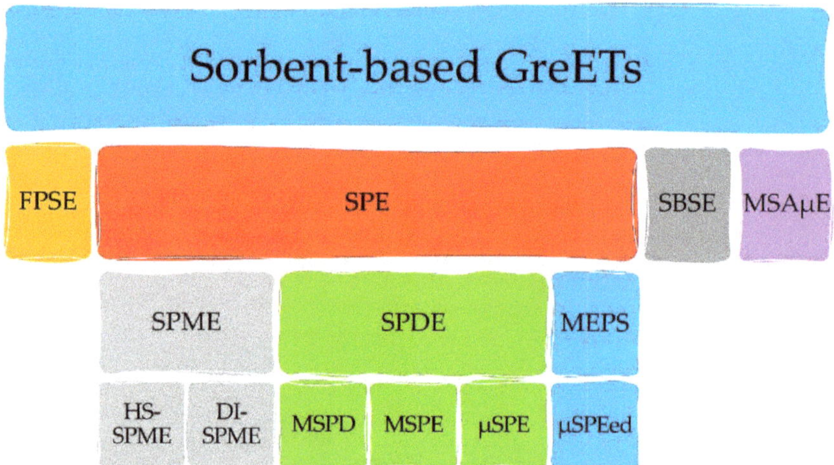

Figure 2. Different sorbent-based GreETs used in several fields of analysis. Legend: DI-SPME: solid-phase microextraction in direct immersion mode; FPSE: fabric-phase solvent extraction; GreETs: green extraction techniques; HS-SPME: solid-phase microextraction in headspace mode; MEPS: microextraction in packed sorbent; MSAμE: multisphere adsorptive microextraction; MSPD: matrix solid-phase dispersion; MSPE: magnetic solid-phase extraction; SBSE: stir-bar sorbent extraction; SPDE: solid-phase dynamic extraction; SPE: solid-phase extraction; SPME: solid-phase microextraction; μSPE: micro-solid-phase extraction.

3.1.1. Fabric-Phase Sorbent Extraction

Introduced in 2014 by Kabir and Furton [5], the solid-phase dynamic extraction (SPDE) format fabric-phase sorptive extraction (FPSE) is a fast, efficient, and versatile sample preparation approach by implementing a natural or synthetic permeable and flexible fabric (e.g., polyester, fiberglass, or cellulose) substrate to host a chemically coated sol–gel organic–inorganic hybrid sorbent in the form of an ultrathin coating. FPSE allows direct extraction of analytes without sample modification, thus minimizing/eliminating the sample pretreatment steps, which are considered the primary source of major analyte loss [6]. A strong covalent interaction between the fabric substrate and sol–gel contributes to improving the extraction efficiency medium, helping expose the FPSE to extreme chemical conditions without compromising the chemical/structural integrity of the microextraction device. The main disadvantages of FPSE are low sample capacity and extensive longer sample preparation time [7].

3.1.2. Solid-Phase Extraction-Based Approaches
Solid-Phase Microextraction (SPME)

A key milestone in the development of microextraction techniques was first achieved by the seminal invention of solvent-free solid-phase microextraction, popularly known as SPME by Arthur and Pawliszyn in the early 1990s [8]. SPME is an equilibrium-based microextraction technique that involves the partitioning of the analytes from the sample solution into the sorbent coating of the SPME fiber owing to the intermolecular interaction or affinity for the sorbent material. Several configurations of SPME integrally optimize the volume of the extraction phase to improve the high surface-area-to-volume ratio of the extraction-phase coating. There are several geometries for SPME, such as planar, spherical, rod, and in-tube or cylinder [9]. The selection of the SPME geometry depends on the target analyte and matrix that will be analyzed. Usually, in the reduced diameter or length of the extraction phase, its higher surface-area-to-volume ratio can result in a smaller extraction period and higher recoveries [9].

SPME is a simple, fast, universal, sensitive, solventless, and economical technique for the preconcentration and sampling of analytes derived from various types of samples [3,10]. This technique combines extraction, enrichment, and sample injection into a single step. Other advantages of SPME are due to the reliability, sensitivity, and selectivity of this technique [4]. SPME allows the detection of semivolatile and nonvolatile compounds [10] and benefits from the constant development of new sorption coatings [3]. This procedure can be performed in different modes: (i) headspace SPME (HS-SPME) mode in which the analytes are adsorbed/absorbed from the gas phase in equilibrium with the samples (as the temperature is a parameter with a significant effect on the kinetics of the process, this is the most adequate for volatiles extraction); (ii) direct mode (DI) in which the SPME fiber is immersed directly into the bulk sample. In this case, the agitation is an important experimental parameter to facilitate the transport of the analytes from the solution to the fiber. (iii) In the third mode, membrane extraction, the extraction of less volatile compounds is facilitated by the use of a protected membrane.

Microextraction in Packed Sorbent (MEPS)

MEPS emerged as a greener alternative to the conventional SPE, consisting of a sample pretreatment technique based on the miniaturization of SPE. This technique uses the same sorbents as SPE but is considered more advantageous since sorbent integration into a liquid-handling syringe results in low void volumes, making sample manipulations easy [4]. MEPS can be applied to smaller samples and requires shorter sample preparation times and lower solvent volumes [2]. Moreover, MEPS can be performed online in a fully automated way using the same syringe for sample extraction and extract injection into the analytical instrument [4]. A typical MEPS application comprises sorbent conditioning, sample loading, washing, and analyte elution. Contrary to SPE, the two-direction flow potential in MEPS provides the duplication of each step and satisfactory sorbent conditioning, enhanced sample–sorbent interaction, sample loading, and improved analyte elution. The elution and washing steps can be performed with 20–50 µL of organic solvent, and 1–4 mg of reused sorbent material is sufficient to extract a target analyte with high efficiency [11]. More recently, µSPEed has been proposed. It represents an advance with respect to MEPS because it has a unidirectional valve that corresponds to a flow in one direction, in addition to the high pressure conferred by the small diameter of the sorbent particles. In this case, the analytes retained in the adsorbent are not altered by the aspiration of solvents, as is the case with MEPS, which results in more efficient extractions. The most remarkable improvement over MEPS procedure is the direct flow through the sorbent bed; therefore, the analytes retained in the sorbent bed are not disturbed by the solvent aspiration as it occurs with MEPS. Moreover, the high pressure and the single direction contribute to obtain more efficient extractions of the target analytes. Figure 3 represents a schematic overview and the most important aspects of µSPEed.

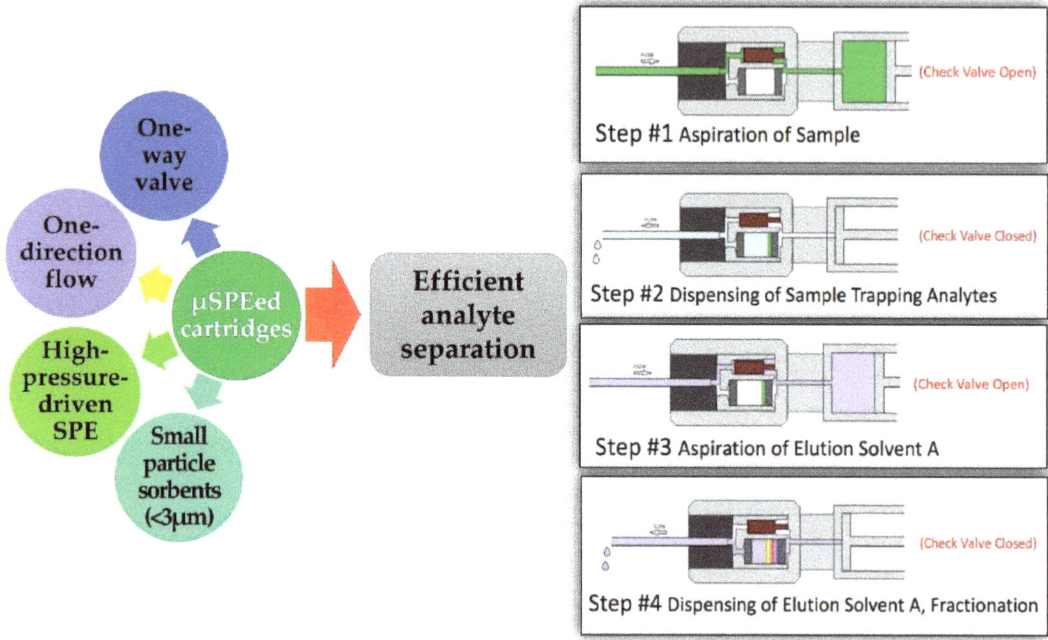

Figure 3. Advantages and schematic overview of μSPEed.

Solid-Phase Dispersion Extraction (SPDE)

In SPDE, the microparticles are dispersed in the solution (liquid sample) until the equilibrium between the two phases is reached. The most popular formats are the matrix solid-phase dispersion extraction (MSPD), magnetic solid-phase extraction (MSPE), and micro-solid-phase extraction (μSPE).

MSPD is an efficient and generic technique for the isolation of a wide range of drugs, pesticides, naturally occurring constituents, and other compounds from a wide variety of complex plant and animal samples. According to Barker [12], the sample is dispersed over the surface of the bonded-phase support material, producing, through hydrophobic and hydrophilic interactions of the various components, a unique mixed-character phase for conducting target analyte isolation [12].

MSPE is based on the use of sorbent materials, such as magnetic nanoparticles, carbon hemimicelles, and molecular imprinted polymers [13]. C18 functionalized magnetic nanoparticles (MNP) are used for preconcentration or cleanup of moderate and nonpolar polar pesticides due to the absence of internal diffusion resistance, the excellent absorption capacity of the target analytes, and the high surface-to-volume ratio [14]. The main advantage of MSPE is that the sorbent is composed of MNPs, often NPs of the most diverse chemistries and geometries, that can be easily recovered from a solution by a simple spin or centrifugation process.

The pipette-tip SPE is the simplest format of μSPE in which the sorbent is placed in a tip and extraction is handled by using a pipette, widely used in preclinical and clinical development programs in addition to the study of metabolomics, genomics, and proteomics. SPE tips, such as the MonoTip®, NuTip®, and ZipTip®, can be used for the purification of peptides or proteins that, using affinity and metal chelation, can be successfully selectively isolated [15].

3.1.3. Stir-Bar Sorbent Extraction (SBSE)

Stir-bar sorbent extraction (SBSE) was introduced by Baltussen, Sandra, David, and Cramers in 1999 as an alternative to SPME and became one of the most powerful microextraction and preconcentration techniques for the enrichment of volatile analytes from aqueous samples due to its simplicity, robustness, cost-effectiveness, and environmental friendliness. After that, its applications have been extended to the analysis of nonvolatile analytes and solid and liquid samples.

This extraction procedure is based on a magnetic stir (0.5–1 mm thickness) coated with polydimethylsiloxane, a nonpolymeric phase used as an extraction phase of target analytes through hydrophilic interaction. SBSE consists of two steps: extraction and desorption. Related to extraction, the coated stir bar can act in immersion mode (immersed in the sample solution) or in headspace mode (stir bar is exposed in the gas phase above the liquid or solid sample). After extraction, the target analytes adsorbed in the stir bar are desorbed by thermal desorption, followed by analysis in a chromatographic system (e.g., GC, HPLC, and CE) [11,16].

Despite that the SBSE principle is similar to SPME, SBSE exhibits higher sensitivity, recovery, and extraction efficiency. This is due to the larger amount of coated phase in SBSE, which is 50–250 times higher than the SPME fiber, making SBSE more suitable to analyze trace levels in complex matrices. On the other hand, a special interface is required for thermal desorption in gas chromatography (GC), and lower recoveries are obtained for target analytes with a logarithm of octanol–water partitioning coefficient (log Ko/w) lower than 3 [11,16].

3.1.4. Multisphere Adsorptive Microextraction (MSAµE)

A new adsorptive microextraction (AµE) technique was proposed by Nogueira et al. [17,18], which represents a great alternative for the enrichment of a wide range of polar analytes at trace levels in aqueous media, selecting appropriate sorbent phases. The new AµE approach can be used through different analytical devices presenting suitable geometry, where specific sorbents are simply sustained through sticking-based technologies. Usually, the sorbent is physically embedded on the substrate and put in the aqueous media. The solution is stirred using a stir bar or vortexed. Since most of the polar targets are nonvolatile and some of them have thermolabile properties, liquid desorption (LD), followed by HPLC is certainly the following combination of choice for analytical purposes. AµE can appear in two geometrical configurations, namely, bar adsorptive microextraction (BAµE) and multisphere adsorptive microextraction (MSAµE). Nevertheless, previous experimental data [17,18] showed that MSAµE devices present much better stability compared with BAµE, especially when they are exposed to an aggressive sample matrix because, in this case, thermal supporting promotes much higher robustness from the fixation point of view. The MSAµE showed several advantages, namely, high recovery for polar analytes, easiness to prepare, economicalness, and selectiveness, as sorbent can be selected based on the target analyte of interest. Nevertheless, the main drawback is the device's stability as should be evaluated on a case-by-case basis.

3.2. Miniaturized LPE-Based Techniques

In addition to sorbent- or solid-based GreETs, several microextraction approaches involving a sorbent phase in a liquid state have been developed in the last decades. Similar to the SPE-based techniques, the major shortcomings of the conventional liquid–liquid extraction technique (LLE), such as emulsion formation, long preparation time, noncompliance with GAC due to the usage of a high volume of toxic organic solvents, and inevitability of solvent evaporation and sample reconstitution, have triggered research into the miniaturized and greener version of LLE. In contrast to SPME, miniaturized LPE techniques include solvent-based extraction techniques that use microliters of organic solvent to accomplish the selective isolation, preconcentration of the analytes, and clean-up of the sample (Figure 4).

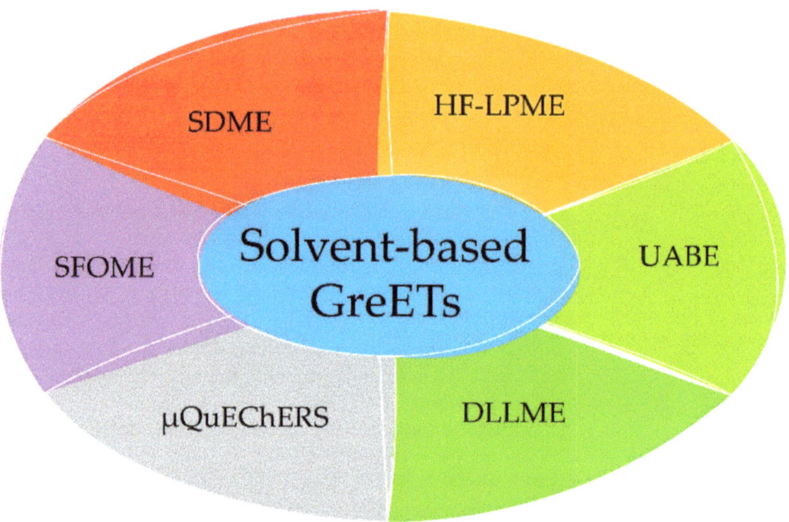

Figure 4. Different liquid-based GreETs used in several fields of analysis. Legend: DLLME: dispersive liquid–liquid microextraction; GreETs: green extraction techniques; HF-LPME: hollow fiber liquid-phase microextraction, SDME: single-drop microextraction; SFOME: solidification of floating organic drop microextraction; UABE: ultrasound-assisted back extraction; μQuEChERS: micro-QuEChERS.

3.2.1. Single-Drop Microextraction (SDME)

The first to be invented for the series of solvent-based microextraction techniques was SDME. SDME is a nearly solvent-free, quick, inexpensive, and easy-to-operate extraction technique. It can be used to highly enrich analytes in a relatively short time and uses simple laboratory equipment, which considerably lowers the cost of analysis [4,10]. This approach implies that a single drop of an extraction solvent is employed for the isolation of the analyte. It was introduced in the mid-1990s by Liu and Dasgupta [19], which used a drop of water to extract ammonia and sulfur dioxide before the spectrophotometric analysis. SDME is based on the principle of the partitioning of the analytes from the sample solution to the extraction solvent with or without mechanical aid. As presented in Figure 5a, SDME can be operated in different modes: direct immersion is employed mostly for nonvolatile analytes, being the extraction solvent immersed in the liquid sample from which the analytes are transferred, subsequently followed by the withdrawal of the drop before instrumental injection; the headspace mode (HS-SDME) is tailored for the isolation of volatile compounds [20]; the bubble-in-drop (BID-SDME) introduced by Williams et al. [21] was designed to enlarge the droplet surface area; continuous-flow microextraction (CFME) proposed by Liu and Lee [22] was designed to increase the contact area between the analyte and the extraction solvent; or the drop-to-drop liquid–liquid microextraction was developed by Wijethunga et al. [23] in which the sample volume required for analysis is reduced. Automation has also been established with SDME coupled with electrothermal atomic absorption spectrometry for the quantitation of Cr (VI) in natural water samples [24].

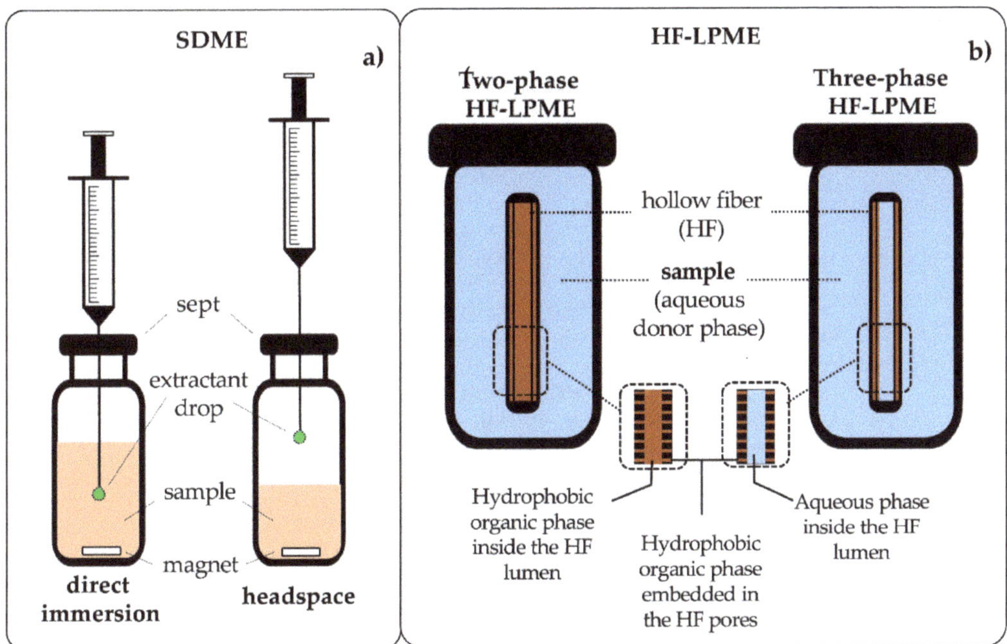

Figure 5. Schematic representation of direct immersion and headspace modes of single-drop microextraction (SDME) (**a**) and two- and three-phase modes of hollow fiber liquid-phase microextraction (HF-LPME) (**b**).

3.2.2. Hollow-Fiber Liquid-Phase Microextraction (HF-LPME)

Pedersen-Bjergaard and Rasmussen [25] first developed the hollow fiber LPME coupled with capillary electrophoresis for methamphetamine in biological samples, such as urine and plasma. It is another mode of solvent-based microextraction technique that is premised on the transfer of the target compounds from the sample (donor) solution via a supported liquid membrane to the acceptor phase. Since its introduction, it has gained wide popularity for the analysis of a wide range of analytes in environmental samples [26], biological samples [27,28], and food samples [29,30]. HF-LPME could be operated in two different modes, a two-phase system and a three-phase system (Figure 5b). Although the two modes share a similarity in principle in that they involve the partitioning of the analytes from the sample (donor phase) solution to other phases (acceptor phase), few lines of demarcation can be observed. In two-phase systems, the analytes are transferred from the aqueous phase to the organic acceptor phase based on their affinity for them. In turn, the three-phase mode involves partitioning from the aqueous donor across the organic solid support liquid membranes (SLMs) into the aqueous acceptor phase in the lumen of the hollow fiber [26]. This results in several advantages that accompany automation, including a lower number of operators to recruit, reduced chemical use, and accelerated analysis time, just to mention a few. Automated HF-LPME has been applied for the analysis of pharmaceutical drugs [31].

3.2.3. Dispersive Liquid–Liquid Microextraction (DLLME)

DLLME has also gained wide popularity over the years. DLLME offers several advantages, including small sample volume, high extraction efficiency, low consumption of solvents, high enrichment factor, good repeatability, and high recovery. Furthermore, this technique is simple and uses small amounts of extraction solvents, and the equilibration between the aqueous phase and extracting solvent is fast [32]. Ultrasounds can be applied to disperse the extraction solvent in the sample, avoiding the reduction of the analyte's partition coefficient between water and the extracting solvent [32]. As a result of efforts to minimize errors incurred in analysis due to intermittent human intervention and to increase the efficiency of the overall process, studies detailing the automation of the LPE techniques have also been reported, for instance, the online sequential injection (SI) DLLME for the isolation and preconcentration of copper and lead using a series of reagents including methanol as disperser solvent mixed with 2.0% (v/v) xylene as the extraction solvent and 0.3% (m/v) ammonium diethyldithiophosphate as the complexing agent. The solvent mixture was merged with the aqueous sample, and 300 µL isobutyl methyl ketone was used to elute the complex of the analytes before the injection into the nebulizer of the flame atomic absorption spectrometry (FAAS) [33]. A similar study involving a modification was reported for the quantitative analysis of cadmium and lead in natural water samples [34].

3.2.4. QuEChERS

At the beginning of this century, Anastassiades, et al. [35] proposed an innovative sample preparation approach with attractive characteristics, a quick, easy, cheap, effective, rugged, and safe (QuEChERS) method for the quantitative measurement of pesticide residues in vegetables and fruits. It is a two-stage process of solid–liquid partitioning with a salting-out effect and a dispersive solid-phase extraction (dSPE). The extraction of the target analytes occurs in the first stage, when a mixture of salts is dispersed in the matrix and mixes thoroughly with an organic phase, often acetonitrile, till an equilibrium is reached. This is followed by a clean-up step (dSPE) using a different combination of porous sorbents and salts according to the matrix interferences that should be removed. Since its invention, QuEChERS has been applied for the analysis of a wide spectrum of analytes in different sample matrices, such as fluoxetine and carbamazepine in benthic invertebrates (*Potamopyrgus antipodarum* and *Valvata piscinalis*) [36], pharmaceuticals and personal care products in sewage and surface waters [37], sulfonamide residues in milk samples [38], or BPA in human urine [39]. More recently, several improvements to the original procedure were reported, notably, its miniaturization applied in different fields of research [40] (additional reports available in Tables S1–S3).

3.2.5. Solidification of Floating Organic Drop Microextraction (SFOME)

The SFOME method was introduced by Khalili Zanjani et al. [41] using polycyclic aromatic hydrocarbons as model compounds. In this technique, the collection of the analytes in a microdrop of an organic extraction solvent under agitation is achieved by the solidification of the suspended microdrop organic layer in ice. The solidified microdrop is allowed to melt before it is injected into the instrument for quantitative assessment. The notable characteristic of the extraction solvent, peculiar for this procedure, is its low melting point usually in the range of 10–30 °C. The use of a little amount of organic solvent indicates the compliance of this simple method with the green analytical chemistry requirements (GAC), and it has been popularly employed either individually or in conjunction with other extraction methods for the analysis of contaminants in environmental [42], food [43], and biological [44] samples. The use of SFOME is not limited to the extraction of organic compounds as it has been deployed for the isolation of inorganic metallic ions, such as lead [45].

3.2.6. UABE

Ultrasound-assisted back extraction (UABE) is another LPE-based extraction method that has been used in tandem with other sample preparation strategies, such as cloud point extraction, for the analysis of brominated flame retardants in water samples (BFRs) [46], heterocyclic aromatic amines in natural water samples [47], and DLLME for the isolation of suvorexant in urine samples [48]. The use of UABE was reported to be a necessity when the extracts contain much extraction solvent that is not compatible with the analytical instrument. Zhou, Gao, Zhang, Li, and Li [46] developed a method for the quantitative determination of BFRs in water samples using a cloud point extraction coupled with UABE before injection into the inlet of HPLC–MS/MS. An amount of 400 µL of an aqueous solution of Triton X-114 and 0.5 M ammonium acetate was added to 40 mL of water sample, which was thoroughly mixed and centrifuged at 5000 rpm for 3 min. After the aqueous phase was decanted, 200 µL of acetonitrile and 2 mL of isooctane were added to the surfactant-rich component and sonicated for 5 min. The isooctane layer was allowed to dry with the aid of N_2 flow in a new centrifuge tube, while the residue was reconstituted in methanol (50 µL) after which it was injected into the HPLC system. This method gave a limit of detection of 0.3 to 3.0 ng L^{-1} and a recovery of 8.7%–54.7%.

3.3. Emergent Green Solvents: Ionic Liquids (ILs), DES, and NADES

ILs have evolved as potential replacements for conventional solvents over the years. They are low-melting organic salts with a combination of an organic cation and an organic or inorganic anion, occurring in the liquid state at a temperature below 100 °C. The striking features of ILs, such as negligible vapor pressure, enhanced synthesis route, fewer by-products, thermal stability, and high hydrophobicity, are factors establishing them as green solvents as they are more environmentally friendly than conventional solvents [49]. Added to the advantages of ILs is the ability to modify functionalities, which enhances the selectivity and specificity of the target molecules [50]. ILs have been shown to improve selectivity and extraction efficiency when used in tandem with a metal organic framework (MOF). For instance, [51] employed imidazole-based ILs as a guest material with Zr-MOFs for the preconcentration of sulfonamides in a water sample in a dSPE-HPLC-DAD method. A limit of detection below 0.03 µg L^{-1} and enrichment factors greater than 270 were obtained. Despite the promising usefulness of ILs in the field of separation science, their notable drawback is that not all ILs are nonvolatile, nonflammable, and stable in air and water as originally considered. In fact, many ILs are volatile, flammable, unstable, and even toxic, particularly to aquatic beings [52]. For this reason, deep eutectic solvents (DES) have emerged as safer alternatives, exhibiting higher stability and lower costs and toxicity [52]. DES are formed by combining different hydrogen bond acceptors and donors, and their classification as ILs is not consensual, mainly because they share more differences than similarities (reviewed in [52]). Among those differences, it is important to highlight in the context of this review that DES are less hazardous and more stable and biodegradable than IL [52]. A specific subclass of DES, composed of components of natural origin, NADES, will be the ultimate green solvent that can be used. For their greener profile and possibilities to fine-tune extraction properties by combining different ILs and, more recently, DES and NADES, the use of these innovative extraction solvents in the most diverse extraction formats is growing exponentially and constitutes one of the forefronts in sample extraction. An exhaustive list of applications using ILs, DES, and NADES is available in Tables S1–S3.

3.4. Other Advanced Extraction Techniques

Other advanced extraction techniques may act at different levels (alone or in combination), including, but not limited to, breaking the overall matrix structure or cell wall, allowing easier penetration of selective solvents with an affinity for the target analytes, fastening mass transfer and extraction kinetics; using more selective and cost-effective extraction solvents, increasing analyte solubility, increasing safety, and decreasing environmental impact (e.g., by changing the type of solvent, decreasing the chemicals needed, reducing energy consumption, or reducing wastes generation). Pulsed electric-field-assisted extraction (PEFAE) is a nonthermal technology that has been primarily applied to disintegrate cells and cellular tissues in food processing and extraction processes. It makes use of very intense electrical pulses with a very short duration. The main objective is to disrupt cell walls and tissues and increase cell permeability without heating the target samples (and analytes), thus increasing extraction efficiency. As the duration of the pulses is designed to avoid thermal effects, it can be used with heat-sensitive compounds. Other electric-field-based approaches are also available. For instance, using moderate electric fields for longer times, combining thermal and nonthermal electric effects, may also be an effective way of extracting the target analytes. This process is also known as ohmic heating. Heat is generated inside the material through the Joule effect, and the matrix is heated almost instantaneously and evenly. Simultaneously, a limited electroporation effect is also expected. It is applicable when heat is needed to achieve an efficient extraction of the target analytes, increasing extraction efficiencies and reducing thermal degradation. Furthermore, energy efficiency in ohmic-heating-based processes is significantly higher than in traditional heating processes [53]. Electric technologies may be used complementary to other more traditional approaches to increase their efficiency and selectivity and decrease the time of sample preparation [54]. Applications of electric fields can be found both in analytes' extraction and in samples' concentration/purification. One example, described by Xu and coworkers [55], is the application of electric fields to enhance SPE extraction of contaminating compounds (tricyclic antidepressants) in environmental waters before their identification by GC–MS [55]. Electro-enhanced solid-phase microextraction can be found applied to several other matrices and analytes, including phthalate esters and bisphenol A from blood and seawater [56] or fluoroquinolones in eggs [57]. Electroextraction of analytes across aqueous–organic phase boundaries has also been described, and membrane-based processes coupled with electric fields are also quite common in the literature [58]. Using greener non-petroleum-based solvents and tuning their properties to increase their efficiency and selectivity by using subcritical or supercritical temperatures and pressures is also an interesting approach to reduce or remove the use of toxic and/or environmental impacting solvents in the extraction step. For instance, when considering analytes with moderate polarity and low thermal sensitivity, subcritical water extraction (SWE) may be considered a viable option. Pressurized solvent extraction (PSE) consists of a liquid–liquid extraction technique where the solvent is used at temperatures higher than its "normal" boiling point (at atmospheric pressure) in pressurized systems. The pressure is kept at values above the boiling point at the selected temperature, but under the critical point, allowing to keep the fluid in the liquid state. Higher temperatures allow increasing the solubility of the analytes and the transfer rate, thus improving extraction efficiencies. Further, viscosity decreases, and the high pressures involved may facilitate the solvent penetration into the matrix from which the analytes are being extracted. SWE is a particular case of pressurized liquid extraction, using water as solvent. Water is a solvent with unique properties in these conditions, not present in other solvents. Besides the above-mentioned advantages, the dielectric constant of water decreases when temperature increases, thus increasing its affinity for less polar analytes. Though this decrease is limited, it is possible to use subcritical water as a greener replacer for "intermediate" polarity organic solvents, such as methanol or ethanol. Further, it is possible to tune water properties to meet the desired affinity for the target analyte by selecting the most appropriate temperature. On top of that, the ionization constant of water also increases, thus liberating more ions, H^+ and OH^-, that may work as a catalyst to break down the matrix, thus improving the solvent accessibility to the analyte [59]. On the other side, for thermolabile nonpolar analytes, supercritical carbon dioxide extraction (SCE) is a relevant option, avoiding the use of solvents, such as hexane. Supercritical fluids have mixed properties between liquids and gases, facilitating extraction processes: diffusion, viscosity, and surface tension similar to gases and density and solvation power as liquids. In the particular case of CO_2, the critical temperature is close to room temperature (31 °C), and working under supercritical conditions is possible at relatively low temperatures. The solvent's low polarity makes it ideal for nonpolar compounds. However, SCE of more polar compounds is possible using a chemical modifier or a cosolvent (such as ethanol), though decreasing the process greenness. Further evaporation or concentration steps are not needed: resuming the atmospheric conditions turns the solvent back into gas that can either return to the environment or be pressurized again to be reused while purifying/concentrating the analyte's sample.

Greener processes tend to use greener solvents with lower environmental impact. Bio-based solvents using renewable sources or water are considered solvents with a lower environmental footprint. However, the utmost target should be using direct analytical methods not requiring reagents or solvents [60]. When the sample's pretreatment is unavoidable, alternatives to reduce or use no solvent at all in the preparation step should be considered [61]. In this context, sample treatments such as simple pressing or extrusion, instant controlled pressure drop, PEFs, or microwaves applied directly in the matrix being analyzed, without adding extra solvents, may cause membrane or cellular structure ruptures enough to free intracellular or structural fluids containing the compounds to be analyzed.

In Table 1 are described the advantages and disadvantages of the most common GreETs used in the analysis of biological, food, and environmental samples.

Table 1. Advantages and disadvantages of some GreETs commonly used in the analysis of biological, food, and environmental matrices.

Extraction Procedure	Advantages	Disadvantages
SPME	Alternative to SPE A limited number of steps Reduced sample amount Reuse of the polymeric phase Environmentally friendly Short extraction time	Potential contamination of the SPME needle
μSPE	Alternative of LLE Simplicity of automation Suitable for large scale Low sorbent Low solvent volume	Requires stirring Possibility of low recoveries
MEPS	Low solvent volume Low sample amount Fast and easy to use Economical Fully automated for online procedure	Requires a wide range of optimization steps
MSPE	Environmentally friendly A limited number of steps Low amount of sorbent material Reuse of sorbent material Short extraction time	Requires vortex/shaker/magnetic stirrerSelection of suitable sorbent
MSDP	Environmentally friendly A limited number of steps Quick Simple	Requires anhydrous sorbents activated at high temperatures
FPSE	Efficient Fast extraction Low volume of solvents High preconcentration factor	Low sorbent capacity Long sample preparation time
DLLME	Economical High recovery Low sample amount Low extraction time Low solvent volume	Low selectivity Requires centrifugation
SFOME	Environmentally friendly High enrichment factor economical Low volume of solvents Simplicity of automation	Requires a wide range of optimization steps

Table 1. *Cont.*

Extraction Procedure	Advantages	Disadvantages
µQuEChERS	Economical Efficient clean-up by dSPE Low solvent consumption	Labor intensive Difficult to automate Emulsion formation
SFE	Environmentally friendly No required solvents Low operating temperatures (40–80 °C) Fast and high yield	Very expensive Complex equipment operating at high pressures High power consumption

Legend: DLLME: dispersive liquid–liquid microextraction; FPSE: fabric-phase sorptive extraction; MEPS: microextraction in packed sorbent; MSPD: matrix solid-phase dispersion; MSPE: magnetic solid-phase extraction; SFE: supercritical fluid extraction; SFOME: solidification of floating organic drop microextraction; SPME: solid-phase microextraction; µQuEChERS: micro-QuEChERS; µSPE: micro-solid-phase extraction.

4. High-Resolution Analytical Techniques

Liquid chromatography (LC) has a great benefit on the efficiency of separating complex matrices, but it is not appropriate to achieve structural information of the target analytes. In this sense, LC combined with mass spectrometry (MS) or MS/MS is certainly the most common analytical approach in the analysis of a diversity of target analytes in environmental, clinical, and food matrices, since it provides higher selectivity, mainly when isomeric mixtures were analyzed. HPLC combined with a traditional detector such as ultraviolet (UV) [62], photodiode array detector (PDA) [63], and fluorescence detector (FLD) [64] have been applied in the determination of several target analytes in environmental, clinical, and food matrices. The benefits of these traditional systems are economical, more accessible in common laboratories, efficient, faster, and easy to use. Despite the lower sensitivity attained using these traditional detectors, excellent results related to validation parameters were achieved, namely, low limit of detection (few µg/L), good accuracy (recoveries higher than 70%), and intra- and interday precisions with relative standard deviation (RSD) lower than 20%. Currently, UV, PDA (DAD), and FLD detectors have been substituted by MS and/or MS/MS detectors since they provide high selectivity, sensitivity (low LODs), and ability to provide information related to molecular mass and structural proprieties. Liquid chromatography–tandem mass spectrometry (LC–MS/MS) [65] is becoming a promising analytical approach to analyzing complex matrices due to its high separation resolution, high sensitivity (low LODs), and capacity to identify compounds and does not require any derivatization step before the analysis. Nevertheless, compared with the traditional systems, LC–MS/MS showed several drawbacks, such as the complexity of the operation, expensiveness, and strong matrix effects that promote in many cases signal suppression or enhancement.

The atmospheric pressure ionization of MS has electrospray ionization (ESI) and atmospheric pressure chemical ionization (APCI). The introduction of ESI and APCI overcame the limitation of previous interfaces by evaporating the mobile phase during the ionization process. This, combined with the orthogonal spray interface, provided a means to stop possibly interfering nonvolatile compounds, such as salts, buffers, and detergents, from entering the MS. For many target analytes, ESI provides high sensitivity, being more used than APCI [65]. APCI is more appropriate for the analysis of nonpolar analytes and volatile organic compounds (VOCs). In addition, the major benefits of the application of ESI for quantitative LC–MS are the production of protonated or deprotonated molecules with slight fragmentation, optimal selection of precursor ions, and maximizing sensitivity, the matrix effect being its main drawback.

Single quadrupole, triple quadrupole (QQQ-MS), ion trap (IT), time of flight (TOF), and quadrupole–time of flight (Q–TOF) are the most common MSs used in tandem mass spectrometry (MS/MS). The IT, TOF, and Q–TOF mass spectrometers usually are used for structural elucidation and the identification of unknown compounds. Nevertheless, achieving structural information of unknown compounds requires higher purity of matrices. More sophisticated analytical approaches, such as ultra-performance liquid chromatography–tandem mass spectrometry (UHPLC–MS/MS) [66], have been recently used in the analysis of compounds in environmental, clinical, and food matrices. UHPLC–MS/MS compared with HPLC–MS/MS provides high pressures, narrow peaks, high chromatographic separation, and lower analysis time and solvent volumes. Moreover, LC coupled to high-resolution mass spectrometry (HRMS) was used for direct determination of glyphosate and its metabolite aminomethylphosphonic acid (AMPA) in human urine by combining cold-induced phase separation (CIPS) with hydrophilic pipette tip solid-phase extraction (PT-SPE) [67]. LC–HRMS compared with LC–MS offers screening for targeted, suspect, and nontargeted analysis in a single run, producing high-resolution accurate masses, their isotopic patterns, and MS2 spectra included in online databases.

Gas chromatography (GC) coupled with a flame ionization detector (FID) [68], MS [69], or MS/MS [70] has also been used in minor extension, when compared with LC–MS and LC–MS/MS, in the analysis of environmental, clinical, and food matrices. This fact could be explained by the derivatization process required for the analysis of some target analytes in GC analysis to promote the volatility and decrease the polarity of the analytes, as well as the time of analysis.

Other analytical approaches have been used in the analysis of environmental, clinical, and food matrices, such as flame atomic absorption spectrometry (FAAS) [71], inductively coupled plasma (ICP) combined with mass spectrometry (MS) [72], or optical emission spectrometry (OES) [73]. Additional details about these and other examples are available in Tables S1–S3, covering the clinical, food, and environmental fields, respectively.

5. Applications of Green Extraction Techniques to Different Fields

As discussed in the previous sections, the use of GreETs spans a wide range of applications, covering the most diverse type of samples, from biofluids to environmental matrices and all type of foods (an exhaustive list of applications reported in the literature since 2016 is available in the Supplementary Material). In modern analytical layouts and to fulfill GAC requirements, the analysis that follows the sample preparation using GreETs should employ fast and efficient analytical instruments able to acquire huge amounts of data. As a consequence, powerful data processing and statistical analysis procedures will be required to produce consistent results (Figure 6).

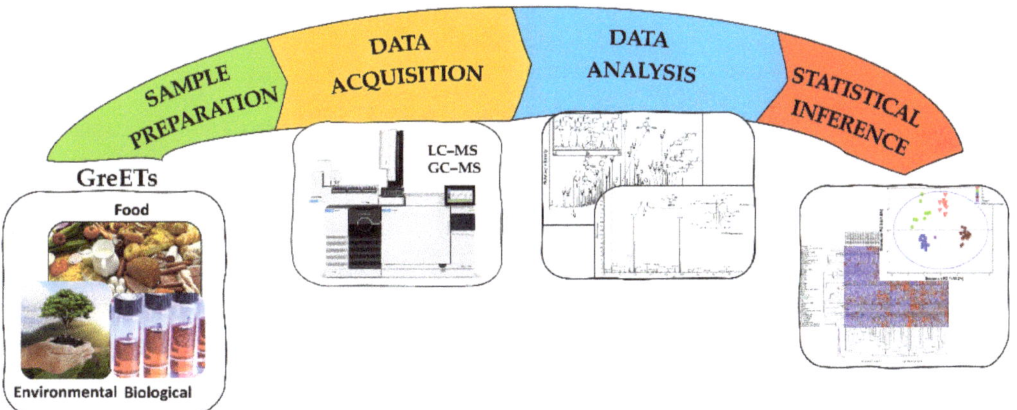

Figure 6. Different steps involved in sample preparation. GreETs: green extraction techniques.

5.1. Biological Samples

The application of GreETs to the clinical field has increased consistently since the beginning of the century [74]. This mostly includes body fluid samples containing lower-molecular-mass organic molecules, less than 500 g/mol, comprising drug analytes, metabolites, environmental exposure contaminants, poisons, tissues, and endogenous substances [74]. These biological samples present great complexity and moderate-to-high levels of protein, thus requiring robust sample preparation approaches able to simplify and isolate the target analytes from the matrix [75]. As discussed in more detail in the previous sections, traditional sample preparation methods are not particularly tailored for clinical applications because they are time-consuming and require various steps and extensive clean-up before analysis. In contrast, most GreETs require low sample amounts, very low or no solvent at all, and simple, fast, and user-friendly systems that can be easily automated [75]. These advantages made SPME, µSPE, MEPS, MSPE, just to name a few GreETs, particularly suitable to process biological samples. Moreover, they also allow spanning a wide range of analytes with different properties, such as drugs for clinical and forensic toxicology assays, pharmacokinetic studies, biochemical analysis, pharmaceuticals, in vivo applications, and metabolomics [75]. SPME and its different formats are particularly efficient in this field of application because they often require minimum sample pretreatment and can be easily coupled to analytical instruments (e.g., CG and LC), providing an enhanced extraction capacity and simultaneous quantification of different compounds with overall sensitivity. This includes the simultaneous identification of drugs of abuse (e.g., amphetamines, barbiturates, methadone), psychoactive substances, pharmaceuticals (e.g., antidepressants, antiepileptic agents, steroids, anorectic agents, corticosteroids, anaesthetics), substances that affect the adrenergic system, nonsteroidal anti-inflammatory substances, and so forth [76]. Examples of such applications are available as Supplementary Material (Table S1). Among the different biological matrices, microextraction of urine samples has the advantage of minimum processing, often not requiring any centrifugation or filtration before extraction. This minimizes sample handling and improves method precision. Additionally, it is suitable for a wide range of sample volumes, including volumes as small as 50 µL, and even for sampling when the volume is not accurately known. Diverse types of GreETs using urine are available in the literature, SPME, µSPE, and MEPS being the most often reported [75] (Table S1). The use of GreETs with blood sampling is also advantageous, particularly when this allows the elimination of blood-withdrawal steps from the analytical workflow, as with SPME. GreET usage also reduces the risk of analyte degradation and matrix changes due to enzymatic conversion, as well as fast sample collection and clean-up. Different examples of applications involving blood sampling using GreETs can be found in the literature, such as VOCs (SPME [77]), polycyclic aromatic hydrocarbons (PAHs, pipette-tip SPE [78]), Ni and Pb (µSPE [79]), opiates (MEPS [70]), and antidepressants (FPSE [80]). SPME has also been reported in in vivo assays with biological matrices like tissues. This can be performed with a removed tissue portion (ex vivo), direct in vivo measurement, exposing the BioSPME needle to the tissue or even inserting the probes directly into the tissue. Regarding this, Musteata [81] observed that microdialysis and SPME were not only appropriate for tissue sampling but also complementary to each other for in vivo sampling and ex situ analysis. By using this approach, the probe extracts only a slight fraction of the free analyte, minimizing disturbances of chemical equilibrium and allowing multiple measurements of analyte concentrations under physiological conditions. Moreover, the accurate determination of analyte concentration is unaffected by the sample volume. Finally, the technique is open to miniaturization, allowing its application within small living systems, sample storage and transportation, and easy coupling to portable instrumentation [75]. An example of such an approach was reported by Cudjoe, et al. [82], which used SPME to monitor neurotransmitter changes in the striatum of a rat brain after dosing antidepressants, variations in serotonin concentrations due to deep-brain stimulations, and distribution of pharmaceuticals in the striatal region and cortex. This elegant experiment shows that SPME can also be very useful in metabolomics assays, particularly at the initial stage of biomarker discovery in medical diagnosis. It is also very relevant to the quantification of different compounds simultaneously, which enables the simultaneous monitoring of drugs in complex treatments. This is possible because GreETs coupling with chromatographic methods, as shown in Section 4, can be easily achieved, allowing the analysis of a whole pharmacopoeia of drugs, such as anticancer, antibiotic, antidepressant, analgesic, anti-inflammatory, steroid, and neurotransmitter drugs. This can help to provide earlier detection of the disease, which is imperative for a successful clinical treatment, especially in some oncologic diseases, where an early diagnosis is crucial for the survival of the patient without suffering severe impacts on health and life quality. FPSE is a very promising GreET having a key advantage regarding other microextraction approaches, allowing a direct analyte extraction with no sample modification [6]. Since its introduction in 2014, many examples of applications involving biological samples have been reported in the literature, such as the cow and human breast milk sample clean-up for screening bisphenol A and residual dental restorative material [83]; the simultaneous monitoring of inflammatory bowel disease treatment drugs [84] and anticancer drugs [85] in whole blood, plasma, and urine; or the assessment of radiation exposure [86] (Table S1). The use of magnetic nanoparticles as microextraction sorbents in MSPE also results in a very simple and efficient extraction procedure because the sorbent can be tailored to extract specific analytes, and the sorbent-retained analyte complex can be easily recovered from the solution using a magnetic field or magnet [87]. MSPE has been used to extract different drugs from urine, such as nonsteroidal anti-inflammatory drugs (NSAIDs) [88], methadone [89], pseudoephedrine [90], fluoxetine [91], and statins [92], as well as antiepileptic drugs [93] or ibuprofen [94] from plasma (Table S1). GreETs involving liquid-phase sorbents, such as DLLME, are also often reported in the literature. This format, mostly assisted by ultrasounds (UA-DLLME), allows the usage of a myriad of extraction solvents, and consequently, the repertoire of applications is very broad. Mabrouk et al. [95], for instance, used UA-DLLME to extract three gliflozins (antidiabetic drugs) from plasma.

5.2. Food Samples

Food analysis is of great importance since ingestion of a growing number of compounds intentionally or not added to food can represent a risk to our health. However, beyond food safety, consumers are also more aware of the nutritional value of food and are also interested in its composition, particularly regarding the presence of bioactive compounds. For these reasons, efficient methodologies for the identification and quantification of all these analytes are required. Accordingly, GreETs have been used in the sample preparation procedures of different food matrices to extract and preconcentrate target analytes to a sufficient level to allow their analysis [96]. The μSPE technique, for instance, has been used in the determination of aflatoxins [96], pesticides [62], trace metals [73], and pollutants, such as bisphenol A [97] and PAHs [68], in a variety of food products. Additionally, it aided in the identification and quantification of rosmarinic acid in medicinal plants [98] and vitamin D3 in bovine milk [99]. MEPS is another GreET that has been employed in the analysis of foodstuffs, including the identification of herbicides in rice [100], insecticides in drinking water [101], pesticides in apple juice and coffee [102], antibiotics [103] and steroids [104] in milk, parabens in vegetable oil [105], PAHs in apple [106], caffeine in drinks [107], and polyphenols in baby food [108]. SPME has been widely used to study the volatile composition of several foods, including walnut oils [109], *hongeo* [110], melon [111], and dairy products [112]. Moreover, this technique has also been used to determine the composition of specific analytes, such as the x-ray induced markers 2-dodecylcyclobutanone and 2-tetradecylcyclobutanone in irradiated dairy products [113], the contaminants 1,4-dioxane and 1,2,3-trichloropropane [114], acrylamide [115], organophosphorus pesticides [116], phthalates [117], synthetic phenolic antioxidants [118], and xanthines [119]. MSPD has been reported in the literature for the extraction of flavonoids [120], polyphenols [121], mangiferin, and hyperoside in mango-processing waste [122], ergosterol in edible fungi [123], and pharmacologically active substances in microalgae [124]. This methodology has also been applied for pesticide [125] and sulfonylurea herbicide [126] extraction in several food matrices. MSPE allowed the extraction of trace metals in food products [127] (additional examples available in Table S2). Moreover, studies have shown that this technique can be used for the determination of acrylamide [78,79], bisphenols [80], PAHs [128], plant growth regulators [129], and caffeine [130]. FPSE is another GreET that has been shown to be very useful for the determination of several classes of pesticides in foods [96]. Other analytes studied using this technique include bisphenol A [131], oligomers [132], PAHs [64], steroid hormone residues [133], and tetracycline residues [134]. DLLME has been vastly applied for the determination of trace metals [71] (additional examples available in Table S2), pesticides [96], chloramphenicol [135], and nonsteroidal anti-inflammatory drugs [136] in different foods. µQuEChERS was employed in the extraction of several analytes from foods, ranging from pesticide residues in wine [137] and PAHs in coffee and tea [138] to polyphenols in baby food [139] and pyrrolizidine alkaloids in oregano [66]. The application of SDME was proved to allow the determination of unfavorable compounds and elements in foods, such as drug metabolites [140], acrylamide [141], ammonia [142], ethyl carbamate [143], formaldehyde [144], tartrazine [145], and Cu(II) [146]. Similarly to SDME, SFOME can be used for the detection of trace metals [147], as well as of β-lactam antibiotic residues [148] and organochlorine pesticides [149]. PEAE has been applied for the extraction of different bioactive compounds [63], including phenolic compounds [43], carotenoids [150], procyanidins [151], and sulforaphane [152]. The use of SFE has been used for the extraction of several antioxidant and antibacterial compounds from feijoa leaf [153], fatty acids and oils from Indian almonds [154], oleoresins from industrial food waste [155], and polar lipid fraction from blackberry and passion fruits [156]. Additionally, SFE was employed for the extraction of phytochemicals from *Terminalia chebula* pulp [157]. Finally, SWE is a technique largely applied to the extraction of several classes of bioactive compounds, including anthocyanins [158], fatty acids [159], hesperidin and narirutin [160], phenolic compounds [161], and scopoletin, alizarin, and rutin [162]. The extraction of antioxidant protein hydrolysates from shellfish waste [163] and pectic polysaccharides from apple pomace has been also previously accomplished by SWE [164].

5.3. Environmental Samples

Most environmental samples have complex matrix compositions and involve the determination of trace and ultra-trace analytes [3]. For instance, the determination of PAHs in water samples or pesticide analysis is challenging due to their very low concentrations [2,3]. This requires efficient clean-up and enrichment procedures before the analytes' analysis [2]. MEPS seems to be tailored for these requirements and has been applied in the analysis of benzene, phenol and their derivates [165], diazinon [166], La^{3+} and Tb^{3+} [167], organophosphorus pesticides [168] in water samples, fipronil and fluazuron residues in wastewater [169], and PAHs in the most diverse samples (see Table S3), including Antarctic snow [170], and in the detection of phthalates in tap and river water [171]. SPME is eventually one of the most used sample extraction procedures and has been applied for the detection of different pesticides in water [172] (additional examples in Table S3), microplastic in coral reef invertebrates [173], PAHs in rainwater [174], and volatile organic compounds (VOCs) in wastewater [69]. Molecularly imprinted polymers (MIPs) have also been employed in the extraction of polychlorinated aromatic compounds from environmental samples. Some applications include the use of MIPs in the analysis of 2-chlorophenol [175], 2,4-dichlorophenoxyacetic acid [176], and endosulfans [177] in water samples and in the determination of organochlorine pesticides in environmental samples [178]. This methodology has also been reported in the preparation of soil samples to increase the extraction efficiency of triazine herbicides [179]. Multisphere adsorptive microextraction (MSAµ) has been applied in the extraction of caffeine, acetaminophen [18], pharmaceuticals, sexual steroid hormones, and antibiotics [17] in water samples. QuEChERS is known as the Swiss knife of extraction. Its µQuEChERS version is even more greener and includes applications such as the detection of insecticides in guttation fluids [65], pesticides in arthropods and gastropods [180], and VOCs in zebrafish [181].

LPME techniques, such as SDME and SLLME, use small volumes of organic solvents to extract the analytes [4]. SDME has gained a lot of interest in the last few years and is mostly used for the determination of trace analysis in environmental matrices, including Cu(II) in tap and seawater [146], PAHs in tap water [182], ranitidine in wastewater [183], and V(V) in water samples [184]. DLLME is another efficient microextraction procedure, and its ultrasound-assisted (UA) DLLME variation has been adopted in several environmental matrices for the analysis of aromatic amines [185], Cd [186], Cr [187], dyes [188], herbicides [189], polybrominated biphenyls [190], pyrethroid insecticides [191], and tetracycline [192] in water samples. SFE was applied to environmental matrices for the analysis of Ag in electronic waste [193], petroleum biomarkers in tar balls and crude oils [194], petroleum hydrocarbons in soil [195], and solanesol in tobacco residues [196]. In turn, SWE has been successfully used for the extraction of Co, Li, and Mn in spent lithium-ion batteries [197], crude oil in soil [198], oil shale in mines [199], and VOCs in sewage sludge [200]. An exhaustive list of GreETs involving environmental samples is available in Table S3.

6. Final Remarks

The sample preparation, including the extraction process, is one of the most important steps for any analysis. It is the step that determines the quality of the measurements of the target analytes, and so it should be critically considered. In recent years, a rapid transition from solvent-based extractions and solid sorbent-based extractions to miniaturized formats promoted a substantial reduction of the solvent consumption. Being simpler, faster, more economical, and user-friendly than CETs, SPE techniques have become more popular. In turn, this has boosted a continuous improvement of innovative techniques respecting the green analytical chemistry principles, such as, FPSE, SPME, SPDE, MEPS, SBSE, and MSAµE. However, there are always opportunities for improvement, and future research should be directed to, for instance, innovative sorbents/nanosorbents able to further improve retention efficiency, loading capacity, and selectivity. Also noticeable will be the use of artificial intelligence, including microfluidics and smartphones, to boost the automation of extraction procedures and the use of alternative detectors. Another challenging aspect in this field in the near future is the integration of the whole analytical procedure in injection loops able to protect operators from harmful solvents and significantly reduce their use.

Supplementary Materials: The following supporting information can be downloaded at https://www.mdpi.com/article/10.3390/molecules27092953/s1: Table S1: Representative applications of GreETs for the analysis of biological samples; Table S2: Representative applications of GreETs for the analysis of food samples; Table S3: Representative applications of GreETs for the analysis of environmental samples.

Author Contributions: J.S.C. and J.A.M.P. planned and managed the manuscript; J.A.M.P. was responsible for Section 1; J.S.C. was responsible for Sections 2 and 6; J.S.C., R.P. and C.V.B. collaborated on Section 3.1; B.O., A.K. and J.A.M.P. collaborated on Section 3.2; C.M.R.R., J.A.T. and R.P. collaborated on Section 3.3; R.P. was responsible for Section 4; T.M.G., C.F.P.A., C.V.B., R.P. and J.A.M.P. collaborated on Section 5; J.S.C. and J.A.M.P. designed all the figures and edited the manuscript; J.S.C. made its submission. All authors have read and agreed to the published version of the manuscript.

Funding: This research was funded by FCT (Fundação para a Ciência e a Tecnologia) through the CQM Base Fund, UIDB/00674/2020, Programmatic Fund, UIDP/00674/2020, and CEB—Centre of Biological Engineering, and by ARDITI (Agência Regional para o Desenvolvimento da Investigação Tecnologia e Inovação) through the project M1420-01-0145-FEDER-000005, Centro de Química da Madeira (CQM+; Madeira 14-20 Program). The authors also acknowledge FCT and the Madeira 14–20 Program to the Portuguese Mass Spectrometry Network (RNEM) through the PROEQUIPRAM program, M14-20 M1420-01-0145-FEDER-000008). ARDITI is also acknowledged for the postdoctoral fellowship granted to J.A.M.P. (Project M1420-09-5369-FSE-000001).

Institutional Review Board Statement: Not applicable.

Informed Consent Statement: Not applicable.

Data Availability Statement: Not applicable.

Acknowledgments: The authors acknowledge Victoria F. Samanidou from Aristotle University of Thessaloniki for the kind invitation to participate in this Special Issue.

Conflicts of Interest: The authors declare no conflict of interest.

Abbreviations

AMPA: aminomethylphosphonic acid; APCI: atmospheric pressure chemical ionization; BIN: barrel insert needle; BID-SDME: bubble-in-drop; BFR: brominated flame retardants; CET: classical extraction techniques; CFME: continuous-flow microextraction; CIPS: cold-induced phase separation; CNFs: carbon nanofibers; CNTs: carbon nanotubes; DAD: diode-array detection; DCM: dichloromethane; DES: deep eutectic solvents; DLLME: dispersive liquid–liquid microextraction; dSPE: dispersive solid-phase extraction; DVB: divinylbenzene; ESI: electrospray ionization; ECD: electrochemical detection; EPT-SPME: effervescent pipette-tip solid-phase microextraction; EtAc: ethyl acetate; EtOH: ethanol; FA: formic acid; FAAS: flame atomic absorption spectrometry; FID: flame ionization detector; FGO-TD-PTFE: functional graphene oxide thermal desorption poly(tetrafluoroethylene); FLD: fluorescence detector; FPSE: fabric-phase sorbent extraction; GA: gallic acid; GAC: green analytical chemistry; GC–MS: gas chromatography–mass spectrometry; GreETs: green extraction techniques; HF-LPME: hollow fiber liquid-phase microextraction; HPLC: high performance liquid chromatography; HS: headspace; HRMS: high-resolution mass spectrometry; ICP: inductively coupled plasma; ILs: ionic liquids; IT: ion trap; LC–MS: liquid chromatography–mass spectrometry; LDH: layered double hydroxide; LLE: liquid–liquid extraction; LLME: liquid–liquid microextraction; LOD: limit of detection; LPE: liquid-phase extraction; M-ILs-SDME: magnetic ionic liquid single-drop microextraction; MAE: microwave-assisted extraction; MIP: molecular imprinted polymer; MIOMS-ir: molecularly imprinted ordered mesoporous silica imprint-removed silica; MISM: molecularly imprinted silica monolithic; MeOH: methanol; MEPS: microextraction by packed sorbent; MMF: multiple monolithic fiber; MNPs: magnetic nanoparticles; MOF: metal organic framework; MS: mass spectrometry; MSAµ: multisphere adsorptive microextraction; MSPD: matrix solid-phase dispersion; MSPE: magnetic solid-phase extraction; MS/MS: tandem mass spectrometry; MWCNTs: multiwalled CNTs; NADES: natural deep eutectic solvents; NPs: nanoparticles; NTD: needle trap device; NTME: needle trap microextraction; NSAIDs: nonsteroidal anti-inflammatory drugs; OES: optical emission spectrometry; PAHs: polycyclic aromatic hydrocarbons; PDA: photodiode array; PDMS: polydimethylsiloxane; PEFAE: pulsed electric-field-assisted extraction; PSE: pressurized solvent extraction; PT: pipette tip; PS/DVB-RP: reverse-phase polystyrene–divinylbenzene sorbent; QQQ: single quadrupole triple quadrupole; Q–TOF: quadrupole–time of flight; SBSE: stir-bar sorptive extraction; SCE: supercritical carbon dioxide extraction; SDME: single-drop microextraction; SFE: supercritical fluid extraction; SFOME: solidification of floating organic drop microextraction; SPE: solid-phase extraction; SPME: solid-phase microextraction; SWE: subcritical water extraction; TOF: quadrupole–time of flight; UA: ultrasound-assisted; UHPLC: ultrahigh performance liquid chromatography; UABE: ultrasound-assisted back extraction; UV: ultraviolet analysis; VOCs: volatile organic compounds; µQuEChERS: micro-QuEChERS; µSPE: micro-solid-phase extraction.

References

1. Gałuszka, A.; Migaszewski, Z.; Namieśnik, J. The 12 principles of green analytical chemistry and the SIGNIFICANCE mnemonic of green analytical practices. *TrAC Trends Anal. Chem.* **2013**, *50*, 78–84. [CrossRef]
2. Fu, S.; Fan, J.; Hashi, Y.; Chen, Z. Determination of polycyclic aromatic hydrocarbons in water samples using online microextraction by packed sorbent coupled with gas chromatography-mass spectrometry. *Talanta* **2012**, *94*, 152–157. [CrossRef] [PubMed]
3. Delińska, K.; Yavir, K.; Kloskowski, A. Ionic liquids in extraction techniques: Determination of pesticides in food and environmental samples. *TrAC Trends Anal. Chem.* **2021**, *143*, 116396. [CrossRef]
4. Casado, N.; Ganan, J.; Morante-Zarcero, S.; Sierra, I. New Advanced Materials and Sorbent-Based Microextraction Techniques as Strategies in Sample Preparation to Improve the Determination of Natural Toxins in Food Samples. *Molecules* **2020**, *25*, 702. [CrossRef]
5. Kabir, A.; Furton, K.G. Fabric Phase Sorptive Extractors. U.S. Patent 20140274660A1, 18 September 2014.

6. Kabir, A.; Samanidou, V. Fabric Phase Sorptive Extraction: A Paradigm Shift Approach in Analytical and Bioanalytical Sample Preparation. *Molecules* **2021**, *26*, 856. [CrossRef]
7. Olia, A.E.A.; Mohadesi, A.; Feizy, J. Ochratoxin Determination in Food Samples by Fabric Phase Sorptive Extraction Coupled with High-Performance Liquid Chromatography Technique. 2022. Available online: https://www.researchsquare.com/article/rs-1478781/v1 (accessed on 1 May 2022).
8. Arthur, C.L.; Pawliszyn, J. Solid phase microextraction with thermal desorption using fused silica optical fibers. *Anal. Chem.* **1990**, *62*, 2145–2148. [CrossRef]
9. Billiard, K.M.; Dershem, A.R.; Gionfriddo, E. Implementing Green Analytical Methodologies Using Solid-Phase Microextraction: A Review. *Molecules* **2020**, *25*, 5297.
10. Ribeiro, C.; Ribeiro, A.R.; Maia, A.S.; Goncalves, V.M.; Tiritan, M.E. New trends in sample preparation techniques for environmental analysis. *Crit. Rev. Anal. Chem.* **2014**, *44*, 142–185. [CrossRef]
11. Filippou, O.; Bitas, D.; Samanidou, V. Green approaches in sample preparation of bioanalytical samples prior to chromatographic analysis. *J. Chromatogr. B Anal. Technol. Biomed. Life Sci.* **2017**, *1043*, 44–62. [CrossRef]
12. Barker, S.A. Matrix solid phase dispersion (MSPD). *J. Biochem. Biophys. Methods* **2007**, *70*, 151–162. [CrossRef]
13. Chang, Y.C.; Chen, D.H. Adsorption kinetics and thermodynamics of acid dyes on a carboxymethylated chitosan-conjugated magnetic nano-adsorbent. *Macromol. Biosci.* **2005**, *5*, 254–261. [CrossRef] [PubMed]
14. Bruzzoniti, M.C.; Sarzanini, C.; Costantino, G.; Fungi, M. Determination of herbicides by solid phase extraction gas chromatography-mass spectrometry in drinking waters. *Anal. Chim. Acta* **2006**, *578*, 241–249. [CrossRef] [PubMed]
15. Hsieh, H.C.; Sheu, C.; Shi, F.K.; Li, D.T. Development of a titanium dioxide nanoparticle pipette-tip for the selective enrichment of phosphorylated peptides. *J. Chromatogr. A* **2007**, *1165*, 128–135. [CrossRef] [PubMed]
16. Berton, P.; Lana, N.B.; Ríos, J.M.; García-Reyes, J.F.; Altamirano, J.C. State of the art of environmentally friendly sample preparation approaches for determination of PBDEs and metabolites in environmental and biological samples: A critical review. *Anal. Chim. Acta* **2016**, *905*, 24–41. [CrossRef] [PubMed]
17. Neng, N.R.; Silva, A.R.M.; Nogueira, J.M.F. Adsorptive micro-extraction techniques—Novel analytical tools for trace levels of polar solutes in aqueous media. *J. Chromatogr. A* **2010**, *1217*, 7303–7310. [CrossRef] [PubMed]
18. Silva, A.R.M.; Neng, N.R.; Nogueira, J.M.F. Multi-Spheres Adsorptive Microextraction (MSAμE)—Application of a Novel Analytical Approach for Monitoring Chemical Anthropogenic Markers in Environmental Water Matrices. *Molecules* **2019**, *24*, 931. [CrossRef]
19. Liu, S.; Dasgupta, P.K. Liquid Droplet. A Renewable Gas Sampling Interface. *Anal. Chem.* **1995**, *67*, 2042–2049. [CrossRef]
20. Jeannot, M.; Przyjazny, A.; Kokosa, J. Single drop microextraction—Development, applications and future trends. *J. Chromatogr. A* **2009**, *1217*, 2326–2336. [CrossRef]
21. Williams, D.B.G.; George, M.J.; Meyer, R.; Marjanovic, L. Bubbles in Solvent Microextraction: The Influence of Intentionally Introduced Bubbles on Extraction Efficiency. *Anal. Chem.* **2011**, *83*, 6713–6716. [CrossRef]
22. Liu, W.; Lee, H.K. Continuous-Flow Microextraction Exceeding 1000-Fold Concentration of Dilute Analytes. *Anal. Chem.* **2000**, *72*, 4462–4467. [CrossRef]
23. Wijethunga, P.A.L.; Nanayakkara, Y.S.; Kunchala, P.; Armstrong, D.W.; Moon, H. On-Chip Drop-to-Drop Liquid Microextraction Coupled with Real-Time Concentration Monitoring Technique. *Anal. Chem.* **2011**, *83*, 1658–1664. [CrossRef] [PubMed]
24. Pena, F.; Lavilla, I.; Bendicho, C. Immersed single-drop microextraction interfaced with sequential injection analysis for determination of Cr(VI) in natural waters by electrothermal-atomic absorption spectrometry. *Spectrochim. Acta Part B At. Spectrosc.* **2008**, *63*, 498–503. [CrossRef]
25. Pedersen-Bjergaard, S.; Rasmussen, K.E. Liquid—liquid—liquid microextraction for sample preparation of biological fluids prior to capillary electrophoresis. *Anal. Chem.* **1999**, *71*, 2650–2656. [CrossRef] [PubMed]
26. Madikizela, L.; Pakade, V.; Ncube, S.; Tutu, H.; Chimuka, L. Application of Hollow Fibre-Liquid Phase Microextraction Technique for Isolation and Pre-Concentration of Pharmaceuticals in Water. *Membranes* **2020**, *10*, 311. [CrossRef] [PubMed]
27. Hrdlička, V.; Navrátil, T.; Barek, J. Application of hollow fibre based microextraction for voltammetric determination of vanillylmandelic acid in human urine. *J. Electroanal. Chem.* **2019**, *835*, 130–136. [CrossRef]
28. Kazakova, J.; Villar-Navarro, M.; Pérez-Bernal, J.L.; Ramos-Payán, M.; Bello-López, M.Á.; Fernández-Torres, R. Urine and saliva biomonitoring by HF-LPME-LC/MS to assess dinitrophenols exposure. *Microchem. J.* **2021**, *166*, 106193. [CrossRef]
29. Dominguez-Tello, A.; Dominguez-Alfaro, A.; Gómez-Ariza, J.L.; Arias-Borrego, A.; García-Barrera, T. Effervescence-assisted spiral hollow-fibre liquid-phase microextraction of trihalomethanes, halonitromethanes, haloacetonitriles, and haloketones in drinking water. *J. Hazard. Mater.* **2020**, *397*, 122790. [CrossRef]
30. Piao, H.; Jiang, Y.; Li, X.; Ma, P.; Wang, X.; Song, D.; Sun, Y. Matrix solid-phase dispersion coupled with hollow fiber liquid phase microextraction for determination of triazine herbicides in peanuts. *J. Sep. Sci.* **2019**, *42*, 2123–2130. [CrossRef]
31. Tajik, M.; Yamini, Y.; Esrafili, A.; Ebrahimpour, B. Automated hollow fiber microextraction based on two immiscible organic solvents for the extraction of two hormonal drugs. *J. Pharm. Biomed. Anal.* **2015**, *107*, 24–31. [CrossRef]
32. Rezaee, M.; Yamini, Y.; Faraji, M. Evolution of dispersive liquid-liquid microextraction method. *J. Chromatogr. A* **2010**, *1217*, 2342–2357. [CrossRef]
33. Anthemidis, A.N.; Ioannou, K.-I.G. On-line sequential injection dispersive liquid–liquid microextraction system for flame atomic absorption spectrometric determination of copper and lead in water samples. *Talanta* **2009**, *79*, 86–91. [CrossRef] [PubMed]

34. Anthemidis, A.N.; Ioannou, K.-I.G. Development of a sequential injection dispersive liquid–liquid microextraction system for electrothermal atomic absorption spectrometry by using a hydrophobic sorbent material: Determination of lead and cadmium in natural waters. *Anal. Chim. Acta* **2010**, *668*, 35–40. [CrossRef] [PubMed]

35. Anastassiades, M.; Lehotay, S.J.; Stajnbaher, D.; Schenck, F.J. Fast and easy multiresidue method employing acetonitrile extraction/partitioning and "dispersive solid-phase extraction" for the determination of pesticide residues in produce. *J. AOAC Int.* **2003**, *86*, 412–431. [CrossRef] [PubMed]

36. Berlioz-Barbier, A.; Baudot, R.; Wiest, L.; Gust, M.; Garric, J.; Cren-Olivé, C.; Buleté, A. MicroQuEChERS–nanoliquid chromatography–nanospray–tandem mass spectrometry for the detection and quantification of trace pharmaceuticals in benthic invertebrates. *Talanta* **2015**, *132*, 796–802. [CrossRef] [PubMed]

37. Kachhawaha, A.S.; Nagarnaik, P.M.; Jadhav, M.; Pudale, A.; Labhasetwar, P.K.; Banerjee, K. Optimization of a Modified QuEChERS Method for Multiresidue Analysis of Pharmaceuticals and Personal Care Products in Sewage and Surface Water by LC-MS/MS. *J. AOAC Int.* **2017**, *100*, 592–597. [CrossRef] [PubMed]

38. Fernandes, F.C.B.; Silva, A.S.; Rufino, J.L.; Pezza, H.R.; Pezza, L. Screening and determination of sulphonamide residues in bovine milk samples using a flow injection system. *Food Chem.* **2015**, *166*, 309–315. [CrossRef]

39. Correia-Sa, L.; Norberto, S.; Delerue-Matos, C.; Calhau, C.; Domingues, V.F. Micro-QuEChERS extraction coupled to GC-MS for a fast determination of Bisphenol A in human urine. *J. Chromatogr. B Anal. Technol. Biomed. Life Sci.* **2018**, *1072*, 9–16. [CrossRef]

40. Rial-Berriel, C.; Acosta-Dacal, A.; Zumbado, M.; Luzardo, O.P. Micro QuEChERS-based method for the simultaneous biomonitoring in whole blood of 360 toxicologically relevant pollutants for wildlife. *Sci. Total Environ.* **2020**, *736*, 139444. [CrossRef]

41. Khalili Zanjani, M.R.; Yamini, Y.; Shariati, S.; Jönsson, J.A. A new liquid-phase microextraction method based on solidification of floating organic drop. *Anal. Chim. Acta* **2007**, *585*, 286–293. [CrossRef]

42. Silva, L.K.; Rangel, J.H.G.; Brito, N.M.; Sousa, E.R.; Sousa, É.M.; Lima, D.L.D.; Esteves, V.I.; Freitas, A.S.; Silva, G.S. Solidified floating organic drop microextraction (SFODME) for the simultaneous analysis of three non-steroidal anti-inflammatory drugs in aqueous samples by HPLC. *Anal. Bioanal. Chem.* **2021**, *413*, 1851–1859. [CrossRef]

43. Moghaddam, T.N.; Elhamirad, A.H.; Asl, M.R.S.; Noghabi, M.S. Pulsed electric field-assisted extraction of phenolic antioxidants from tropical almond red leaves. *Chem. Pap.* **2020**, *74*, 3957–3961. [CrossRef]

44. Golpayegani, M.R.; Akramipour, R.; Fattahi, N. Sensitive determination of deferasirox in blood of patients with thalassemia using dispersive liquid-liquid microextraction based on solidification of floating organic drop followed by HPLC–UV. *J. Pharm. Biomed. Anal.* **2021**, *193*, 113735. [CrossRef] [PubMed]

45. Dadfarnia, S.; Salmanzadeh, A.M.; Shabani, A.M.H. A novel separation/preconcentration system based on solidification of floating organic drop microextraction for determination of lead by graphite furnace atomic absorption spectrometry. *Anal. Chim. Acta* **2008**, *623*, 163–167. [CrossRef] [PubMed]

46. Zhou, X.; Gao, Y.; Zhang, Q.; Li, X.; Li, H. Cloud point extraction coupled with ultrasound-assisted back-extraction for determination of trace legacy and emerging brominated flame retardants in water using isotopic dilution high-performance liquid chromatography-atmospheric pressure chemical ionization-tandem mass spectrometry. *Talanta* **2021**, *224*, 121713. [CrossRef]

47. Canales, R.; Guiñez, M.; Bazán, C.; Reta, M.; Cerutti, S. Determining heterocyclic aromatic amines in aqueous samples: A novel dispersive liquid-liquid micro-extraction method based on solidification of floating organic drop and ultrasound assisted back extraction followed by UPLC-MS/MS. *Talanta* **2017**, *174*, 548–555. [CrossRef] [PubMed]

48. Iqbal, M.; Ezzeldin, E.; Khalil, N.Y.; Alam, P.; Al-Rashood, K.A. UPLC-MS/MS determination of suvorexant in urine by a simplified dispersive liquid-liquid micro-extraction followed by ultrasound assisted back extraction from solidified floating organic droplets. *J. Pharm. Biomed. Anal.* **2019**, *164*, 1–8. [CrossRef] [PubMed]

49. Trujillo-Rodriguez, M.J.; Nan, H.; Varona, M.; Emaus, M.N.; Souza, I.D.; Anderson, J.L. Advances of Ionic Liquids in Analytical Chemistry. *Anal. Chem.* **2019**, *91*, 505–531. [CrossRef]

50. Soares da Silva Burato, J.; Vargas Medina, D.A.; de Toffoli, A.L.; Vasconcelos Soares Maciel, E.; Mauro Lancas, F. Recent advances and trends in miniaturized sample preparation techniques. *J. Sep. Sci.* **2020**, *43*, 202–225. [CrossRef]

51. Lu, D.; Liu, C.; Qin, M.; Deng, J.; Shi, G.; Zhou, T. Functionalized ionic liquids-supported metal organic frameworks for dispersive solid phase extraction of sulfonamide antibiotics in water samples. *Anal. Chim. Acta* **2020**, *1133*, 88–98. [CrossRef]

52. Płotka-Wasylka, J.; De la Guardia, M.; Andruch, V.; Vilková, M. Deep eutectic solvents vs ionic liquids: Similarities and differences. *Microchem. J.* **2020**, *159*, 105539. [CrossRef]

53. Rocha, C.M.R.R.; Genisheva, Z.; Ferreira-Santos, P.; Rodrigues, R.; Vicente, A.A.; Teixeira, J.A.; Pereira, R.N. Electric field-based technologies for valorization of bioresources. *Bioresour. Technol.* **2018**, *254*, 325–339. [CrossRef]

54. Wuethrich, A.; Haddad, P.R.; Quirino, J.P. The electric field—An emerging driver in sample preparation. *TrAC Trends Anal. Chem.* **2016**, *80*, 604–611. [CrossRef]

55. Xu, R.; Lee, H.K. Application of electro-enhanced solid phase microextraction combined with gas chromatography–mass spectrometry for the determination of tricyclic antidepressants in environmental water samples. *J. Chromatogr. A* **2014**, *1350*, 15–22. [CrossRef] [PubMed]

56. Mousa, A.; Basheer, C.; Al-Arfaj, A.R. Application of electro-enhanced solid-phase microextraction for determination of phthalate esters and bisphenol A in blood and seawater samples. *Talanta* **2013**, *115*, 308–313. [CrossRef] [PubMed]

57. Ribeiro, C.C.; Orlando, R.M.; Rohwedder, J.J.; Reyes, F.G.; Rath, S. Electric field-assisted solid phase extraction and cleanup of ionic compounds in complex food matrices: Fluoroquinolones in eggs. *Talanta* **2016**, *152*, 498–503. [CrossRef]

58. Huang, C.; Seip, K.F.; Gjelstad, A.; Pedersen-Bjergaard, S. Electromembrane extraction for pharmaceutical and biomedical analysis-Quo vadis. *J. Pharm. Biomed. Anal.* **2015**, *113*, 97–107. [CrossRef]
59. Gomes-Dias, J.S.; Romaní, A.; Teixeira, J.A.; Rocha, C.M.R.R. Valorization of Seaweed Carbohydrates: Autohydrolysis as a Selective and Sustainable Pretreatment. *ACS Sustain. Chem. Eng.* **2020**, *8*, 17143–17153. [CrossRef]
60. Nowak, P.M.; Wietecha-Posłuszny, R.; Pawliszyn, J. White Analytical Chemistry: An approach to reconcile the principles of Green Analytical Chemistry and functionality. *TrAC Trends Anal. Chem.* **2021**, *138*, 116223. [CrossRef]
61. Chemat, F.; Fabiano-Tixier, A.S.; Vian, M.A.; Allaf, T.; Vorobiev, E. Solvent-free extraction of food and natural products. *TrAC, Trends Anal. Chem.* **2015**, *71*, 157–168. [CrossRef]
62. Khiltash, S.; Heydari, R.; Ramezani, M. Graphene oxide/polydopamine-polyacrylamide nanocomposite as a sorbent for dispersive micro-solid phase extraction of diazinon from environmental and food samples and its determination by HPLC-UV detection. *Int. J. Environ. Anal. Chem.* **2021**. [CrossRef]
63. Barbosa-Pereira, L.; Guglielmetti, A.; Zeppa, G. Pulsed Electric Field Assisted Extraction of Bioactive Compounds from Cocoa Bean Shell and Coffee Silverskin. *Food Bioprocess Technol.* **2018**, *11*, 818–835. [CrossRef]
64. Gazioglu, I.; Zengin, O.S.; Tartaglia, A.; Locatelli, M.; Furton, K.G.; Kabir, A. Determination of polycyclic aromatic hydrocarbons in nutritional supplements by fabric phase sorptive extraction (FPSE) with high-performance liquid chromatography (HPLC) with fluorescence detection. *Anal. Lett.* **2021**, *54*, 1683–1696. [CrossRef]
65. Hrynko, I.; Łozowicka, B.; Kaczyński, P. Development of precise micro analytical tool to identify potential insecticide hazards to bees in guttation fluid using LC-ESI-MS/MS. *Chemosphere* **2021**, *263*, 128143. [CrossRef] [PubMed]
66. Izcara, S.; Casado, N.; Morante-Zarcero, S.; Sierra, I. A miniaturized QuEChERS method combined with ultrahigh liquid chromatography coupled to tandem mass spectrometry for the analysis of pyrrolizidine alkaloids in oregano samples. *Foods* **2020**, *9*, 1319. [CrossRef]
67. Chen, D.; Miao, H.; Zhao, Y.; Wu, Y. A simple liquid chromatography-high resolution mass spectrometry method for the determination of glyphosate and aminomethylphosphonic acid in human urine using cold-induced phase separation and hydrophilic pipette tip solid-phase extraction. *J. Chromatogr. A* **2019**, *1587*, 73–78. [CrossRef]
68. Atirah Mohd Nazir, N.; Raoov, M.; Mohamad, S. Spent tea leaves as an adsorbent for micro-solid-phase extraction of polycyclic aromatic hydrocarbons (PAHs) from water and food samples prior to GC-FID analysis. *Microchem. J.* **2020**, *159*, 105581. [CrossRef]
69. Moufid, M.; Hofmann, M.; El Bari, N.; Tiebe, C.; Bartholmai, M.; Bouchikhi, B. Wastewater monitoring by means of e-nose, VE-tongue, TD-GC-MS, and SPME-GC-MS. *Talanta* **2021**, *221*, 121450. [CrossRef]
70. Prata, M.; Ribeiro, A.; Figueirinha, D.; Rosado, T.; Oppolzer, D.; Restolho, J.; Araújo, A.R.T.S.; Costa, S.; Barroso, M.; Gallardo, E. Determination of opiates in whole blood using microextraction by packed sorbent and gas chromatography-tandem mass spectrometry. *J. Chromatogr. A* **2019**, *1602*, 1–10. [CrossRef]
71. Altunay, N.; Elik, A.; Gürkan, R. Monitoring of some trace metals in honeys by flame atomic absorption spectrometry after ultrasound assisted-dispersive liquid liquid microextraction using natural deep eutectic solvent. *Microchem. J.* **2019**, *147*, 49–59. [CrossRef]
72. Shirkhanloo, H.; Khaleghi Abbasabadi, M.; Hosseini, F.; Faghihi Zarandi, A. Nanographene oxide modified phenyl methanethiol nanomagnetic composite for rapid separation of aluminum in wastewaters, foods, and vegetable samples by microwave dispersive magnetic micro solid-phase extraction. *Food Chem.* **2021**, *347*, 129042. [CrossRef]
73. Nyaba, L.; Nomngongo, P.N. Determination of trace metals in vegetables and water samples using dispersive ultrasound-assisted cloud point-dispersive μ-solid phase extraction coupled with inductively coupled plasma optical emission spectrometry. *Food Chem.* **2020**, *322*, 126749. [CrossRef] [PubMed]
74. Ulrich, S. Solid-phase microextraction in biomedical analysis. *J. Chromatogr. A* **2000**, *902*, 167–194. [CrossRef]
75. Souza-Silva, E.A.; Jiang, R.F.; Rodriguez-Lafuente, A.; Gionfriddo, E.; Pawliszyn, J. A critical review of the state of the art of solid-phase microextraction of complex matrices I. Environmental analysis. *TrAC Trends Anal. Chem.* **2015**, *71*, 224–235. [CrossRef]
76. Roszkowska, A.; Miękus, N.; Bączek, T. Application of solid-phase microextraction in current biomedical research. *J. Sep. Sci.* **2019**, *42*, 285–302. [CrossRef] [PubMed]
77. Silva, C.L.; Perestrelo, R.; Capelinha, F.; Tomás, H.; Câmara, J.S. An integrative approach based on GC–qMS and NMR metabolomics data as a comprehensive strategy to search potential breast cancer biomarkers. *Metabolomics* **2021**, *17*, 72. [CrossRef]
78. Zhang, Y.; Zhao, Y.-G.; Chen, W.-S.; Cheng, H.-L.; Zeng, X.-Q.; Zhu, Y. Three-dimensional ionic liquid-ferrite functionalized graphene oxide nanocomposite for pipette-tip solid phase extraction of 16 polycyclic aromatic hydrocarbons in human blood sample. *J. Chromatogr. A* **2018**, *1552*, 1–9. [CrossRef]
79. Lari, A.; Esmaeili, N.; Ghafari, H. Ionic liquid functionlized on multiwall carbon nanotubes for nickel and lead determination in human serum and urine samples by micro solid-phase extraction. *Anal. Methods Environ. Chem. J.* **2021**, *4*, 72–85. [CrossRef]
80. Lioupi, A.; Kabir, A.; Furton, K.G.; Samanidou, V. Fabric phase sorptive extraction for the isolation of five common antidepressants from human urine prior to HPLC-DAD analysis. *J. Chromatogr. B Anal. Technol. Biomed. Life Sci.* **2019**, *1118–1119*, 171–179. [CrossRef]
81. Musteata, F.M. Recent progress in in-vivo sampling and analysis. *TrAC Trends Anal. Chem.* **2013**, *45*, 154–168. [CrossRef]
82. Cudjoe, E.; Bojko, B.; de Lannoy, I.; Saldivia, V.; Pawliszyn, J. Solid-phase microextraction: A complementary in vivo sampling method to microdialysis. *Angew. Chem. Int. Ed. Engl.* **2013**, *52*, 12124–12126. [CrossRef]

83. Samanidou, V.; Filippou, O.; Marinou, E.; Kabir, A.; Furton, K.G. Sol-gel-graphene-based fabric-phase sorptive extraction for cow and human breast milk sample cleanup for screening bisphenol A and residual dental restorative material before analysis by HPLC with diode array detection. *J. Sep. Sci.* **2017**, *40*, 2612–2619. [CrossRef] [PubMed]
84. Kabir, A.; Furton, K.G.; Tinari, N.; Grossi, L.; Innosa, D.; Macerola, D.; Tartaglia, A.; Di Donato, V.; D'Ovidio, C.; Locatelli, M. Fabric phase sorptive extraction-high performance liquid chromatography-photo diode array detection method for simultaneous monitoring of three inflammatory bowel disease treatment drugs in whole blood, plasma and urine. *J. Chromatogr. B Anal. Technol. Biomed. Life Sci.* **2018**, *1084*, 53–63. [CrossRef] [PubMed]
85. Locatelli, M.; Tinari, N.; Grassadonia, A.; Tartaglia, A.; Macerola, D.; Piccolantonio, S.; Sperandio, E.; D'Ovidio, C.; Carradori, S.; Ulusoy, H.I.; et al. FPSE-HPLC-DAD method for the quantification of anticancer drugs in human whole blood, plasma, and urine. *J. Chromatogr. B Anal. Technol. Biomed. Life Sci.* **2018**, *1095*, 204–213. [CrossRef] [PubMed]
86. Taraboletti, A.; Goudarzi, M.; Kabir, A.; Moon, B.H.; Laiakis, E.C.; Lacombe, J.; Ake, P.; Shoishiro, S.; Brenner, D.; Fornace, A.J., Jr.; et al. Fabric Phase Sorptive Extraction-A Metabolomic Preprocessing Approach for Ionizing Radiation Exposure Assessment. *J. Proteome Res.* **2019**, *18*, 3020–3031. [CrossRef]
87. Manousi, N.; Plastiras, O.E.; Deliyanni, E.A.; Zachariadis, G.A. Green Bioanalytical Applications of Graphene Oxide for the Extraction of Small Organic Molecules. *Molecules* **2021**, *26*, 2790. [CrossRef]
88. Asgharinezhad, A.A.; Ebrahimzadeh, H. Poly(2-aminobenzothiazole)-coated graphene oxide/magnetite nanoparticles composite as an efficient sorbent for determination of non-steroidal anti-inflammatory drugs in urine sample. *J. Chromatogr. A* **2016**, *1435*, 18–29. [CrossRef]
89. Lamei, N.; Ezoddin, M.; Ardestani, M.S.; Abdi, K. Dispersion of magnetic graphene oxide nanoparticles coated with a deep eutectic solvent using ultrasound assistance for preconcentration of methadone in biological and water samples followed by GC–FID and GC–MS. *Anal. Bioanal. Chem.* **2017**, *409*, 6113–6121. [CrossRef]
90. Taghvimi, A.; Hamishehkar, H.; Ebrahimi, M. Magnetic nano graphene oxide as solid phase extraction adsorbent coupled with liquid chromatography to determine pseudoephedrine in urine samples. *J. Chromatogr. B-Anal. Technol. Biomed. Life Sci.* **2016**, *1009*, 66–72. [CrossRef]
91. Barati, A.; Kazemi, E.; Dadfarnia, S.; Shabani, A.M.H. Synthesis/characterization of molecular imprinted polymer based on magnetic chitosan/graphene oxide for selective separation/preconcentration of fluoxetine from environmental and biological samples. *J. Ind. Eng. Chem.* **2017**, *46*, 212–221. [CrossRef]
92. Peng, J.; Tian, H.R.; Du, Q.Z.; Hui, X.H.; He, H. A regenerable sorbent composed of a zeolite imidazolate framework (ZIF-8), Fe_3O_4 and graphene oxide for enrichment of atorvastatin and simvastatin prior to their determination by HPLC. *Mikrochim. Acta* **2018**, *185*, 141. [CrossRef]
93. Zhang, J.; Liu, D.; Meng, X.; Shi, Y.; Wang, R.; Xiao, D.; He, H. Solid phase extraction based on porous magnetic graphene oxide/beta-cyclodextrine composite coupled with high performance liquid chromatography for determination of antiepileptic drugs in plasma samples. *J. Chromatogr. A* **2017**, *1524*, 49–56. [CrossRef] [PubMed]
94. Yuvali, D.; Narin, I.; Soylak, M.; Yilmaz, E. Green synthesis of magnetic carbon nanodot/graphene oxide hybrid material (Fe3O4@C-nanodot@GO) for magnetic solid phase extraction of ibuprofen in human blood samples prior to HPLC-DAD determination. *J. Pharm. Biomed. Anal.* **2020**, *179*, 113001. [CrossRef] [PubMed]
95. Mabrouk, M.M.; Soliman, S.M.; El-Agizy, H.M.; Mansour, F.R. Ultrasound-assisted dispersive liquid–liquid microextraction for determination of three gliflozins in human plasma by HPLC/DAD. *J. Chromatogr. B* **2020**, *1136*, 121932. [CrossRef] [PubMed]
96. Ghoraba, Z.; Aibaghi, B.; Soleymanpour, A. Ultrasound-assisted dispersive liquid-liquid microextraction followed by ion mobility spectrometry for the simultaneous determination of bendiocarb and azinphos-ethyl in water, soil, food and beverage samples. *Ecotoxicol. Environ. Saf.* **2018**, *165*, 459–466. [CrossRef]
97. Kaykhaii, M.; Yavari, E.; Sargazi, G.; Ebrahimi, A.K. Highly Sensitive Determination of Bisphenol A in Bottled Water Samples by HPLC after Its Extraction by a Novel Th-MOF Pipette-Tip Micro-SPE. *J. Chromatogr. Sci.* **2020**, *58*, 373–382. [CrossRef]
98. Alipanahpour Dil, E.; Asfaram, A.; Goudarzi, A.; Zabihi, E.; Javadian, H. Biocompatible chitosan-zinc oxide nanocomposite based dispersive micro-solid phase extraction coupled with HPLC-UV for the determination of rosmarinic acid in the extracts of medical plants and water sample. *Int. J. Biol. Macromol.* **2020**, *154*, 528–537. [CrossRef]
99. Sereshti, H.; Toloutehrani, A.; Nodeh, H.R. Determination of cholecalciferol (vitamin D3) in bovine milk by dispersive micro-solid phase extraction based on the magnetic three-dimensional graphene-sporopollenin sorbent. *J. Chromatogr. B* **2020**, *1136*, 121907. [CrossRef]
100. Mousavi, K.Z.; Yamini, Y.; Karimi, B.; Seidi, S.; Khorasani, M.; Ghaemmaghami, M.; Vali, H. Imidazolium-based mesoporous organosilicas with bridging organic groups for microextraction by packed sorbent of phenoxy acid herbicides, polycyclic aromatic hydrocarbons and chlorophenols. *Microchim. Acta* **2019**, *186*, 239. [CrossRef]
101. Teixeira, R.A.; Dinali, L.A.F.; Silva, C.F.; de Oliveira, H.L.; da Silva, A.T.M.; Nascimento, C.S.; Borges, K.B. Microextraction by packed molecularly imprinted polymer followed by ultra-high performance liquid chromatography for determination of fipronil and fluazuron residues in drinking water and veterinary clinic wastewater. *Microchem. J.* **2021**, *168*, 106405. [CrossRef]
102. Dinali, L.A.F.; de Oliveira, H.L.; Teixeira, L.S.; de Souza Borges, W.; Borges, K.B. Mesoporous molecularly imprinted polymer core@shell hybrid silica nanoparticles as adsorbent in microextraction by packed sorbent for multiresidue determination of pesticides in apple juice. *Food Chem.* **2021**, *345*, 128745. [CrossRef]

103. Aresta, A.; Cotugno, P.; Zambonin, C. Determination of ciprofloxacin, enrofloxacin, and marbofloxacin in bovine urine, serum, and milk by microextraction by a packed sorbent coupled to ultra-high performance liquid chromatography. *Anal. Lett.* **2019**, *52*, 790–802. [CrossRef]
104. Florez, D.H.Â.; de Oliveira, H.L.; Borges, K.B. Polythiophene as highly efficient sorbent for microextraction in packed sorbent for determination of steroids from bovine milk samples. *Microchem. J.* **2020**, *153*, 104521. [CrossRef]
105. Jiang, Y.; Qin, Z.; Song, X.; Piao, H.; Li, J.; Wang, X.; Song, D.; Ma, P.; Sun, Y. Facile preparation of metal organic framework-based laboratory semi-automatic micro-extraction syringe packed column for analysis of parabens in vegetable oil samples. *Microchem. J.* **2020**, *158*, 105200. [CrossRef]
106. Paris, A.; Gaillard, J.L.; Ledauphin, J. Rapid extraction of polycyclic aromatic hydrocarbons in apple: Ultrasound-assisted solvent extraction followed by microextraction by packed sorbent. *Food Anal. Methods* **2019**, *12*, 2194–2204. [CrossRef]
107. Teixeira, L.S.; Silva, C.F.; de Oliveira, H.L.; Dinali, L.A.F.; Nascimento, C.S.; Borges, K.B. Microextraction by packed molecularly imprinted polymer to selectively determine caffeine in soft and energy drinks. *Microchem. J.* **2020**, *158*, 105252. [CrossRef]
108. Casado, N.; Perestrelo, R.; Silva, C.L.; Sierra, I.; Câmara, J.S. Comparison of high-throughput microextraction techniques, MEPS and μ-SPEed, for the determination of polyphenols in baby food by ultrahigh pressure liquid chromatography. *Food Chem.* **2019**, *292*, 14–23. [CrossRef]
109. Kalogiouri, N.P.; Manousi, N.; Rosenberg, E.; Zachariadis, G.A.; Paraskevopoulou, A.; Samanidou, V. Exploring the volatile metabolome of conventional and organic walnut oils by solid-phase microextraction and analysis by GC-MS combined with chemometrics. *Food Chem.* **2021**, *363*, 130331. [CrossRef]
110. Zhao, C.C.; Eun, J.B. Characterization of volatile compounds and physicochemical properties of hongeo using headspace solid-phase microextraction and gas chromatography-mass spectrometry during fermentation. *Food Biosci.* **2021**, *44*, 101379. [CrossRef]
111. Majithia, D.; Metrani, R.; Dhowlaghar, N.; Crosby, K.M.; Patil, B.S. Assessment and classification of volatile profiles in melon breeding lines using headspace solid-phase microextraction coupled with gas chromatography-mass spectrometry. *Plants* **2021**, *10*, 2166. [CrossRef]
112. Thomas, C.F.; Zeh, E.; Dörfel, S.; Zhang, Y.; Hinrichs, J. Studying dynamic aroma release by headspace-solid phase microextraction-gas chromatography-ion mobility spectrometry (HS-SPME-GC-IMS): Method optimization, validation, and application. *Anal. Bioanal. Chem.* **2021**, *413*, 2577–2586. [CrossRef]
113. Zianni, R.; Mentana, A.; Campaniello, M.; Chiappinelli, A.; Tomaiuolo, M.; Chiaravalle, A.E.; Marchesani, G. An investigation using a validated method based on HS-SPME-GC-MS detection for the determination of 2-dodecylcyclobutanone and 2-tetradecylcyclobutanone in X-ray irradiated dairy products. *LWT* **2022**, *153*, 112466. [CrossRef]
114. He, X.; Majid, B.; Zhang, H.; Liu, W.; Limmer, M.A.; Burken, J.G.; Shi, H. Green analysis: Rapid-throughput analysis of volatile contaminants in plants by freeze-thaw-equilibration sample preparation and SPME-GC-MS analysis. *J. Agric. Food. Chem.* **2021**, *69*, 5428–5434. [CrossRef] [PubMed]
115. Passos, C.P.; Petronilho, S.; Serodio, A.F.; Neto, A.C.M.; Torres, D.; Rudnitskaya, A.; Nunes, C.; Kukurova, K.; Ciesarova, Z.; Rocha, S.M.; et al. HS-SPME Gas Chromatography Approach for Underivatized Acrylamide Determination in Biscuits. *Foods* **2021**, *10*, 2183. [CrossRef] [PubMed]
116. Pang, L.; Yang, P.; Pang, R.; Lu, X.; Xiao, J.; Li, S.; Zhang, H.; Zhao, J. Ionogel-based ionic liquid coating for solid-phase microextraction of organophosphorus pesticides from wine and juice samples. *Food Anal. Methods* **2018**, *11*, 270–281. [CrossRef]
117. Perestrelo, R.; Silva, C.L.; Algarra, M.; Câmara, J.S. Evaluation of the occurrence of phthalates in plastic materials used in food packaging. *Appl. Sci.* **2021**, *11*, 2130. [CrossRef]
118. Chen, Y.; Zhang, Y.; Xu, L. A rapid method for analyzing synthetic phenolic antioxidants in food grade lubricant samples based on headspace solid-phase microextracion coupled with gas chromatography-mass spectrometer. *Food Anal. Methods* **2021**, *14*, 2524–2533. [CrossRef]
119. Mejía-Carmona, K.; Lanças, F.M. Modified graphene-silica as a sorbent for in-tube solid-phase microextraction coupled to liquid chromatography-tandem mass spectrometry. Determination of xanthines in coffee beverages. *J. Chromatogr. A* **2020**, *1621*, 461089. [CrossRef]
120. Peng, L.Q.; Zhang, Y.; Yan, T.C.; Gu, Y.X.; Zi, X.; Cao, J. Carbonized biosorbent assisted matrix solid-phase dispersion microextraction for active compounds from functional food. *Food Chem.* **2021**, *365*, 130545. [CrossRef]
121. Gomez-Mejia, E.; Mikkelsen, L.H.; Rosales-Conrado, N.; Leon-Gonzalez, M.E.; Madrid, Y. A combined approach based on matrix solid-phase dispersion extraction assisted by titanium dioxide nanoparticles and liquid chromatography to determine polyphenols from grape residues. *J. Chromatogr. A* **2021**, *1644*, 462128. [CrossRef]
122. Segatto, M.L.; Zanotti, K.; Zuin, V.G. Microwave-assisted extraction and matrix solid-phase dispersion as green analytical chemistry sample preparation techniques for the valorisation of mango processing waste. *Curr. Res. Chem. Biol.* **2021**, *1*, 100007. [CrossRef]
123. Qian, Z.; Wu, Z.; Li, C.; Tan, G.; Hu, H.; Li, W. A green liquid chromatography method for rapid determination of ergosterol in edible fungi based on matrix solid-phase dispersion extraction and a core-shell column. *Anal. Methods* **2020**, *12*, 3327–3343. [CrossRef] [PubMed]
124. Martín-Girela, I.; Albero, B.; Tiwari, B.K.; Miguel, E.; Aznar, R. Screening of contaminants of emerging concern in microalgae food supplements. *Separations* **2020**, *7*, 28. [CrossRef]

125. Souza, M.R.R.; Jesus, R.A.; Costa, J.A.S.; Barreto, A.S.; Navickiene, S.; Mesquita, M.E. Applicability of metal–organic framework materials in the evaluation of pesticide residues in egg samples of chicken (Gallus gallus domesticus). *J. Consum. Prot. Food Saf.* **2021**, *16*, 83–91. [CrossRef]
126. Liang, T.; Gao, L.; Qin, D.; Chen, L. Determination of sulfonylurea herbicides in grain samples by matrix solid-phase dispersion with mesoporous structured molecularly imprinted polymer. *Food Anal. Methods* **2019**, *12*, 1938–1948. [CrossRef]
127. Narimani-Sabegh, S.; Noroozian, E. Magnetic solid-phase extraction and determination of ultra-trace amounts of antimony in aqueous solutions using maghemite nanoparticles. *Food Chem.* **2019**, *287*, 382–389. [CrossRef]
128. Boon, Y.H.; Mohamad Zain, N.N.; Mohamad, S.; Osman, H.; Raoov, M. Magnetic poly(beta-cyclodextrin-ionic liquid) nanocomposites for micro-solid phase extraction of selected polycyclic aromatic hydrocarbons in rice samples prior to GC-FID analysis. *Food Chem.* **2019**, *278*, 322–332. [CrossRef]
129. Chen, J.Y.; Cao, S.R.; Xi, C.X.; Chen, Y.; Li, X.L.; Zhang, L.; Wang, G.M.; Chen, Y.L.; Chen, Z.Q. A novel magnetic β-cyclodextrin modified graphene oxide adsorbent with high recognition capability for 5 plant growth regulators. *Food Chem.* **2018**, *239*, 911–919. [CrossRef]
130. Rahimi, A.; Zanjanchi, M.A.; Bakhtiari, S.; Dehsaraei, M. Selective determination of caffeine in foods with 3D-graphene based ultrasound-assisted magnetic solid phase extraction. *Food Chem.* **2018**, *262*, 206–214. [CrossRef]
131. Mesa, R.; Kabir, A.; Samanidou, V.; Furton, K.G. Simultaneous determination of selected estrogenic endocrine disrupting chemicals and bisphenol A residues in whole milk using fabric phase sorptive extraction coupled to HPLC-UV detection and LC-MS/MS. *J. Sep. Sci.* **2019**, *42*, 598–608. [CrossRef]
132. Ubeda, S.; Aznar, M.; Nerín, C.; Kabir, A. Fabric phase sorptive extraction for specific migration analysis of oligomers from biopolymers. *Talanta* **2021**, *233*, 122603. [CrossRef]
133. Guedes-Alonso, R.; Sosa-Ferrera, Z.; Santana-Rodríguez, J.J.; Kabir, A.; Furton, K.G. Fabric phase sorptive extraction of selected steroid hormone residues in commercial raw milk followed by ultra-high-performance liquid chromatography–tandem mass spectrometry. *Foods* **2021**, *10*, 343. [CrossRef] [PubMed]
134. Agadellis, E.; Tartaglia, A.; Locatelli, M.; Kabir, A.; Furton, K.G.; Samanidou, V. Mixed-mode fabric phase sorptive extraction of multiple tetracycline residues from milk samples prior to high performance liquid chromatography-ultraviolet analysis. *Microchem. J.* **2020**, *159*, 105437. [CrossRef]
135. Campone, L.; Celano, R.; Piccinelli, A.L.; Pagano, I.; Cicero, N.; Sanzo, R.D.; Carabetta, S.; Russo, M.; Rastrelli, L. Ultrasound assisted dispersive liquid-liquid microextraction for fast and accurate analysis of chloramphenicol in honey. *Food Res. Int.* **2019**, *115*, 572–579. [CrossRef] [PubMed]
136. Qiao, L.Z.; Sun, R.T.; Yu, C.M.; Tao, Y.; Yan, Y. Novel hydrophobic deep eutectic solvents for ultrasound-assisted dispersive liquid-liquid microextraction of trace non-steroidal anti-inflammatory drugs in water and milk samples. *Microchem. J.* **2021**, *170*, 106686. [CrossRef]
137. Bernardi, G.; Kemmerich, M.; Adaime, M.B.; Prestes, O.D.; Zanella, R. Miniaturized QuEChERS method for determination of 97 pesticide residues in wine by ultra-high performance liquid chromatography coupled with tandem mass spectrometry. *Anal. Methods* **2020**, *12*, 2682–2692. [CrossRef] [PubMed]
138. Kamal El-Deen, A.; Shimizu, K. Modified μ-QuEChERS coupled to diethyl carbonate-based liquid microextraction for PAHs determination in coffee, tea, and water prior to GC–MS analysis: An insight to reducing the impact of caffeine on the GC–MS measurement. *J. Chromatogr. B Anal. Technol. Biomed. Life Sci.* **2021**, *1171*, 122555. [CrossRef]
139. Casado, N.; Perestrelo, R.; Silva, C.L.; Sierra, I.; Câmara, J.S. An improved and miniaturized analytical strategy based on μ-QuEChERS for isolation of polyphenols. A powerful approach for quality control of baby foods. *Microchem. J.* **2018**, *139*, 110–118. [CrossRef]
140. Abreu, D.C.P.; Botrel, B.M.C.; Bazana, M.J.F.; e Rosa, P.V.; Sales, P.F.; Marques, M.d.S.; Saczk, A.A. Development and comparative analysis of single-drop and solid-phase microextraction techniques in the residual determination of 2-phenoxyethanol in fish. *Food Chem.* **2019**, *270*, 487–493. [CrossRef]
141. Saraji, M.; Javadian, S. Single-drop microextraction combined with gas chromatography-electron capture detection for the determination of acrylamide in food samples. *Food Chem.* **2019**, *274*, 55–60. [CrossRef]
142. Jain, A.; Soni, S.; Verma, K.K. Combined liquid phase microextraction and fiber-optics-based cuvetteless micro-spectrophotometry for sensitive determination of ammonia in water and food samples by the indophenol reaction. *Food Chem.* **2021**, *340*, 128156. [CrossRef]
143. Ma, Z.; Zhao, T.; Cui, S.; Zhao, X.; Fan, Y.; Song, J. Determination of ethyl carbamate in wine by matrix modification-assisted headspace single-drop microextraction and gas chromatography—mass spectrometry technique. *Food Chem.* **2021**, *373*, 131573. [CrossRef] [PubMed]
144. Qi, T.; Xu, M.; Yao, Y.; Chen, W.; Xu, M.; Tang, S.; Shen, W.; Kong, D.; Cai, X.; Shi, H.; et al. Gold nanoprism/Tollens' reagent complex as plasmonic sensor in headspace single-drop microextraction for colorimetric detection of formaldehyde in food samples using smartphone readout. *Talanta* **2020**, *220*, 121388. [CrossRef] [PubMed]
145. Tiwari, S.; Deb, M.K. Modified silver nanoparticles-enhanced single drop microextraction of tartrazine in food samples coupled with diffuse reflectance Fourier transform infrared spectroscopic analysis. *Anal. Methods* **2019**, *11*, 3552–3562. [CrossRef]
146. Neri, T.S.; Rocha, D.P.; Munoz, R.A.A.; Coelho, N.M.M.; Batista, A.D. Highly sensitive procedure for determination of Cu(II) by GF AAS using single-drop microextraction. *Microchem. J.* **2019**, *147*, 894–898. [CrossRef]

147. Tavakoli, M.; Jamali, M.R.; Nezhadali, A. Ultrasound-Assisted Dispersive Liquid–Liquid Microextraction (DLLME) Based on Solidification of Floating Organic Drop Using a Deep Eutectic Solvent for Simultaneous Preconcentration and Determination of Nickel and Cobalt in Food and Water Samples. *Anal. Lett.* **2021**, *54*, 2863–2873. [CrossRef]
148. Shirani, M.; Akbari-adergani, B.; Shahdadi, F.; Faraji, M.; Akbari, A. A Hydrophobic Deep Eutectic Solvent-Based Ultrasound-Assisted Dispersive Liquid–Liquid Microextraction for Determination of β-Lactam Antibiotics Residues in Food Samples. *Food Anal. Methods* **2021**, *15*, 391–400. [CrossRef]
149. Mardani, A.; Torbati, M.; Farajzadeh, M.A.; Mohebbi, A.; Alizadeh, A.A.; Afshar Mogaddam, M.R. Development of temperature-assisted solidification of floating organic droplet-based dispersive liquid–liquid microextraction performed during centrifugation for extraction of organochlorine pesticide residues in cocoa powder prior to GC-ECD. *Chem. Pap.* **2021**, *75*, 1691–1700. [CrossRef]
150. Pataro, G.; Carullo, D.; Ferrari, G. Effect of PEF pre-treatment and extraction temperature on the recovery of carotenoids from tomato wastes. *Chem. Eng. Trans.* **2019**, *75*, 139–144. [CrossRef]
151. Dong, Z.Y.; Wang, H.H.; Li, M.Y.; Liu, W.; Zhang, T.H. Optimization of high-intensity pulsed electric field-assisted extraction of procyanidins from Vitis amurensis seeds using response surface methodology. *E3S Web Conf.* **2020**, *189*, 02029. [CrossRef]
152. Mahn, A.; Comett, R.; Segura-Ponce, L.A.; Díaz-Álvarez, R.E. Effect of pulsed electric field-assisted extraction on recovery of sulforaphane from broccoli florets. *J. Food Process Eng.* **2021**, e13837. [CrossRef]
153. Santos, P.H.; Kammers, J.C.; Silva, A.P.; Oliveira, J.V.; Hense, H. Antioxidant and antibacterial compounds from feijoa leaf extracts obtained by pressurized liquid extraction and supercritical fluid extraction. *Food Chem.* **2021**, *344*, 128620. [CrossRef] [PubMed]
154. Santos, O.V.; Lorenzo, N.D.; Souza, A.L.G.; Costa, C.E.F.; Conceição, L.R.V.; Lannes, S.C.d.S.; Teixeira-Costa, B.E. CO_2 supercritical fluid extraction of pulp and nut oils from Terminalia catappa fruits: Thermogravimetric behavior, spectroscopic and fatty acid profiles. *Food Res. Int.* **2021**, *139*, 109814. [CrossRef] [PubMed]
155. Fornereto Soldan, A.C.; Arvelos, S.; Watanabe, É.O.; Hori, C.E. Supercritical fluid extraction of oleoresin from Capsicum annuum industrial waste. *J. Clean. Prod.* **2021**, *297*, 126593. [CrossRef]
156. Arturo-Perdomo, D.; Mora, J.P.J.; Ibáñez, E.; Cifuentes, A.; Hurtado-Benavides, A.; Montero, L. Extraction and Characterization of the Polar Lipid Fraction of Blackberry and Passion Fruit Seeds Oils Using Supercritical Fluid Extraction. *Food Anal. Methods* **2021**, *14*, 2026–2037. [CrossRef]
157. Jha, A.K.; Sit, N. Comparison of response surface methodology (RSM) and artificial neural network (ANN) modelling for supercritical fluid extraction of phytochemicals from Terminalia chebula pulp and optimization using RSM coupled with desirability function (DF) and genetic. *Ind. Crops Prod.* **2021**, *170*, 113769. [CrossRef]
158. Wang, Y.; Ye, Y.; Wang, L.; Yin, W.; Liang, J. Antioxidant activity and subcritical water extraction of anthocyanin from raspberry process optimization by response surface methodology. *Food Bioscience* **2021**, *44*, 101394. [CrossRef]
159. Pangestuti, R.; Haq, M.; Rahmadi, P.; Chun, B.-s. Nutritional Value and Biofunctionalities of Two Edible Green Seaweeds (Ulva lactuca and Caulerpa racemosa) from Indonesia by Subcritical Water Hydrolysis. *Mar. Drugs* **2021**, *19*, 578. [CrossRef]
160. Hwang, H.J.; Kim, H.J.; Ko, M.J.; Chung, M.S. Recovery of hesperidin and narirutin from waste Citrus unshiu peel using subcritical water extraction aided by pulsed electric field treatment. *Food Sci. Biotechnol.* **2021**, *30*, 217–226. [CrossRef]
161. Pinto, D.; Vieira, E.F.; Peixoto, A.F.; Freire, C.; Freitas, V.; Costa, P.; Delerue-Matos, C.; Rodrigues, F. Optimizing the extraction of phenolic antioxidants from chestnut shells by subcritical water extraction using response surface methodology. *Food Chem.* **2021**, *334*, 127521. [CrossRef]
162. Jamaludin, R.; Kim, D.S.; Salleh, L.M.; Lim, S.B. Kinetic study of subcritical water extraction of scopoletin, alizarin, and rutin from morinda citrifolia. *Foods* **2021**, *10*, 2260. [CrossRef]
163. Rodrigues, L.A.; Matias, A.A.; Paiva, A. Recovery of antioxidant protein hydrolysates from shellfish waste streams using subcritical water extraction. *Food Bioprod. Process.* **2021**, *130*, 154–163. [CrossRef]
164. Zhang, F.; Zhang, L.; Chen, J.; Du, X.; Lu, Z.; Wang, X.; Yi, Y.; Shan, Y.; Liu, B.; Zhou, Y.; et al. Systematic evaluation of a series of pectic polysaccharides extracted from apple pomace by regulation of subcritical water conditions. *Food Chem.* **2022**, *368*, 130833. [CrossRef] [PubMed]
165. Darvishnejad, M.; Ebrahimzadeh, H. Graphitic carbon nitride-reinforced polymer ionic liquid nanocomposite: A novel mixed-mode sorbent for microextraction in packed syringe. *Int. J. Environ. Anal. Chem.* **2020**, 1–14. [CrossRef]
166. Saraji, M.; Jafari, M.T.; Amooshahi, M.M. Sol-gel/nanoclay composite as a sorbent for microextraction in packed syringe combined with corona discharge ionization ion mobility spectrometry for the determination of diazinon in water samples. *J. Sep. Sci.* **2018**, *41*, 493–500. [CrossRef]
167. Moradi, E.; Mehrani, Z.; Ebrahimzadeh, H. Gelatin/sodium triphosphate hydrogel electrospun nanofiber mat as a novel nanosorbent for microextraction in packed syringe of La3+ and Tb3+ ions prior to their determination by ICP-OES. *React. Funct. Polym.* **2020**, *153*, 104627. [CrossRef]
168. Taghani, A.; Goudarzi, N.; Bagherian, G.A.; Arab Chamjangali, M.; Amin, A.H. Application of nanoperlite as a new natural sorbent in the preconcentration of three organophosphorus pesticides by microextraction in packed syringe coupled with gas chromatography and mass spectrometry. *J. Sep. Sci.* **2018**, *41*, 2245–2252. [CrossRef]
169. Matin, P.; Ayazi, Z.; Jamshidi-Ghaleh, K. Montmorillonite reinforced polystyrene nanocomposite supported on cellulose as a novel layered sorbent for microextraction by packed sorbent for determination of fluoxetine followed by spectrofluorimetry based on multivariate optimisation. *Int. J. Environ. Anal. Chem.* **2020**, 1–16. [CrossRef]

170. Arcoleo, A.; Bianchi, F.; Careri, M. A sensitive microextraction by packed sorbent-gas chromatography-mass spectrometry method for the assessment of polycyclic aromatic hydrocarbons contamination in Antarctic surface snow. *Chemosphere* **2021**, *282*, 131082. [CrossRef]

171. Amiri, A.; Chahkandi, M.; Targhoo, A. Synthesis of nano-hydroxyapatite sorbent for microextraction in packed syringe of phthalate esters in water samples. *Anal. Chim. Acta* **2017**, *950*, 64–70. [CrossRef]

172. Vera, J.; Fernandes, V.C.; Correia-Sá, L.; Mansilha, C.; Delerue-Matos, C.; Domingues, V.F. Occurrence of Selected Known or Suspected Endocrine-Disrupting Pesticides in Portuguese Surface Waters Using SPME-GC-IT/MS. *Separations* **2021**, *8*, 81. [CrossRef]

173. Saliu, F.; Montano, S.; Hoeksema, B.W.; Lasagni, M.; Galli, P. A non-lethal SPME-LC/MS method for the analysis of plastic-associated contaminants in coral reef invertebrates. *Anal. Methods* **2020**, *12*, 1935–1942. [CrossRef]

174. Terzaghi, E.; Falakdin, P.; Fattore, E.; Di Guardo, A. Estimating temporal and spatial levels of PAHs in air using rain samples and SPME analysis: Feasibility evaluation in an urban scenario. *Sci. Total Environ.* **2021**, *762*, 144184. [CrossRef] [PubMed]

175. El-Sheikh, A.H.; Al-Quse, R.W.; El-Barghouthi, M.I.; Al-Masri, F.S. Derivatization of 2-chlorophenol with 4-amino-anti-pyrine: A novel method for improving the selectivity of molecularly imprinted solid phase extraction of 2-chlorophenol from water. *Talanta* **2010**, *83*, 667–673. [CrossRef] [PubMed]

176. Anirudhan, T.S.; Alexander, S. Multiwalled carbon nanotube based molecular imprinted polymer for trace determination of 2,4-dichlorophenoxyaceticacid in natural water samples using a potentiometric method. *Appl. Surf. Sci.* **2014**, *303*, 180–186. [CrossRef]

177. Shaikh, H.; Memon, N.; Bhanger, M.I.; Nizamani, S.M.; Denizli, A. Core-shell molecularly imprinted polymer-based solid-phase microextraction fiber for ultra trace analysis of endosulfan I and II in real aqueous matrix through gas chromatography-micro electron capture detector. *J. Chromatogr. A* **2014**, *1337*, 179–187. [CrossRef] [PubMed]

178. Gao, X.; Pan, M.; Fang, G.; Jing, W.; He, S.; Wang, S. An ionic liquid modified dummy molecularly imprinted polymer as a solid-phase extraction material for the simultaneous determination of nine organochlorine pesticides in environmental and food samples. *Anal. Methods* **2013**, *5*, 6128–6134. [CrossRef]

179. Zhao, F.; Wang, S.; She, Y.; Zhang, C.; Zheng, L.; Jin, M.; Shao, H.; Jin, F.; Du, X.; Wang, J. Subcritical water extraction combined with molecular imprinting technology for sample preparation in the detection of triazine herbicides. *J. Chromatogr. A* **2017**, *1515*, 17–22. [CrossRef]

180. Stoeckelhuber, M.; Müller, C.; Vetter, F.; Mingo, V.; Lötters, S.; Wagner, N.; Bracher, F. Determination of Pesticides Adsorbed on Arthropods and Gastropods by a Micro-QuEChERS Approach and GC–MS/MS. *Chromatographia* **2017**, *80*, 825–829. [CrossRef]

181. Kurth, D.; Krauss, M.; Schulze, T.; Brack, W. Measuring the internal concentration of volatile organic compounds in small organisms using micro-QuEChERS coupled to LVI-GC-MS/MS. *Anal. Bioanal. Chem.* **2017**, *409*, 6041–6052. [CrossRef]

182. Mehravar, A.; Feizbakhsh, A.; Sarafi, A.H.M.; Konoz, E.; Faraji, H. Deep eutectic solvent-based headspace single-drop microextraction of polycyclic aromatic hydrocarbons in aqueous samples. *J. Chromatogr. A* **2020**, *1632*, 461618. [CrossRef]

183. Kiszkiel-Taudul, I.; Starczewska, B. Single drop microextraction coupled with liquid chromatography-tandem mass spectrometry (SDME-LC-MS/MS) for determination of ranitidine in water samples. *Microchem. J.* **2019**, *145*, 936–941. [CrossRef]

184. Nunes, L.S.; Korn, M.G.A.; Lemos, V.A. A novel direct-immersion single-drop microextraction combined with digital colorimetry applied to the determination of vanadium in water. *Talanta* **2021**, *224*, 121893. [CrossRef] [PubMed]

185. Werner, J. Low Density Ionic Liquid-Based Ultrasound-Assisted Dispersive Liquid–Liquid Microextraction for the Preconcentration of Trace Aromatic Amines in Waters. *J. Anal. Chem.* **2021**, *76*, 1182–1188. [CrossRef]

186. Yang, S.; Liu, H.; Hu, K.; Deng, Q.; Wen, X. Investigation of thermospray flame furnace atomic absorption spectrometric determination of cadmium combined with ultrasound-assisted dispersive liquid-liquid microextraction. *Int. J. Environ. Anal. Chem.* **2022**, *102*, 443–455. [CrossRef]

187. Ali, J.; Tuzen, M.; Citak, D.; Uluozlu, O.D.; Mendil, D.; Kazi, T.G.; Afridi, H.I. Separation and preconcentration of trivalent chromium in environmental waters by using deep eutectic solvent with ultrasound-assisted based dispersive liquid-liquid microextraction method. *J. Mol. Liq.* **2019**, *291*, 111299. [CrossRef]

188. Shojaei, S.; Shojaei, S.; Nouri, A.; Baharinikoo, L. Application of chemometrics for modeling and optimization of ultrasound-assisted dispersive liquid–liquid microextraction for the simultaneous determination of dyes. *npj Clean Water* **2021**, *4*, 23. [CrossRef]

189. Xizhi, S.; Sun, A.-l.; Wang, Q.-h.; Hengel, M.; Shibamoto, T. Rapid Multi-Residue Analysis of Herbicides with Endocrine-Disrupting Properties in Environmental Water Samples Using Ultrasound-Assisted Dispersive Liquid–Liquid Microextraction and Gas Chromatography–Mass Spectrometry. *Chromatographia* **2018**, *81*, 1071–1083. [CrossRef]

190. Wang, X.M.; Du, T.T.; Wang, J.; Kou, H.X.; Du, X.Z. Determination of polybrominated biphenyls in environmental water samples by ultrasound-assisted dispersive liquid-liquid microextraction followed by high-performance liquid chromatography. *Microchem. J.* **2019**, *148*, 85–91. [CrossRef]

191. Xinya, L.; Liu, C.; Qian, H.; Qu, Y.; Zhang, S.; Lu, R.; Gao, H.; Zhou, W. Ultrasound-assisted dispersive liquid-liquid microextraction based on a hydrophobic deep eutectic solvent for the preconcentration of pyrethroid insecticides prior to determination by high-performance liquid chromatography. *Microchem. J.* **2019**, *146*, 614–621. [CrossRef]

192. Rahimi Moghadam, M.; Zargar, B.; Rastegarzadeh, S. Determination of Tetracycline Using Ultrasound-Assisted Dispersive Liquid-Liquid Microextraction Based on Solidification of Floating Organic Droplet Followed by HPLC-UV System. *J. AOAC Int.* **2021**, *104*, 999–1004. [CrossRef]
193. Fayaz, S.M.; Abdoli, M.A.; Baghdadi, M.; Karbasi, A. Ag removal from e-waste using supercritical fluid: Improving efficiency and selectivity. *Int. J. Environ. Stud.* **2021**, *78*, 459–473. [CrossRef]
194. Falsafi, Z.; Raofie, F.; Kazemi, H.; Ariya, P.A. Simultaneous extraction and fractionation of petroleum biomarkers from tar balls and crude oils using a two-step sequential supercritical fluid extraction. *Mar. Pollut. Bull.* **2020**, *159*, 111484. [CrossRef] [PubMed]
195. Meskar, M.; Sartaj, M.; Infante Sedano, J.A. Assessment and comparison of PHCs removal from three types of soils (sand, silt loam and clay) using supercritical fluid extraction. *Environ. Technol.* **2019**, *40*, 3040–3053. [CrossRef] [PubMed]
196. Tita, G.J.; Navarrete, A.; Martin, A.; Cocero, M.J. Model assisted supercritical fluid extraction and fractionation of added-value products from tobacco scrap. *J. Supercrit. Fluids* **2021**, *167*, 105046. [CrossRef]
197. Lie, J.; Tanda, S.; Liu, J.-C. Subcritical Water Extraction of Valuable Metals from Spent Lithium-Ion Batteries. *Molecules* **2020**, *25*, 2166. [CrossRef]
198. Taki, G.; Islam, M.N.; Park, S.-J.; Park, J.-H. Optimization of operating parameters to remove and recover crude oil from contaminated soil using subcritical water extraction process. *Environ. Eng. Res.* **2018**, *23*, 175–180. [CrossRef]
199. Kang, S.J.; Sun, Y.H.; Qiao, M.Y.; Li, S.L.; Deng, S.H.; Guo, W.; Li, J.S.; He, W.T. The enhancement on oil shale extraction of $FeCl_3$ catalyst in subcritical water. *Energy* **2022**, *238*, 121763. [CrossRef]
200. Zohar, M.; Matzrafi, M.; Abu-Nassar, J.; Khoury, O.; Gaur, R.Z.; Posmanik, R. Subcritical water extraction as a circular economy approach to recover energy and agrochemicals from sewage sludge. *J. Environ. Manag.* **2021**, *285*, 112111. [CrossRef]

Review

Factors Affecting Preparation of Molecularly Imprinted Polymer and Methods on Finding Template-Monomer Interaction as the Key of Selective Properties of the Materials

Aliya Nur Hasanah [1,2,*], Nisa Safitri [1], Aulia Zulfa [1], Neli Neli [1] and Driyanti Rahayu [1,2]

1   Department of Pharmaceutical Analysis and Medicinal Chemistry, Faculty of Pharmacy,
    Universitas Padjadjaran, Jl. Raya Bandung Sumedang KM 21.5, Sumedang 45363, Indonesia;
    nisa18005@mail.unpad.ac.id (N.S.); aulia18007@mail.unpad.ac.id (A.Z.); neli18001@mail.unpad.ac.id (N.N.);
    driyanti.rahayu@unpad.ac.id (D.R.)
2   Drug Development Study Center, Faculty of Pharmacy, Universitas Padjadjaran,
    Jl. Raya Bandung Sumedang KM 21.5, Sumedang 45363, Indonesia
*   Correspondence: aliya.n.hasanah@unpad.ac.id; Tel.: +62-812-2346-382

Citation: Hasanah, A.N.; Safitri, N.; Zulfa, A.; Neli, N.; Rahayu, D. Factors Affecting Preparation of Molecularly Imprinted Polymer and Methods on Finding Template-Monomer Interaction as the Key of Selective Properties of the Materials. *Molecules* **2021**, *26*, 5612. https://doi.org/10.3390/molecules26185612

Academic Editors: Victoria Samanidou and Irene Panderi

Received: 22 August 2021
Accepted: 14 September 2021
Published: 16 September 2021

Publisher's Note: MDPI stays neutral with regard to jurisdictional claims in published maps and institutional affiliations.

Abstract: Molecular imprinting is a technique for creating artificial recognition sites on polymer matrices that complement the template in terms of size, shape, and spatial arrangement of functional groups. The main advantage of Molecularly Imprinted Polymers (MIP) as the polymer for use with a molecular imprinting technique is that they have high selectivity and affinity for the target molecules used in the molding process. The components of a Molecularly Imprinted Polymer are template, functional monomer, cross-linker, solvent, and initiator. Many things determine the success of a Molecularly Imprinted Polymer, but the Molecularly Imprinted Polymer component and the interaction between template-monomers are the most critical factors. This review will discuss how to find the interaction between template and monomer in Molecularly Imprinted Polymer before polymerization and after polymerization and choose the suitable component for MIP development. Computer simulation, UV-Vis spectroscopy, Fourier Transform Infrared Spectroscopy (FTIR), Proton-Nuclear Magnetic Resonance (^{1}H-NMR) are generally used to determine the type and strength of intermolecular interaction on pre-polymerization stage. In turn, Suspended State Saturation Transfer Difference High Resolution/Magic Angle Spinning (STD HR/MAS) NMR, Raman Spectroscopy, and Surface-Enhanced Raman Scattering (SERS) and Fluorescence Spectroscopy are used to detect chemical interaction after polymerization. Hydrogen bonding is the type of interaction that is becoming a focus to find on all methods as this interaction strongly contributes to the affinity of molecularly imprinted polymers (MIPs).

Keywords: molecular imprinted polymer; interaction mechanism; template-monomer interaction; MIP-template interaction

1. Introduction

Molecular imprinting is a technique for creating artificial recognition sites on polymer matrices that complement the template in terms of size, shape, and spatial arrangement of functional groups. This molecular imprinting technique uses target molecules in a synthetic polymer matrix by selectively binding [1,2]. Recently, molecular imprinting technology has been used to create biometric surfaces in biosensors. There are many molecular imprinting technologies, including bulk printing, surface printing, and epitope printing.

In the bulk printing method, the template molecules are printed on the entire polymer matrix, and at the end of the method, the template needs to be removed from the polymer. The polymer produced in this method is large or bulk, so grinding must be carried out on the polymer to obtain template-specific binding sites on the polymer [1]. Due to the thick morphology of the polymer in the bulk printing, it causes low access for the target molecule to bind to its specific site. Therefore, another method was developed to overcome this limitation, namely the surface printing method and the epitope printing method.

In surface printing, the removal of template molecules will result in specific binding sites on the polymer surface [3]. The binding site on the polymer surface causes this type of polymer to provide greater access to the binding target molecule than bulk imprinting [1]. This technique has been widely used for various types of analytes such as proteins [4], cells [5], and micro-organisms [6]. While in epitope printing, the target molecule is a protein and uses only a small portion or fragment of the macromolecule is printed to represent the whole molecule (epitope) as a template [1]. In this method, the peptide epitope is covalently bonded to the silicon surface where the monomer is polymerized. In epitope printing, more specific and strong interactions can be obtained. The polymer has the ability to recognize templates as well as whole proteins very well [7].

Molecularly Imprinted Polymer (MIP), which contains specific bonds between template molecules and polymers [8], is an example of materials that use molecular imprinting techniques. MIP is a unique recognition system resulting from templates and functional monomers that are polymerized, enabling molecular recognition utilizing principles similar to those underlying the action of enzymes and their substrates [9,10]. The main advantage of MIP is high selectivity and affinity for the target molecules used in the molding process. Compared to biological systems such as proteins and nucleic acids, imprinted polymers have higher physical strength, high temperature, pressure resistance, and inertia to acids, alkalis, metal ions, and organic solvents. In addition, the synthesis cost is low, the storage life of the polymer can be very high, and the recognition capability can be maintained for several years at room temperature.

Templates, functional monomers, solvent, initiator, and cross-linker are MIP components [11]. MIP is based on the formation of complexes between analytes (templates) and monomers of functional compounds. In the presence of excess cross-linking agents, three-dimensional polymers are formed [12]. In the process of making MIP, the selection of the constituent components will affect the performance of the resulting imprinted polymer. A functional monomer is preferred over an ordinary monomer because a functional monomer contains a Y functional group that can interact with template molecule via hydrogen bonding, dipole-dipole, and ionic interaction to produce a template-monomer complex. The complex is then fixed in the presence of a large excess of a cross-linking agent, and a three-dimensional polymer network is formed. After the polymerization process, template molecules are removed from the polymer using a solvent, resulting in selective complementary polymer-template bonds [11,13]. The scheme of the molecular imprinting process can be seen in Figure 1.

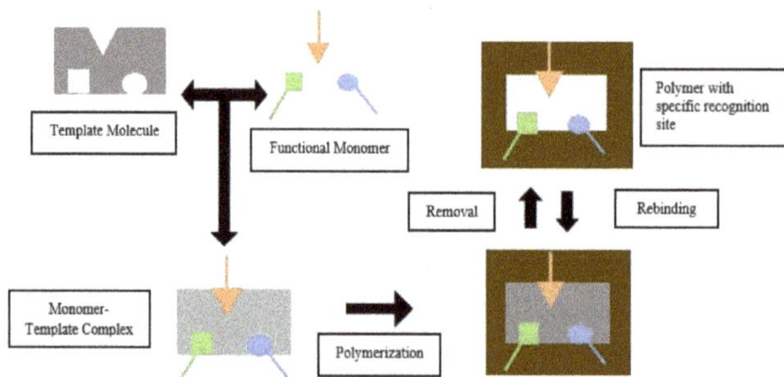

Figure 1. Molecular Imprinting Process (green, orange and blue arrow indicated different type of functional groups that could be appeared in the functional monomer).

Several interactions occur in the molecular imprinting process, like hydrogen bonds, dipole-dipole, and ionic interactions. The interactions between the template molecule and functional groups present in the polymer matrix drive the molecular recognition phenomena. Thus, the resultant polymer recognizes and binds selectively to the template molecules. As interaction is the driving force for molecular recognition, finding the chemical interaction that happened while developing MIP is crucial to be known. According to a literature search, no review has been made on methods to find chemical interaction between template-functional monomers before and after polymerization. For that reason, this review will discuss the factors that influence the success of making MIP, especially on choosing the suitable component to form MIP and methods to determine the type of interactions happening in MIP. We will divide the method on finding the type of interaction in MIP into two stages: pre-polymerization and post-polymerization [14–17]. Thus, we will discuss computer simulation, UV-Vis Spectroscopy, Fourier transform infrared spectroscopy (FTIR), Proton Nuclear Magnetic Resonance (^{1}H-NMR) for methods on finding the type of interaction in the pre-polymerization stage. In contrast, as a method for detecting the type of interaction after polymerization, we will discuss Suspended State Saturation Transfer Difference High Resolution/Magic Angle Spinning (STD HR/MAS) NMR, Raman Spectroscopy, Surface-Enhanced Raman scattering (SERS), and Fluorescence Spectroscopy.

2. MIP Application

2.1. Environmental Monitoring

The increasing contaminants in the environmental water is an alarming issue [18]. Environmental monitoring can be carried out on environmentally hazardous materials such as dyes [19], persistent organic pollutants [20], pesticide residues [21], mycotoxins [22], heavy metals, and antibiotics [23]. The presence of these contaminants can cause harmful effects on public health, either directly or indirectly [24]. Environmental matrices that can be polluted by pollutants include air, soil, atmosphere, sediment, and flora and fauna [25,26]. The pre-treatment sample aims to eliminate matrix interference so that only the target analyte is obtained [27]. MIP is commonly used as a sorbent in pre-treatment samples in samples with complex matrices because of its selected properties compared to conventional sorbents [26].

One of the applications was presented by Song et al. [28] that analyzed ten macrolide drugs (spiramycin, clarithromycin, erythromycin, tulathromycin, midecamycin, roxithromycin, josamycin, kitasamycin, tilmicosin, and azithromycin) in environmental water. MIP with functional monomer methacrylic acid and tylosin as a template was used as a sorbent in the molecular imprinted solid-phase extraction (MI-SPE) pre-treatment technique and then analyzed using the LC-MS/MS method. As a result, MI-SPE showed a high recognition ability of macrolides. The mean percentage recovery of macrolides at four spiked concentration levels was 62.6–100.9%, with intra-day and inter-day relative standard deviations below 12.6%.

2.2. Food Analysis

Contaminants from the environment can enter the body through food or drinking water and have the potential to cause harmful effects on health [29]. For example, insecticide and herbicide contaminants used in agriculture include fruits, vegetables, and cereals. Effective analytical methods and technologies are needed to ensure food safety. Food is a complex matrix sample, so MIP can be used as a sorbent in the analysis process [30].

Garcia et al. [31] assessed dimethoate spiked in olive oil using core-shell magnetic-photonic dual responsive molecularly imprinted polymers (magnetic-photonic DR-MIP) as sorbent. Sample preparation using a DR-MIP-based sorbent is superior to the MI-SPE technique in terms of the procedure and minimized processing time. As a result, DR-MIP was declared a promising sorbent for the analysis of spiked dimethoate olive oil in sample preparation because it produced a high percentage recovery of 83.5% ± 0.3% with a low detection limit of 0.03 μg/mL [31].

Besides olive oil, MIP is also used for the analysis of macrolide antibiotics in honey, milk, and drinking water samples [32], porcine serum albumin in raw meat extract [33], lincomycin residue in pasteurized milk [34], triazine pesticides on cereals [35], fluoro-quinolones in fish samples [36], and so on.

2.3. Biomedical Diagnostic

In biomedical applications, MIP is known as an artificial receptor that has basic capabilities like natural receptors, one of which is the ability to recognize cells [37]. Compared to natural receptors, MIP has advantages such as high affinity and selectivity, more temperature stability, rapid preparation, and low costs [38]. MIPs are designed to have specific recognition sites for target molecules such as antibodies and enzymes [38,39]. In the immune system, antibodies must be specific in recognizing certain antigens. MIP as an alternative to antibodies is often used as an analytical method for diagnostic purposes [40], such as breast cancer diagnostics [41], cardiovascular disease [42], and dengue fever [43]. Therefore, MIP acts as a biomarker [44] because it is able to show certain physical characteristics or measurable biologically generated changes in the body associated with certain diseases or health conditions [45].

2.4. Drug Delivery

Drug-delivery systems (DDS) are a method of delivering drugs to the desired place before drug release and absorption so as to increase the pharmacological activity of the drug and achieve the desired therapeutic effect [46,47]. DDS must be able to control the amount and speed of drug release [46]. MIP as a DDS agent offers several advantages, namely long shelf life, easy preparation, high chemical, and physical stability, and low cost [39].

Suksuwan et al. [48] evaluated the ability of drug delivery to cancer cells using an enantioselective receptor for (R)-thalidomide enantiomer designed in nanoparticles using methacrylic acid, a fluorescently active 2,6-bis(acrylamido)pyridine and N,N_0-methylene-bis-acrylamide, via both a covalent approach and a physical approach. The results of his research show that MIP nanoparticles, through a physical approach, have the potential to make effective drugs to attack multidrug-resistant cells with the right temperature at the target location.

3. Choosing Right Component for MIP

Many factors can influence the success of making molecularly imprinted polymers. These factors include the properties of monomers, cross-linkers, and solvents [17,49,50]. The properties of monomers, cross-linker, and solvents will affect the morphology and size of the polymer formed and template-monomer interactions.

3.1. Functional Monomers

Functional monomers are one of the components of MIP whose properties will influence the success of making a MIP. Functional monomers are an essential factor for binding interactions in molecular imprinting technology [51]. The functional monomer will influence the binding site affinity of a MIP [52]. In the MIP pre-polymerization, the functional monomers are going to interact with the template molecules [11]. The formation of a stable template-monomer complex is critical to the success of MIP [53,54]. It is commonly accepted that stronger interaction between the template and functional monomer will result in a more stable template-monomer complex prior to polymerization, and consequently, the better imprinting efficiency of the polymer resulting will be [55–57]. Monomers are positioned spatially around the template, and due to cross-linking, the position of the individual repeat units that interact with template molecules will become fixed. Polymer microporous matrix possessing microcavities will be formed with a three-dimensional structure complementary to that of the template [11,58].

Research conducted by Zhong M. et al. [57] on the preparation of magnetic molecularly imprinted polymers for isolation of chelerythrine (CHE) showed that stable complex between CHE as a template (T) and MAA as a functional monomer (FM) in a 1:4 ratio gave the lowest binding energy in computational calculation compared to other ratios (1:1; 1:2; 1:3; and 1:5). The lowest binding energy indicated the highest complex stability. When the imprinting ratio (template:monomer) was 1:5, the excess monomers would increase their own association. At the same time, the spatial structure of the complex was unstable due to the space steric hindrance between the template and the functional monomer molecules. The result of the wet laboratory experiment also showed that the ratio (T:FM = 1:4) gave higher selectivity and adsorption capacity for template molecules toward analogous compounds.

Three types of functional monomers are most commonly used: (i) acidic compounds, such as methacrylic acid, (ii) alkaline compounds, such as 4-vinyl pyridine, and (iii) pH neutral compounds, such as styrene [59]. Some typical functional monomers are listed in Figure 2.

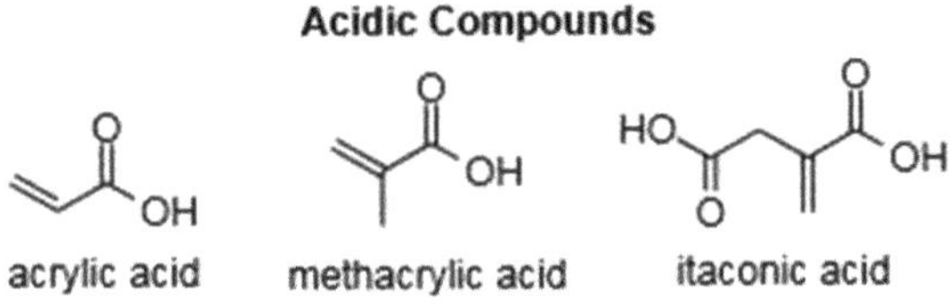

Figure 2. Chemical Structure of Functional Monomers.

A sufficiently high number of functional monomers is required to ensure a high binding capacity for the target molecule [60]. If the amount of monomer is too large, it usually results in a more non-specific interaction site [50]. The use of correct proportions of monomers and templates in the pre-polymerization process will lead to the formation of many high-affinity sites [60]. The effect of monomer amount on MIP was studied by Zhao et al. [61]. They studied the effect of amount variation of methyl methacrylate (MMA) as a functional monomer on the adsorption performance of a Solasenol MIP (SSO-MIP). Their research proved that the adsorption performance of a MIP would reach its highest point at a certain monomer concentration, and increasing the concentration after reaching the optimum point caused a decreased adsorption capacity value (Q). The highest point of SSO-MIP adsorption capacity, 43 mg/g, occurred when the MMA concentration was 0.2 mmol. When the MMA concentration was increased to 0.25 mmol, the adsorption capacity decreased to 41 mg/g, as well as at concentrations of 0.30 mmol (Q = 37 mg/g) and 0.35 mmol (Q = 36 mg/g) [61].

The interaction between template and functional monomer must be interdependent, i.e., based on the nature of the template/analyte and its functional monomer (acid analyte:

alkaline functional monomers; alkaline analyte: functional acid monomer) [62], like the study of Xu et al. [63] who synthesized MIP Hexamethylenetetramine. Hexamethylenetetramine is an alkaline compound, and they used methacrylic acid as the monomer. Meanwhile, in another study by Zunngu et al. [64], they succeeded in synthesizing MIP with ketoprofen (acid) as a template and using 2-vinyl pyridine (alkaline) as a monomer.

The choice of functional monomer in MIP synthesis should also be based on the interactions formed by the template based on the functional groups in the selected template. Like the study of Barros et al. [50], who synthesized MIP using hydrochlorothiazide as a template. Hydrochlorothiazide has three interaction sites, one site on the thiazide ring and two sites on the sulfonamide group. These groups can form hydrogen bonds with carboxyl, hydroxyl, amine, and even amide groups. Therefore, various monomers (methacrylic acid, allylamine, 4-vinyl pyridine, methacrylamide, acrylamide, acrylic acid, and 2-(trifluoromethyl)acrylic acid) were investigated, which may interact through hydrogen bonding with hydrochlorothiazide. The choice of monomer is based on the interaction energy formed between the monomer and hydrochlorothiazide through computational simulations. It is known that methacrylic acid is the best monomer for synthesizing MIP hydrochlorothiazide because it has the highest interaction energy compared to other monomers [50].

3.2. Cross-Linker

The cross-linker, as a reagent of MIP, has a significant role in the formation of MIP. The role of the cross-linker is to participate in the formation of the physical characteristics of polymers in MIP. The physical characteristics of the polymer affected by the presence of the cross-linker are the morphological characteristics of the absorbent structure, three-dimensional structure, optimal stiffness, and the durability of the MIP [59,65].

There are several important factors in the use of cross-linkers, including the type and number of cross-linkers used for polymerization. The higher number of cross-linkers makes it possible to obtain a stable porous material [17]. A study was carried out by Zhao et al. [61] to see the effect of cross-linker amount on adsorption performances of a MIP. Their research proved that the adsorption performance of a MIP would increase to a certain concentration of cross-linker. After reaching optimum conditions, the further addition of the cross-linker will not increase the adsorption performance; it will even decrease the adsorption performance. The chemical compounds most often used as cross-linkers are trimethylolpropane trimethacrylate (TRIM), divinylbenzene (DVB), and ethylene glycol dimethacrylate (EGDMA) [59,66].

The presence of a cross-linker also affects the morphology of MIP that will affect the binding capacity of MIP. Studies conducted by Holland et al. [67] and Rosengren et al. [68] showed similar results that lower cross-linker concentrations provide greater binding capacity. This phenomenon occurs because the mean pore diameter decreased with increasing EGDMA, which may have resulted in a greater sieving ability of these polymers, which leads to lower overall capacity. Several studies have shown a comparison of the cross-linkers commonly used in MIP. Esfandyari-Manesh et al. [66] compared the use of TRIM and EGDMA in MIP with methacrylic acid (MAA) as functional monomer and carbamazepine (CBZ) as the template. Their study showed that MIP using TRIM (MAA-TRIM) as a cross-linker showed better binding capacity and had more carbamazepine recognition sites compared to MAA-EGDMA, with MAA-TRIM binding capacity > 300 (around 300–500) micrograms CBZ while MAA-EGDMA binding capacity was around 300 micrograms CBZ, both measured at 180 min. Another study from Pangkamta et al. [69] also, MIP with TRIM as a cross-linker showed a higher binding capacity percentage of 45–50% than MIP with EGDMA as cross-linker, which only shows 40–45%. The structure of TRIM and EGDMA possibly influences this. In the TRIM molecule, three-branched chains contain three vinyl groups, making TRIM more preferable in the polymerization process and resulting in more rigid complementary recognition sites for the template than the EGDMA molecule only has two vinyl groups. The contribution of the cross-linker, of course, also depends on other components, such as using the suitable functional monomer for the template or selecting the right solvent in the polymerization process [66,69].

3.3. Solvents

Solvents act as a medium for reactions and have a significant effect on template-monomer interactions, which are certain components that must be considered to make a MIP. Solvents in the making of MIP must interact and dissolve all the starting materials but should not interfere too much during the polymerization reaction process [60]. When a high solvation value is used in the synthesis process of a MIP, the solvent will protect the molecular interaction site and weaken the strong interaction between the monomer and template, thus making the molecular recognition ability of the MIP relatively poor [14]. A study by Dong et al. [70] examined the effect of solvents on the adsorption selectivity of MIP with theophylline (THO) as a template and MAA as a functional monomer. They compared three solvents, namely chloroform, tetrahydrofuran (THF), and dimethyl sulfoxide (DMSO), and found that DMSO had the highest affinity for THO and MAA but had the lowest imprinting factor (IF), 1.0533, compared to that of THF (IF = 1.1076) and chloroform (IF = 3.3197). The Imprinting Factor (IF) value shows a particular analyte's distribution ratio on the imprinted polymer as well as under the same conditions as the non-imprinted polymer (NIP). An IF value larger than one indicates good imprinting [71]. MIP with chloroform as solvent has the highest imprinting factor value because it provides the weakest interference on the template-monomer interaction, which results in the formation of the strongest hydrogen bonds between them [70].

MIP is usually synthesized in the organic solvent to increase the hydrogen bonding and electrostatic interactions between the template and monomer [72]. The polarity of the solvent greatly influences the template-monomer interactions. Less polar solvents promote the formation of the template-monomer functional complex, whereas more polar solvents interfere with the interactions in the template-monomer functional complex that form [17]. The statement is supported by the results of research conducted by Song et al. [73] on the effect of porogenic solvent in the manufacture of MIP for quercetin. They used four organic solvents with different polarities. The solvents are 1,4-dioxane, tetrahydrofuran (THF), acetone, and acetonitrile, from lowest to highest polarity. MIP with THF as the solvent showed a higher IF value of 1.2 compared to MIP using other solvents: 1,4-dioxane (IF = 1.05), acetone (IF = 1.07), and acetonitrile (IF = 1.03). This clearly shows that the medium polar solvent (THF) provides a better imprinting factor and the polar solvent (acetonitrile) provides the worst imprinting factor. Using a medium polar solvent (THF) will form an optimum interaction between the template and monomer and develop a uniform printing site on the MIP. When using a polar solvent (acetonitrile), the solvent interacts strongly with quercetin (template) and acrylamide (functional monomer), making it difficult for them to interact with each other. While using a less polar solvent (1,4-dioxane), a strong interaction will be formed between the template-monomer, but because the polymer's solubility in the less polar solvent is low, the MIP formed will quickly form a precipitate [73].

4. Template-Monomer Interaction

A study has shown that physical properties and recognition of MIP depend on the successful interaction between template and monomer. This process occurs in the pre-polymerization stage [55]. The selected functional monomer will interact with the template, producing a stable template-monomer complex [17]. Therefore, functional monomers and templates must complement each other [59].

There are currently two strategies used in MIP technology based on the nature of the template-monomer interaction. Two types of molecular imprinting strategies have been set by covalent or non-covalent interactions between the template and functional monomer [74]. An example of interaction can be seen in Figure 3. These two strategies are:

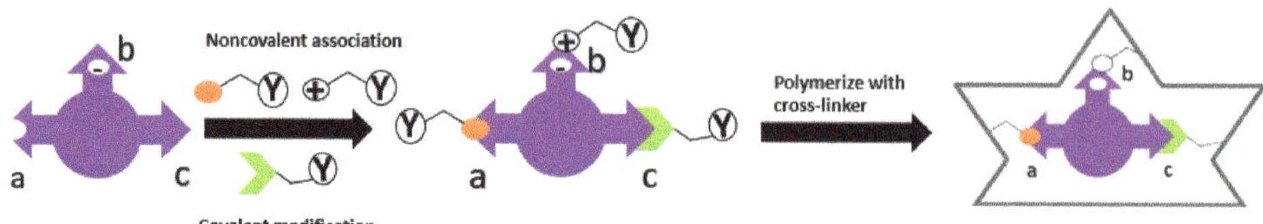

Figure 3. Type of interaction between template and monomer: (**a**) Noncovalent (nonionic) (**b**) noncovalent (electrostatic/ionic) (**c**) covalent.

1. Self-assembly approach, which uses non-covalent bonds between monomer and template, such as hydrogen bonds, Van der Waals forces, ionic or hydrophobic interactions, and others [11]. The functional monomers are regularly positioned around the template molecules during the self-assembling process due to different exchanges [75];

2. Pre-organized approach, which uses reversible covalent bonds between the functional monomer and template. This strategy will reduce non-specific sites on MIP [11,76].

The technique most widely used in the manufacture of MIP is the self-assembly approach. In this technique, template-monomer complexes are formed in situ by non-covalent interaction [77]. Hydrogen bond, hydrophobic, and electrostatic interactions are the most widely used bonds for manufacturing MIP due to their excellent adaptability [65].

MIP manufacturing techniques using covalent and non-covalent bonds have their respective advantages and disadvantages, as listed in Table 1.

Table 1. Comparison Between Covalent and Non-Covalent Imprinting Techniques.

Imprinting Type	Covalent	Non-Covalent	References
Interaction	• Reversible condensation reactions (ketal, acetal, esters, boronate, Schiff's bases)	• Ionic interaction, hydrogen bonding, Van Der Waals forces	[17,78,79]
Advantage	• More durable and rigid types of interactions • The template-monomer complex is usually stable during the polymerization process • Forming a polymer with a more homogeneous binding cavity	• Flexible, fast, and straightforward binding interactions • Easy template molecule removal • Easy template-monomer complex preparation • Resulting in a MIP with high-affinity binding, greater affinity, and selectivity to the site	[17,49,51,78,80]
Disadvantage	• Strong covalent interactions result in slow binding and re-binding • Difficult to remove the template	• Non-covalent interactions are sensitive to disruptions	[17,79]

The interaction between the template molecules and functional monomer and the formation of the polymerization mixture can lead to three different results [81]:

1. If the template is added in the pre-polymerization step, at the same time as the monomer, solvent, cross-linker, and initiator, the reaction medium will be very rich in functional monomers. The polymerization process will occur rapidly due to the dominant role of the cross-linker. A weak complex will be formed between the template and the functional monomer, producing MIP with a relatively low imprinting effect;

2. If the template is added after the start of the polymerization process, a few minutes after adding the monomer, cross-linker, initiator, and solvent, the reaction medium will contain functional macromonomers and a few monomers. Due to the nature of functional macromonomers, namely their high flexibility, they will freely re-arrange around the template to produce a MIP with a better imprinting effect;

3. If the template is added too late in the polymerization process, minutes after the optimum time of adding the template (point b), the reaction medium will be rich in pre-formed nanogel particles, which will further form a cross-linked macrogel. The condition is particularly unfavorable because the cross-linked nanogel particles are more rigid than the functional macromonomers. The MIP formed is predicted to have a low imprinting effect.

5. Analysis of Template-Monomer Functional Interactions

Intermolecular interactions between molecular templates and functional monomers will affect the selectivity and affinity of MIP [82]. These intermolecular interactions can be analyzed in the pre-polymerization process using computer simulation, UV-Vis spectroscopy, FTIR, and ^{1}H-NMR. Meanwhile, Suspended-State STD HR/MAS NMR, Raman spectroscopy, SERS, and fluorescence spectroscopy were used after MIP formation [14,16,83–86]. Hydrogen bonding interactions are the kind of interaction that focuses on all methods as this interaction strongly contributes to the affinity of molecularly imprinted polymers (MIPs), especially for low molecular weight compounds in organic, aprotic solvents [87].

5.1. Computer Simulation

The development of in silico-based technology makes it easy to select MIP components such as templates, functional monomers, and suitable porogens [88,89]. In addition, it can be used to determine intermolecular interactions that occur in the pre-polymerization mixture [88]. In the pre-polymerization process, a strong template-monomer interaction will result in a suitable MIP [90]. A computational approach can be used to evaluate hydrogen bond interactions between functional template-monomer [91]. In making MIP, computational studies play a role in determining the best monomer type and ratio in a shorter time than doing experimental trials to get high selectivity [92]. Predictable functional template-monomer interactions can be found using Density Functional Theory (DFT) [83]. The DFT method can check intermolecular interactions based on the distance between the template and functional monomers and become the most extensively utilized approach in the design of MIP [70,93–96].

Wungu et al. [93] used the DFT method based on Becke three-parameter Lee-Yang-Parr (B3LYP)/6-311 + G (d) to investigate the interaction between MAA as a template molecule with D-glucose as a functional monomer. This study used Gaussian 09 software to calculate electronic properties. Before optimization, the initial structure of the MAA-D-glucose complex had a distance of 6.62 Å between the O2 atoms of MAA and the C4 atoms of D-glucose. After optimization, the distance between the MAA complexes and D-glucose was reduced to 2.81 Å, indicating non-covalent interactions with hydrogen bonds [93].

The type of interaction between the monomer templates also can be predicted using computer simulations based on the change in charge on the natural bond orbitals (NBO). NBO analysis allows for the calculation of the number of atoms in the molecule, the molecular structure, and the intermolecular or intramolecular interactions [97]. Huang and Zhu [98] conducted a computational model study to see the interaction between spermidine (template) and methacrylic acid (monomer). In theory, the two compounds can interact through hydrogen bonds. From the observation of the NBO charge, there was an interaction between the N13 atom in spermidine and the H5 atom in methacrylic acid, which was seen from the change in the NBO charge. Before complex formation, the NBO charge of N13 was −0.671, and H5 was 0.485. Meanwhile, after complex shape, the NBO charge of N13 became −0.707 and H5 became 0.501. This change in NBO charge indicates a charge transfer between the proton donor (H5) and proton acceptor (N13) in both molecules (template and monomer) and demonstrates that spermidine and methacrylic acid interact through hydrogen bonds [98].

Intermolecular interactions can also be seen through binding energy (ΔE), even though we can not always determine the interaction type [93]. The initial confirmation of the respective molecular templates and functional monomers was optimized to obtain the molecule's Gibbs free energy (ΔG). The molecular templates and functional monomers are combined to produce stable complex conformations without imaginary frequencies. Functional monomers can be selected depending on the size of the ΔE [97]. The formation of the complex will be more stable as the binding energy value decreases; the lower the value of the binding energy, indicated by the negative ΔE value, the more likely the complex formed will exist in its complex form [99–101]. ΔE can also be used to determine the extent of the reaction, while ΔG is used to determine the spontaneity of the reaction [97]. The ΔE values were calculated using the following equation:

$$\Delta E = E\ (\text{complex}) - E\ (\text{template}) - \Sigma E\ (\text{monomer}) \tag{1}$$

where E (complex) is the total energy of the template-monomer complex, E (template) is the energy of the template, and ΣE (monomer) is the energy of the functional monomer [102].

5.2. UV-Vis Spectroscopy

The UV-Vis spectroscopy analytical method aims to determine the strength of the intermolecular interaction between the template and the functional monomer. The strength and affinity of the template-monomer will affect the selectivity and affinity of the polymer. Therefore, it is crucial to determine the suitable functional monomer which will interact strongly with the template to form a stable complex [52].

The interaction of template and monomer can be investigated using UV titration with nothing changes in absorbance [103]. The titration method using UV-Vis spectroscopy can evaluate the template-monomer association constant (Ka) so that the intramolecular interaction between template and functional monomer can be determined during the pre-polymerization process and also the specificity and selectivity of the polymer [104,105]. The Ka is calculated using the Benesi-Hildebrand equation [103] as follows:

$$\frac{1}{\Delta Y} = \frac{1}{Y\Delta HGKa[G]} + \frac{1}{Y\Delta HG} \tag{2}$$

where ΔY is the change in absorbance, $Y\Delta HG$ is the change in absorbance at the end of the titration, and [G] is the concentration of the monomer added [106].

Hasanah et al. [103] used UV titration to determine the interaction between atenolol (template) and itaconic acid (functional monomer). Based on the calculation results, the association constant was $542.67\ \text{M}^{-1}$ in methanol solvent. [103]. In another study, Hasanah et al. [107] used UV spectroscopy to determine the value of the association constant between atenolol template and Itaconic acid as a functional monomer in a mixture of methanol: acetonitrile (1:1) resulted in a constant association value of $6.277 \times 10^2\ \text{M}^{-1}$. The interaction between atenolol and itaconic acid is a hydrogen bond from the amine groups, and the carboxylic group of atenolol and itaconic acid as the Ka value was increased in a more polar aprotic solvent [103,105].

Hasanah et al. [108] used UV Vis spectroscopy to see the interaction between itaconic acid monomer and diazepam as a template. The value of Ka obtained using the Benesi–Hildebrand formula was $381.9\,\text{M}^{-1} \pm 0.4$. As stated by Wang and Yu [109], a weak interaction has a Ka value less than $25\ \text{M}^{-1}$, and a strong interaction has a Ka value more than $100\ \text{M}^{-1}$. So the interaction between itaconic acid and diazepam includes a strong interaction [108]. Fu et al. [52] also used UV-Vis spectroscopy to see the strength of the interaction between template luteolin and three functional monomers, namely acrylamide (AM), 4-vinyl pyridine (4-VP), and 1-aryl piperazine (1-ALPP) with different concentrations. Interaction strength was detected from significant changes of absorbance on maximum wavelength. 1-ALPP was found to cause reducing absorbance on maximum wavelength due to the π-π transition of luteolin when added in higher concentration. Meanwhile, no changes were found on AM and slight changes on 4-VP [52].

Based on the literature search, UV-Vis spectroscopy is usually used to see the strength of the interaction (from Ka value) or binding experiment. In Lulu Wang's research in 2019 [110], a critical study was conducted to evaluate the adsorption performance of MIP and NIP for gossypol (an acidic organic compound) made with 2-(Dimethylamino)ethyl methacrylate (DMAEMA) monomer using UV spectroscopy at 373 nm by comparing absorbance values before and after absorption. More significant differences in the value represented higher binding. The gossypol molecule has a pKa value of 6.5 [110,111] and six phenolic hydroxyl functional groups -OH in its structure; it can form an acid-base ionic pair interaction with the basic amino group -NH$_2$ of DMAEMA monomer. Theoretically, functional monomers containing basic groups can interact strongly with phenolic acid hydroxyl groups through acid-base interactions [112]. When we want to know the type of interaction, experiments in a few different solvents need to be run.

5.3. Fourier Transform Infrared Spectroscopy Analysis (FTIR)

FTIR spectroscopy can analyze samples in many forms, including liquids, solutions, gases, powders, and pastes [113]. FTIR is used in many MIP developing to identify the formation of chemical bonds, especially hydrogen bonds between molecular templates and functional monomers, due to absorption of the infrared spectrum by the sample and the peak shifting [83,114].

Xie et al. [14], in a theoretical spectrum, investigated the interaction between template chloramphenicol (CAP) and MAA functional monomer. There are peaks characteristic of strain vibrations O-H (3354 cm^{-1}) and C=O (1669 cm^{-1}) in the MAA spectrum. In the CAP spectrum, there are distinct peaks of O-H (3481 cm^{-1}), C=O (1165 cm^{-1}), and N-H stretch (3705 cm^{-1}). In the spectrum of the CAP-MAA complex, which was compared with the respective MAA and CAP spectra, there was a slight shift in the stretching peaks of N-H and C=O to shorter wavenumbers, respectively 3705 to 3684 cm^{-1} and 1669 to 1657 cm^{-1}. The peak shift to the lower wavenumber in the spectrum of the CAP-MAA complex compared to the pure CAP and MAA spectra indicates that hydrogen bonds are formed between CAP-MAA [14].

Analysis of the functional template-monomer intramolecular interaction of hydrogen bonds using FTIR was carried out in the study of Xie et al. [83] with deltamethrin (DM) as a template and AM as a functional monomer. In the AM spectrum, there is a peak characteristic of the NH stretching group at wavenumbers 3354 and 3184 cm^{-1}. Meanwhile, In the DM spectrum, there is a peak characteristic of the C=O group at wave number 1735 cm^{-1}. Compared to the single spectrum of AM, the DM-AM complex shows a shift in the peak of the NH stretching to shorter wave numbers 3346 and 3167 cm^{-1}. This peak shift indicates the formation of hydrogen bonds. The hydrogen bond formation is due to the reaction between the C=O group from DM with -NH from AM [83].

Tadi and Motghare [115] also examined the hydrogen bond interaction between oxalic acid (OA) and AM as a functional monomer at the OA: AM (1:4) ratio. At a wavelength of 2640 cm^{-1}, a shift occurred, which is thought to result from NH stretching of four AM monomer units. In addition, there are hydrogen bonds in the OH group (3421 cm^{-1}) and hydrogen bonds in the C=O group (1650 cm^{-1}), which form broad peaks [115]. Therefore, it can be concluded that the analysis of hydrogen bonding interactions between templates and functional monomers using FTIR is influenced by the electron distribution, which will cause a shift of the peak to a lower wavenumber. The formation of hydrogen bonds can be determined based on the results of MIP analysis before and after extraction [97,103,116].

5.4. Proton-Nuclear Magnetic Resonance (^{1}H-NMR)

^{1}H-NMR can be used for the analysis of complex formation during pre-polymerization; this method is commonly used to obtain optimal MIP preparation with the appearance of non-covalent compounds (hydrogen interactions, electrostatic interactions, and hydrophobic interactions) [117–119]. Studies by Quaglia et al. [120] have demonstrated the use of ^{1}H-NMR on the significance of hydrogen bonding in achieving imprinting effects, while work by Whitcombe et al. [121] showed how NMR investigations of monomer–template dissociation constants can predict MIP binding capacities.

During the formation of hydrogen bonds, there is a downfield shift caused by the reduced electron density around the hydrogen nucleus [83]. The reduction in ^{1}H-shielding correlates with the shortening of the distance between the donor and acceptor of the hydrogen bond. The ^{1}H-shielding value of hydrogen bonding protons is determined by the reduced electron charge density around the hydrogen atom. The greater the deshielding proton, the shorter the hydrogen bond distance [119]. The downfield shift of the NMR signal of the bridging proton in a hydrogen bond is divided into two phenomena. First, the formation of hydrogen bonds due to polarization and charge transfer leading to deshielding. Second is the presence of proton-accepting groups, where the response to an external magnetic field and electron density causes an effect on the position of the proton bridge, exclusive of any H-bonding phenomenon [122]. The interaction between template molecules and functional monomers will be stronger along with the significant chemical shift [83]

The research of Xie et al. [83] reveals the hydrogen bond interaction between and AM from the chemical shifts in the single AM spectra compared to the DM-AM complex spectra. In the AM spectra, the peak of the amino hydrogen atom, which was at 3.10 ppm, was shifted to 2.89 ppm in the DM-AM complex spectra. The hydrogen bond of the carbonyl group in DM with the amino group in AM causes the electron density around the AM amino hydrogen atom to decrease so that a downfield shift occurs. This indicates that the DM template and AM functional monomer are successfully conjugated [83,123]

Sánchez-González et al. [55] investigated the interaction between template cocaine hydrochloride (COCH) and functional monomers methacrylic acid (MAA) and ethylene dimethacrylate (EDMA). The investigation was carried out by varying the molar ratio in the range 0.3–3.0 between COCH: MAA and COCH: EDMA, then comparing the chemical shift disturbance caused by certain functional groups present in the template and functional monomers. The result is a chemical shift at 2.84 ppm COCH (H from the protonated COCH), 1.89 ppm MAA (H close to the acid group in MAA), and 4.36 ppm EDMA (H close to the ester groups in EDMA). This shift occurs due to the interaction of the amino group hydrogen bonds in COCH with oxygen in the acid group (MAA) and oxygen in the ester group (EDMA) [55].

Wang et al. [99] also investigated the intermolecular interactions between the oblongifolin C (OC) template derived from the fruit extract of Garcinia yunnanensis Hu and the functional monomer AM. The molar ratio used is OC:AM = 1:4. When the single spectrum of OC was compared to the mixed spectrum of OC:AM, the proton -OH peak at 9.912 ppm experienced a downfield shift to 9.916 ppm. When the single spectrum of AM was compared to the mixed spectrum of OC:AM, the $-NH_2$ proton peak at 7.509 ppm underwent a downfield shift to 7.513 ppm. When the spectrum of the ^{13}C-NMR mixture (OC:AM = 1:4) was compared with the AM spectrum, the -C=O group at 166.33 ppm experienced a downfield shift to 166.35 ppm. The results of this chemical shift confirm the occurrence of intermolecular interactions between OC and AM in the form of hydrogen bonds [99].

As for MIP, which is formed non-covalently with electrostatic (ionic) interactions, for example, 2,4-D herbicide, with 4-vinyl pyridine as a functional monomer, and synthesized using methanol/water solvent so that the nature of the interaction that will form is impossible for hydrogen bonds to form but possible for the electrostatic interaction. An NMR titration study of the system revealed that the primary interaction is based on the electrostatic effect. The basic pyridine molecule and the 2,4-D acid molecule form an ionic pair, and a solvent is also used. Controls use a non-polar solvent (deuterated chloroform); this solvent is used because observing interactions in the solvent (methanol) is complicated due to the exchange process between the labile protons of the analyte and the deuterium atom of the solvent takes place rapidly [124]. The acidic proton signal of the carboxylic acid migrates strongly upwards as the titration progresses for a higher concentration of 4-VP, as the 4-VP in solution is progressively protonated. The electrostatic interactions supported by the imprinted polymer show the most excellent affinity for the structural analogs of phenoxy acetic acid and 2,4-dichlorobenzoic acid. Both compounds can interact electrostatically with their carboxylic substituents, and both contain an aromatic group so that the affinity for 2,4-dichlorobenzoic acid shows steric [125].

5.5. Suspended-State High Resolution/Magic Angle Spinning Nuclear Magnetic Resonance Spectroscopy (STD HR/MAS NMR)

Analysis using Suspended-State STD HR/MAS NMR involves solids (MIP) suspended or dispersed in a liquid. Suspended state NMR can be used to investigate the interaction between templates/analytes and MIP [85]. In this technique, the protons in the MIP are saturated. Energy will be transferred from the MIP protons to the protons in the analyte, and an increased signal in the analyte protons will occur. This increase in the signal can provide information about the possible interactions of MIP with the analyte [126].

Like the study of Courtois et al., who researched the interaction of MIP and template (bupivacaine) using suspended-state STD HR/MAS NMR. The MIP-bupivacaine interaction was observed from the increase in proton signaling of bupivacaine. Interaction observations were carried out on monolithic MIP (mMIP), and monolithic non imprinted polymer (mNIP). The mMIP showed significantly stronger interactions with aromatic protons and twice as many with nitrogen-bound protons than mNIP, thus indicating that mMIP-bupivacaine interacts via a polar interaction. While mNIP showed a signal twice as strong as the methyl group on the alkyl chain (non-polar interaction). The analysis results using STD NMR were following the chromatography results, namely bupivacaine in mMIP formed a specific binding site with the polar group of the functional monomer (MAA) [85].

Another study was from Skogsberg et al. [127], who used STD HR/MAS NMR to investigate the interaction of 9-ethyladenine (9EA) as a template/analyte and MIP. From the STD HR/MAS NMR spectrum, it was known that the signal increase occurs in several hydrogen atoms (H-2, H-8, H-10, and H-12), indicating that H-12 has the most prominent signal increase. From the observations, it was also known that the primary interaction of EA and MIP occurs through hydrogen bonds, namely through the amino group in 9EA. From the spectrum, there may be multiple hydrogen bonds between 9EA and MIP, including N-1 and N-3 (affecting H-2); N-7 (affecting H-8) [127]. Therefore, STD HR/MAS NMR is useful to see the type of interaction, whether polar or non-polar interaction between MIP and its template.

5.6. Raman Spectroscopy

Raman spectroscopy can be used to detect and characterize the binding of template molecules to MIP by seeing the shifting of the Raman spectrum [84,128]. Raman spectroscopy has been extensively used to study polymerization processes and characterize polymers [129]. Raman spectroscopy is one of the promising techniques for understanding hydrogen bonds [130]. With Raman spectroscopy, laser photons will be scattered by the sample molecules resulting in a loss of energy during the process. The amount of energy lost will be detected as a change in the energy (wavelength) of the irradiating photon and represent certain bonds in molecules [128].

In the study of Xi et al. [131], investigations of intra and intermolecular interactions using Raman spectroscopy were carried out on non-covalent MIPs with three minor analgesics (aspirin, caffeine, and acetaminophen) as templates using MAA as a monomer. The investigation was carried out through changes in vibrational mode frequency shifts induced at engineered binding cavities [131]. Vibrational frequencies are sensitive to small chemical environmental changes such as temperature [132], phase transitions [133], and pressure [134]. Collective Intermolecular interactions between analgesic and MIP were assumed to influence the vibrational band higher than non-specific interactions that happened in the surfaces of the polymer particles. Xi et al. [131] measured Raman spectra of analgesics and polymer in solution, with non imprinted and MIP-bound, then examined the vibrational bands sensitive to intermolecular interactions, such as $C-N-CH_3$ deformation phenyl bending and CH_3 rocking. Higher shifting of the vibrational frequencies is related to stronger hydrogen bonding. When non-specific binding exists, vibrational frequencies will shift minorly or not shift at all [131,135].

5.7. Surface-Enhanced Raman Scattering (SERS)

SERS is a surface-sensitive technique that enhances Raman scattering during the binding process by incorporating the incident laser and local surface plasmons on the polymer structure [86]. SERS has a vibrational spectrum that becomes characteristic of the adsorbed substance and is used to identify compounds by comparison with a reference spectrum [136]. SERS has increased the scattering signal several times from Raman by using a rough metal surface when the molecule is adsorbed because Raman has a scattering signal that is too weak [137].

Activating SERS on the target molecule in MIP with the active metal surface placed at the recognition site is essential for enhancing the adhesion of molecularly imprinted xerogels to the SERS substrate [138]. In this study, when analyte molecules selectively bound to the recognition cavity, they spontaneously adsorb onto the surface of the surrounding silver nanoparticles. Adsorption on the surface of silver nanoparticles will produce an enhanced Raman signal from the target molecule; therefore, re-binding can be seen by measuring the intensity of SERS. For example, in Liu's study, THO (Theophylline) -MIP polymers without silver nanoparticles lacked theophylline Raman bands even though they contained as much theophylline as Ag-THO-MIPs. This proves that the silver nanoparticles embedded around the recognition cavity will increase the Raman signal from the template theophylline molecule, resulting in a clear return of the SERS band, indicating the presence of hydrogen bonds that cause the theophylline molecule to re-bind to Ag-THO-MIPs. Because of this, SERS are widely used in identifying extraction and re-binding of template molecules due to sensitivity and specificity [137].

5.8. Fluorescence Spectroscopy

Fluorescence methods are already known can be applied for the characterization of polymeric materials. As an example, the fluorescence of the dansyl group known can be used to investigate the polymer structure due to sensitivity to media polarity changes [139,140]. A polymer that already formed and still contained the template was suspended in dansyl glycine solution in water then fluorescence spectrofluorometer measurements were used to measure the fluorescence signal. A study by Piletsky et al. [141] on the development of MIP for D- and L-phenylalanine used a fluorescence spectrofluorometer to see the hydrophobic and electrostatic interaction that happened between the functional group of the polymer formed with the template. The size of differences in the emission wavelength spectra of dansyl glycine in water alone and after polymer being suspended can be used to indicate hydrophobic interaction and polar interaction. More enormous differences in the emission wavelength mean hydrophobic and electrostatic interactions appeared instead of polar interaction [141].

Another study by Takeuchi et al. [142] used methacrylic acid (MAA) and 2-(trifluoromethyl) acrylic acid (TFMAA) for the synthesis of MIPs of cinchona alkaloids. The emission maximum of free cinchonidine in chloroform/acetonitrile solution is 360 nm and was shifted to 390 nm upon binding to the TFMAA-based polymers, suggested hydrogen bond interaction between TFMAA and cinchonidine molecules [142].

5.9. Method Comparison for Detecting Interactions Template-Monomer and MIP-Template

Here, we summarize the advantages and disadvantages of each method for detecting interactions between template-monomer and MIP-template listed in Table 2.

Table 2. Method Comparison for detecting interactions template-monomer and MIP-template.

Tools	Advantages	Disadvantages	References
Computer simulation	• Able to determine the type and ratio of suitable monomers in less time than experiments • Able to predict the occurrence of unwanted interactions such as the formation of crosslinker-monomer or crosslinker-template complexes • Fast, simple, economic sensing system	• The results of computer simulations must still be confirmed through experiments	[90,97,143]
UV-Vis Spectroscopy	• Able to determine the strength of intermolecular interactions between templates and functional monomers	• It takes a long time and needs to run an experiment in a few different solvents to see the type of chemical interaction	[52,144]
FTIR	• Able to analyze liquid samples, gas solutions, powders, and pastes • Able to identify bond formation from the absorption of the infrared spectrum by the sample • Can also be used to see quantification of the degree of polymerization and reactivity for each type of polymerizable group on the monomers	• Less selective	[113,114,145,146]
^{1}H-NMR	• Able to detects hydrogen bonds by observing the downfield shift that occurs due to the reduced electron density around the protons during hydrogen bond formation • A good tool for studying the power and geometry of hydrogen bonds	• The study of hydrogen bond interactions of small molecules by NMR is often hampered by the fast exchange of species in solution.	[83,147]
Suspended-State STD HR/MAS NMR	• Simple preparation • Fast • Quantitative data may be obtained conveniently • Able to see polar and non-polar interaction between MIP and its template	• A signal from MIP/analyte may overlap with solvent	[85,127,148]
Raman Spectroscopy	• Provides fingerprint information on the chemical structure of the analyte based on the shift in the Raman scattering frequency • Requires little or no sample preparation • In the final Raman spectrum, water is negligible	• Raman scattering cross-section is tiny • Only a small fraction of the photons are inelastically scattered (about 1 in 10 million) • Limited by low signal-to-noise because only 1 photon out of 109 incident photons undergoes Raman scattering	[131,149,150]
SERS	• Highly specific, sensitive, and accurate • Fast and nondestructive method	• Sample decomposition is often encountered in the measurement of SERS • Sensitivity of SERS to the changes of the local environment of the molecules-MIP including H+ concentration, sensing, potential, and other factors influencing accuracy of SERS result on detection of intermolecular binding • SERS mechanisms require that the molecular chromophores interact with the metal interface at short distances • Molecules with poor affinity to metal often require nanoparticle functionalization or molecule derivatization	[137,151–156]
Fluorescence Spectroscopy	• Highly sensitive • It can be used to visualize interaction on biological systems when MIP is used as a sensor	• Need fluorescence monomer or other fluorescence compounds for interaction detection • Background signals usually exist caused either by strong scattering of the particles, strong absorption, or background luminescence of the polymeric system.	[140,157–161]

6. Conclusions

The molecular imprinting technique uses target molecules in a synthetic polymer matrix by performing selective binding. Functional monomer and cross-linker selection in type and ratio, type of solvents, will determine the chemical interaction in pre-polymerization solution and after the formation of the imprinted polymer. This type of interaction can be predicted and identify using computer simulations, UV-Vis spectroscopy, FTIR, ^{1}H-NMR for a pre-polymerization solution. While suspended-state STD HR/MAS NMR, Raman spectroscopy, SERS, and fluorescence spectroscopy were used after the polymer being formed, ^{1}H-NMR is more valuable and sensitive to find chemical interaction between template-monomers besides hydrogen bonds in pre-polymerization solution. Computer simulation, especially the DFT method, is the most extensive method used in MIP design. Computer simulation can certainly predict all types of interaction happening in MIP but still need a laboratory experiment for confirmation. After polymers are formed, the suspended-state STD HR/MAS NMR and SERS is a highly sensitive method to see hydrogen or electrostatic interaction. Even the last method needs metal to interact with the molecular chromophores. Other chemical interactions besides hydrogen bonds still need to be explored from the study as this is not the only interaction happening in MIP.

The development of molecularly imprinted polymers (MIPs) has grown fast. MIPs have been used widely in various fields using single and multi templates, such as pharmaceuticals analysis, food, and beverage quality, environmental pollutant analysis, doping analysis, and therapeutic drug monitoring. However, there are still some problems to be explored and solved:

1. Nanostructured MIP materials often suffer from problems such as the hardenest of template removal. The condition often results in MIP nanomaterials with non-uniform recognition sites;

2. Different efficient synthesis methods to produce MIP with high binding ability and selectivity are still needed to explore to maximize commercial conversion;

3. Knowing chemical interaction that happened between MIP and template with the affordable instrument still needs to be explored shortly as an instrument available now expensive for some laboratories.

Author Contributions: A.N.H. Conceptualization, draft preparation, editing draft, review, and funding acquisition; N.S. original draft preparation and editing draft; A.Z. original draft preparation, and editing draft; N.N. original draft preparation and editing draft; D.R. review and methodology. All authors have read and agreed to the published version of the manuscript.

Funding: The APC was funded by Directorat of Research and Community Services, Universitas Padjadjaran.

Institutional Review Board Statement: Not applicable.

Informed Consent Statement: Not applicable.

Data Availability Statement: Data sharing not applicable.

Conflicts of Interest: The authors declare no conflict of interest.

References

1. Ertürk, G.; Mattiasson, B. Molecular imprinting techniques used for the preparation of biosensors. *Sensors* **2017**, *17*, 288. [CrossRef]
2. Sanbe, H.; Haginaka, J. Uniformly sized molecularly imprinted polymers for bisphenol A and β-estradiol: Retention and molecular recognition properties in hydro-organic mobile phases. *J. Pharm. Biomed. Anal.* **2003**, *30*, 1835–1844. [CrossRef]
3. Abo Dena, A.S.; Ali, A.M.; El-Sherbiny, I. Surface-Imprinted Polymers (SIPs): Advanced Materials for Bio-Recognition. *J. Nanotechnol. Adv. Mater.* **2020**, *8*, 1–19. [CrossRef]
4. Kalecki, J.; Iskierko, Z.; Cieplak, M.; Sharma, P.S. Oriented Immobilization of Protein Templates: A New Trend in Surface Imprinting. *ACS Sens.* **2020**, *5*, 3710–3720. [CrossRef] [PubMed]
5. Yongabi, D.; Khorshid, M.; Losada-Pérez, P.; Eersels, K.; Deschaume, O.; D'Haen, J.; Bartic, C.; Hooyberghs, J.; Thoelen, R.; Wübbenhorst, M.; et al. Cell detection by surface imprinted polymers SIPs: A study to unravel the recognition mechanisms. *Sens. Actuators B Chem.* **2018**, *255*, 907–917. [CrossRef]

6. Idil, N.; Bakhshpour, M.; Perçin, I.; Mattiasson, B. Whole Cell Recognition of Staphylococcus aureus Using Biomimetic SPR Sensors. *Biosensors* **2021**, *11*, 140. [CrossRef]
7. Pasquardini, L.; Bossi, A.M. Molecularly imprinted polymers by epitope imprinting: A journey from molecular interactions to the available bioinformatics resources to scout for epitope templates. *Anal. Bioanal. Chem.* **2021**, *413*, 6101–6115. [CrossRef]
8. Trends, T.; Chemistry, A.; Merko, A. New materials for electrochemical sensing IV. Molecular imprinted polymers. *TrAC Trends Anal. Chem.* **2016**, *21*, 717–725.
9. Sellergren, B.; Allender, C.J. Molecularly imprinted polymers: A bridge to advanced drug delivery. *Adv. Drug Deliv. Rev.* **2005**, *57*, 1733–1741. [CrossRef] [PubMed]
10. Chen, L.; Wang, X.; Lu, W.; Wu, X.; Li, J. Molecular imprinting: Perspectives and applications. *Chem. Soc. Rev.* **2016**, *45*, 2137–2211. [CrossRef]
11. Vasapollo, G.; Del Sole, R.; Mergola, L.; Lazzoi, M.R.; Scardino, A.; Scorrano, S.; Mele, G. Molecularly imprinted polymers: Present and future prospective. *Int. J. Mol. Sci.* **2011**, *12*, 5908–5945. [CrossRef]
12. Guć, M.; Schroeder, G. Application of Molecularly Imprinted Polymers (MIP) and Magnetic Molecularly Imprinted Polymers (mag-MIP) to Selective Analysis of Quercetin in Flowing Atmospheric-Pressure Afterglow Mass Spectrometry (FAPA-MS) and in Electrospray Ionization Mass Spectrom. *Molecules* **2019**, *24*, 2364. [CrossRef]
13. Arifuzzaman, M.; Zhao, Y. Water-Soluble Molecularly Imprinted Nanoparticle Receptors with Hydrogen-Bond-Assisted Hydrophobic Binding. *J. Org. Chem.* **2016**, *81*, 7518–7526. [CrossRef]
14. Xie, L.; Xiao, N.; Li, L.; Xie, X.; Li, Y. Theoretical Insight into the Interaction between Chloramphenicol and Functional Monomer (Methacrylic Acid) in Molecularly Imprinted Polymers. *Int. J. Mol. Sci.* **2020**, *21*, 4139. [CrossRef]
15. Amin, S.; Damayanti, S.; Ibrahim, S.; Riset, A. Synthesis and Characterization Molecularly Imprinted Polymers for Analysis of Dimethylamylamine Using Acrylamide as Monomer Functional Sintesis dan Karakterisasi Molecularly Imprinted Polymers untuk Analisis Dimetilalamin Menggunakan Akrilamid sebagai Mon. *J. Kefarmasian Indones.* **2018**, *8*, 76–84. [CrossRef]
16. Bitas, D.; Samanidou, V. Molecularly imprinted polymers as extracting media for the chromatographic determination of antibiotics in milk. *Molecules* **2018**, *23*, 316. [CrossRef]
17. Włoch, M.; Datta, J. Synthesis and polymerisation techniques of molecularly imprinted polymers. In *Comprehensive Analytical Chemistry*; Elsevier: Amsterdam, The Netherlands, 2019; Volume 86, pp. 17–40.
18. Rebelo, P.; Costa-Rama, E.; Seguro, I.; Pacheco, J.G.; Nouws, H.P.A.; Cordeiro, M.N.D.S.; Delerue-Matos, C. Molecularly imprinted polymer-based electrochemical sensors for environmental analysis. *Biosens. Bioelectron.* **2021**, *172*, 112719. [CrossRef]
19. Okutucu, B. Wastewater Treatment Using Imprinted Polymeric Adsorbents. In *Waste in Textile and Leather Sectors*; IntechOpen: London, UK, 2020. [CrossRef]
20. Ren, X.; Zeng, G.; Tang, L.; Wang, J.; Wan, J.; Liu, Y.; Yu, J.; Yi, H.; Ye, S.; Deng, R. Sorption, transport and biodegradation—An insight into bioavailability of persistent organic pollutants in soil. *Sci. Total Environ.* **2018**, *610–611*, 1154–1163. [CrossRef] [PubMed]
21. Farooq, S.; Nie, J.; Cheng, Y.; Yan, Z.; Li, J.; Bacha, S.A.S.; Mushtaq, A.; Zhang, H. Molecularly imprinted polymers' application in pesticide residue detection. *Analyst* **2018**, *143*, 3971–3989. [CrossRef] [PubMed]
22. Capcarova, M.; Zbynovska, K.; Kalafova, A.; Bulla, J.; Bielik, P. Environment contamination by mycotoxins and their occurrence in food and feed: Physiological aspects and economical approach. *J. Environ. Sci. Health Part B Pestic. Food Contam. Agric. Wastes* **2016**, *51*, 236–244. [CrossRef] [PubMed]
23. Naseri, M.; Mohammadniaei, M.; Sun, Y.; Ashley, J. The use of aptamers and molecularly imprinted polymers in biosensors for environmental monitoring: A tale of two receptors. *Chemosensors* **2020**, *8*, 32. [CrossRef]
24. Yang, Q.; Li, J.; Wang, X.; Peng, H.; Xiong, H.; Chen, L. Strategies of molecular imprinting-based fluorescence sensors for chemical and biological analysis. *Biosens. Bioelectron.* **2018**, *112*, 54–71. [CrossRef]
25. Azizi, A.; Bottaro, C.S. A critical review of molecularly imprinted polymers for the analysis of organic pollutants in environmental water samples. *J. Chromatogr. A* **2020**, *1614*, 460603. [CrossRef]
26. Ndunda, E.N.; Mizaikoff, B. Molecularly imprinted polymers for the analysis and removal of polychlorinated aromatic compounds in the environment: A review. *Analyst* **2016**, *141*, 3141–3156. [CrossRef] [PubMed]
27. Gao, M.; Gao, Y.; Chen, G.; Huang, X.; Xu, X.; Lv, J.; Wang, J.; Xu, D.; Liu, G. Recent Advances and Future Trends in the Detection of Contaminants by Molecularly Imprinted Polymers in Food Samples. *Front. Chem.* **2020**, *8*, 1142. [CrossRef] [PubMed]
28. Song, X.; Zhou, T.; Li, J.; Zhang, M.; Xie, J.; He, L. Determination of ten macrolide drugs in environmental water using molecularly imprinted solid-phase extraction coupled with liquid chromatography-tandem mass spectrometry. *Molecules* **2018**, *23*, 1172. [CrossRef] [PubMed]
29. Garcia, R.; Cabrita, M.J.; Costa Freitas, A.M. Application of Molecularly Imprinted Polymers for the Analysis of Pesticide Residues in Food—A Highly Selective and Innovative Approach. *Am. J. Anal. Chem.* **2011**, *2*, 16–25. [CrossRef]
30. Song, X.; Xu, S.; Chen, L.; Wei, Y.; Xiong, H. Recent advances in molecularly imprinted polymers in food analysis. *J. Appl. Polym. Sci.* **2014**, *131*, 40766. [CrossRef]
31. Garcia, R.; Carreiro, E.P.; Lima, J.C.; da Silva, M.G.; Freitas, A.M.C.; Cabrita, M.J. Assessment of dimethoate in olive oil samples using a dual responsive molecularly imprinting-based approach. *Foods* **2020**, *9*, 618. [CrossRef] [PubMed]

32. Liu, Y.; Yang, Q.; Chen, X.; Song, Y.; Wu, Q.; Yang, Y.; He, L. Sensitive analysis of trace macrolide antibiotics in complex food samples by ambient mass spectrometry with molecularly imprinted polymer-coated wooden tips. *Talanta* **2019**, *204*, 238–247. [CrossRef] [PubMed]

33. Cheubong, C.; Yoshida, A.; Mizukawa, Y.; Hayakawa, N.; Takai, M.; Morishita, T.; Kitayama, Y.; Sunayama, H.; Takeuchi, T. Molecularly Imprinted Nanogels Capable of Porcine Serum Albumin Detection in Raw Meat Extract for Halal Food Control. *Anal. Chem.* **2020**, *92*, 6401–6407. [CrossRef]

34. Negarian, M.; Mohammadinejad, A.; Mohajeri, S.A. Preparation, evaluation and application of core–shell molecularly imprinted particles as the sorbent in solid-phase extraction and analysis of lincomycin residue in pasteurized milk. *Food Chem.* **2019**, *288*, 29–38. [CrossRef]

35. Zhao, Q.; Li, H.; Xu, Y.; Zhang, F.; Zhao, J.; Wang, L.; Hou, J.; Ding, H.; Li, Y.; Jin, H.; et al. Determination triazine pesticides in cereal samples based on single-hole hollow molecularly imprinted microspheres. *J. Chromatogr. A* **2015**, *1376*, 26–34. [CrossRef]

36. Sun, X.; Wang, J.; Li, Y.; Yang, J.; Jin, J.; Shah, S.M.; Chen, J. Novel dummy molecularly imprinted polymers for matrix solid-phase dispersion extraction of eight fluoroquinolones from fish samples. *J. Chromatogr. A* **2014**, *1359*, 1–7. [CrossRef]

37. Pan, J.; Chen, W.; Ma, Y.; Pan, G. Molecularly imprinted polymers as receptor mimics for selective cell recognition. *Chem. Soc. Rev.* **2018**, *47*, 5574–5587. [CrossRef]

38. Zhang, H. Molecularly Imprinted Nanoparticles for Biomedical Applications. *Adv. Mater.* **2019**, *32*, 1806328. [CrossRef]

39. El-Schich, Z.; Zhang, Y.; Feith, M.; Beyer, S.; Sternbæk, L.; Ohlsson, L.; Stollenwerk, M.; Wingren, A.G. Molecularly imprinted polymers in biological applications. *Biotechniques* **2020**, *69*, 407–420. [CrossRef] [PubMed]

40. Takeuchi, T.; Sunayama, H. Beyond natural antibodies—A new generation of synthetic antibodies created by post-imprinting modification of molecularly imprinted polymers. *Chem. Commun.* **2018**, *54*, 6243–6251. [CrossRef] [PubMed]

41. Wang, H.Y.; Cao, P.P.; He, Z.Y.; He, X.W.; Li, W.Y.; Li, Y.H.; Zhang, Y.K. Targeted imaging and targeted therapy of breast cancer cells: Via fluorescent double template-imprinted polymer coated silicon nanoparticles by an epitope approach. *Nanoscale* **2019**, *11*, 17018–17030. [CrossRef] [PubMed]

42. Regan, B.; Boyle, F.; O'Kennedy, R.; Collins, D. Evaluation of molecularly imprinted polymers for point-of-care testing for cardiovascular disease. *Sensors* **2019**, *19*, 3485. [CrossRef] [PubMed]

43. Arshad, R.; Rhouati, A.; Hayat, A.; Nawaz, M.H.; Yameen, M.A.; Mujahid, A.; Latif, U. MIP-Based Impedimetric Sensor for Detecting Dengue Fever Biomarker. *Appl. Biochem. Biotechnol.* **2020**, *191*, 1384–1394. [CrossRef] [PubMed]

44. Selvolini, G.; Marrazza, G. MIP-based sensors: Promising new tools for cancer biomarker determination. *Sensors* **2017**, *17*, 718. [CrossRef]

45. Mayeux, R. Biomarkers: Potential Uses and Limitations. *NeuroRx* **2004**, *1*, 182–188. [CrossRef]

46. Zaidi, S.A. Molecular imprinting: A useful approach for drug delivery. *Mater. Sci. Energy Technol.* **2020**, *3*, 72–77. [CrossRef]

47. Li, C.; Wang, J.; Wang, Y.; Gao, H.; Wei, G.; Huang, Y.; Yu, H.; Gan, Y.; Wang, Y.; Mei, L.; et al. Recent progress in drug delivery. *Acta Pharm. Sin. B* **2019**, *9*, 1145–1162. [CrossRef] [PubMed]

48. Suksuwan, A.; Lomlim, L.; Rungrotmongkol, T.; Nakpheng, T.; Dickert, F.L.; Suedee, R. The composite nanomaterials containing (R)-thalidomide-molecularly imprinted polymers as a recognition system for enantioselective-controlled release and targeted drug delivery. *J. Appl. Polym. Sci.* **2015**, *132*, 41930. [CrossRef]

49. Urraca, J.L.; Carbajo, M.C.; Torralvo, M.J.; González-Vázquez, J.; Orellana, G.; Moreno-Bondi, M.C. Effect of the template and functional monomer on the textural properties of molecularly imprinted polymers. *Biosens. Bioelectron.* **2008**, *24*, 155–161. [CrossRef] [PubMed]

50. Barros, L.A.; Custodio, R.; Rath, S. Design of a New Molecularly Imprinted Polymer Selective for Hydrochlorothiazide Based on Theoretical Predictions Using Gibbs Free Energy. *J. Braz. Chem. Soc.* **2016**, *27*, 2300–2311. [CrossRef]

51. Joke Chow, A.L.; Bhawani, S.A. Synthesis and Characterization of Molecular Imprinting Polymer Microspheres of Cinnamic Acid: Extraction of Cinnamic Acid from Spiked Blood Plasma. *Int. J. Polym. Sci.* **2016**, *2016*, 2418915. [CrossRef]

52. Fu, X.; Yang, Q.; Zhou, Q.; Lin, Q.; Wang, C. Template-Monomer Interaction in Molecular Imprinting: Is the Strongest the Best? *Open J. Org. Polym. Mater.* **2015**, *5*, 58–68. [CrossRef]

53. Singh, M.; Singh, S.; Singh, S.P.; Patel, S.S. Recent advancement of carbon nanomaterials engrained molecular imprinted polymer for environmental matrix. *Trends Environ. Anal. Chem.* **2020**, *27*, e00092. [CrossRef]

54. Yu, H.; He, Y.; She, Y.; Wang, M.; Yan, Z.; Ren, J.H.; Cao, Z.; Shao, Y.; Wang, S.; Abd El-Aty, A.M.; et al. Preparation of molecularly imprinted polymers coupled with high-performance liquid chromatography for the selective extraction of salidroside from Rhodiola crenulata. *J. Chromatogr. B* **2019**, *1118–1119*, 180–186. [CrossRef]

55. Sánchez-González, J.; Peña-Gallego, Á.; Sanmartín, J.; Bermejo, A.M.; Bermejo-Barrera, P.; Moreda-Piñeiro, A. NMR spectroscopy for assessing cocaine-functional monomer interactions when preparing molecularly imprinted polymers. *Microchem. J.* **2019**, *147*, 813–817. [CrossRef]

56. Wu, H.; Tian, Q.; Zheng, W.; Jiang, Y.; Xu, J.; Li, X.; Zhang, W.; Qiu, F. Non-enzymatic glucose sensor based on molecularly imprinted polymer: A theoretical, strategy fabrication and application. *J. Solid State Electrochem.* **2019**, *23*, 1379–1388. [CrossRef]

57. Zhong, M.; Wang, Y.-H.; Wang, L.; Long, R.-Q.; Chen, C.-L. Preparation and application of magnetic molecularly imprinted polymers for the isolation of chelerythrine from Macleaya cordata. *J. Sep. Sci.* **2018**, *41*, 3318–3327. [CrossRef] [PubMed]

58. Nikoleli, G.P.; Nikolelis, D.P.; Siontorou, C.G.; Karapetis, S.; Varzakas, T. Novel Biosensors for the Rapid Detection of Toxicants in Foods. *Adv. Food Nutr. Res.* **2018**, *84*, 57–102. [CrossRef] [PubMed]

59. Marć, M.; Wieczorek, P.P. Chapter One—Introduction to MIP synthesis, characteristics and analytical application. In *Comprehensive Analytical Chemistry*; Elsevier: Amsterdam, The Netherlands, 2019; Volume 86, pp. 1–15.

60. Anene, A.; Kalfat, R.; Chevalier, Y.; Hbaieb, S. Design of Molecularly Imprinted Polymeric Materials: The Crucial Choice of Functional Monomers. *Chem. Afr.* **2020**, *3*, 769–781. [CrossRef]

61. Zhao, G.; Liu, J.; Liu, M.; Han, X.; Peng, Y.; Tian, X.; Liu, J.; Zhang, S. Synthesis of molecularly imprinted polymer via emulsion polymerization for application in solanesol separation. *Appl. Sci.* **2020**, *10*, 2868. [CrossRef]

62. Ibarra, I.S.; Miranda, J.M.; Pérez-Silva, I.; Jardinez, C.; Islas, G. Sample treatment based on molecularly imprinted polymers for the analysis of veterinary drugs in food samples: A review. *Anal. Methods* **2020**, *12*, 2958–2977. [CrossRef] [PubMed]

63. Xu, X.; Duhoranimana, E.; Zhang, X. Preparation and characterization of magnetic molecularly imprinted polymers for the extraction of hexamethylenetetramine in milk samples. *Talanta* **2017**, *163*, 31–38. [CrossRef]

64. Zunngu, S.S.; Madikizela, L.M.; Chimuka, L.; Mdluli, P.S. Synthesis and application of a molecularly imprinted polymer in the solid-phase extraction of ketoprofen from wastewater. *Comptes Rendus Chim.* **2017**, *20*, 585–591. [CrossRef]

65. Cui, F.; Zhou, Z.; Zhou, H.S. Molecularly Imprinted Polymers and Surface Imprinted Polymers Based Electrochemical Biosensor for Infectious Diseases. *Sensors* **2020**, *20*, 996. [CrossRef]

66. Esfandyari-Manesh, M.; Javanbakht, M.; Shahmoradi, E.; Dinarvand, R.; Atyabi, F. The control of morphological and size properties of carbamazepine-imprinted microspheres and nanospheres under different synthesis conditions. *J. Mater. Res.* **2013**, *28*, 2677–2686. [CrossRef]

67. Holland, N.; Frisby, J.; Owens, E.; Hughes, H.; Duggan, P.; McLoughlin, P. The influence of polymer morphology on the performance of molecularly imprinted polymers. *Polymer* **2010**, *51*, 1578–1584. [CrossRef]

68. Rosengren, A.M.; Karlsson, B.C.G.; Nicholls, I.A. Consequences of Morphology on Molecularly Imprinted Polymer-Ligand Recognition. *Int. J. Mol. Sci* **2013**, *14*, 1207–1217. [CrossRef]

69. Pengkamta, T.; Mala, M.; Klakasikit, C.; Kanawuttikorn, P.; Boonkorn, P.; Chuaejedton, A.; Karuehanon, W. Synthesis and Evaluation of Molecularly Imprinted Polymer as a Selective Material for Vanillin. *Suan Sunandha Sci. Technol. J.* **2020**, *7*, 1–6.

70. Dong, W.; Yan, M.; Liu, Z.; Wu, G.; Li, Y. Effects of solvents on the adsorption selectivity of molecularly imprinted polymers: Molecular simulation and experimental validation. *Sep. Purif. Technol.* **2007**, *53*, 183–188. [CrossRef]

71. Ansell, R.J. Characterization of the Binding Properties of Molecularly Imprinted Polymers. *Adv. Biochem. Eng. Biotechnol.* **2015**, *150*, 51–93. [CrossRef] [PubMed]

72. Shen, F.; Zhang, Q.; Ren, X. A triple-function zwitterion for preparing water compatible diclofenac imprinted polymers. *Chem. Commun.* **2015**, *51*, 183–186. [CrossRef]

73. Song, X.; Wang, J.; Zhu, J. Effect of Porogenic Solvent on Selective Performance of Molecularly Imprinted Polymer for Quercetin. *Mater. Res.* **2009**, *12*, 299–304. [CrossRef]

74. Hashim, S.N.N.S.; Boysen, R.I.; Schwarz, L.J.; Danylec, B.; Hearn, M.T.W. A comparison of covalent and non-covalent imprinting strategies for the synthesis of stigmasterol imprinted polymers. *J. Chromatogr. A* **2014**, *1359*, 35–43. [CrossRef]

75. Li, S.; Zhu, M.; Whitcombe, M.J.; Piletsky, S.A.; Turner, A.P.F. Molecularly Imprinted Polymers for Enzyme-like Catalysis: Principle, Design, and Applications. In *Molecularly Imprinted Catalysts*; Elsevier: Amsterdam, The Netherlands, 2016; pp. 1–17. [CrossRef]

76. Iacob, B.-C.; Bodoki, A.E.; Oprean, L.; Bodoki, E. Metal–Ligand Interactions in Molecular Imprinting. *Ligand* **2018**, *23*, 1875–1895. [CrossRef]

77. Bakhtiar, S.; Ahmad Bhawani, S.; Rizwan Shafqat, S. Synthesis and characterization of molecular imprinting polymer for the removal of 2-phenylphenol from spiked blood serum and river water. *Chem. Biol. Technol. Agric.* **2019**, *6*, 15. [CrossRef]

78. Yi, L.X.; Fang, R.; Chen, G.H. Molecularly imprinted solid-phase extraction in the analysis of agrochemicals. *J. Chromatogr. Sci.* **2013**, *51*, 608–618. [CrossRef] [PubMed]

79. Torres-Cartas, S.; Catalá-Icardo, M.; Meseguer-Lloret, S.; Simó-Alfonso, E.F.; Herrero-Martínez, J.M. Recent advances in molecularly imprinted membranes for sample treatment and separation. *Separations* **2020**, *7*, 69. [CrossRef]

80. BelBruno, P.P. Molecularly Imprinted Polymers. *Chem. Rev.* **2019**, *119*, 94–119. [CrossRef] [PubMed]

81. Anfossi, L.; Cavalera, S.; Di Nardo, F.; Spano, G.; Giovannoli, C.; Baggiani, C. Delayed Addition of Template Molecules Enhances the Binding Properties of Diclofenac-Imprinted Polymers. *Polymers* **2020**, *12*, 1178. [CrossRef] [PubMed]

82. Sikiti, P.; Msagati, T.A.M.; Mamba, B.B.; Mishra, A.K. Synthesis and characterization of molecularly imprinted polymers for the remediation of PCBs and dioxins in aqueous environments. *J. Environ. Health Sci. Eng.* **2014**, *12*, 82. [CrossRef]

83. Xie, L.; Xiao, N.; Li, L.; Xie, X.; Li, Y. An investigation of the intermolecular interactions and recognition properties of molecular imprinted polymers for deltamethrin through computational strategies. *Polymers* **2019**, *11*, 1872. [CrossRef]

84. McStay, D.; Al-Obaidi, A.H.; Hoskins, R.; Quinn, P.J. Raman spectroscopy of molecular imprinted polymers. *J. Opt. A Pure Appl. Opt.* **2005**, *7*, s340–s345. [CrossRef]

85. Courtois, J.; Fischer, G.; Schauff, S.; Albert, K.; Irgum, K. Interactions of bupivacaine with a molecularly imprinted polymer in a monolithic format studied by NMR. *Anal. Chem.* **2006**, *78*, 580–584. [CrossRef]

86. Bompart, M.; De Wilde, Y.; Haupt, K. Chemical nanosensors based on composite molecularly imprinted polymer particles and surface-enhanced Raman scattering. *Adv. Mater.* **2010**, *22*, 2343–2348. [CrossRef]

87. Verheyen, E.; Schillemans, J.P.; Van Wijk, M.; Demeniex, M.A.; Hennink, W.E.; Van Nostrum, C.F. Challenges for the effective molecular imprinting of proteins. *Biomaterials* **2011**, *32*, 3008–3020. [CrossRef]

88. El Nashar, R.M.; Abdel Ghani, N.T.; El Gohary, N.A.; Barhoum, A.; Madbouly, A. Molecularly imprinted polymers based biomimetic sensors for mosapride citrate detection in biological fluids. *Mater. Sci. Eng. C* **2017**, *76*, 123–129. [CrossRef]

89. Suryana, S.; Mutakin; Rosandi, Y.; Hasanah, A.N. An Update on Molecularly Imprinted Polymer Design through a Computational Approach to Produce Molecular Recognition Material with Enhanced Analytical Performance. *Molecules* **2021**, *26*, 1891. [CrossRef] [PubMed]

90. Cheong, W.J.; Yang, S.H.; Ali, F. Molecular imprinted polymers for separation science: A review of reviews. *J. Sep. Sci.* **2013**, *36*, 609–628. [CrossRef]

91. Hasanah, A.N.; Soni, D.; Pratiwi, R.; Rahayu, D.; Megantara, S.; Mutakin. Synthesis of Diazepam-Imprinted Polymers with Two Functional Monomers in Chloroform Using a Bulk Polymerization Method. *J. Chem.* **2020**, *2020*, 7282415. [CrossRef]

92. Abdel Ghani, N.T.; Mohamed El Nashar, R.; Abdel-Haleem, F.M.; Madbouly, A. Computational Design, Synthesis and Application of a New Selective Molecularly Imprinted Polymer for Electrochemical Detection. *Electroanalysis* **2016**, *28*, 1530–1538. [CrossRef]

93. Wungu, T.D.K.; Marsha, S.E.; Widayani; Suprijadi. Density Functional Theory (DFT) Study of Molecularly Imprinted Polymer (MIP) Methacrylic Acid (MAA) with D-Glucose. *IOP Conf. Ser. Mater. Sci. Eng.* **2017**, *214*, 012004. [CrossRef]

94. Pardeshi, S.; Patrikar, R.; Dhodapkar, R.; Kumar, A. Validation of computational approach to study monomer selectivity toward the template gallic acid for rational molecularly imprinted polymer design. *J. Mol. Model.* **2012**, *18*, 4797–4810. [CrossRef] [PubMed]

95. Dong, W.; Yan, M.; Zhang, M.; Liu, Z.; Li, Y. A computational and experimental investigation of the interaction between the template molecule and the functional monomer used in the molecularly imprinted polymer. *Anal. Chim. Acta* **2005**, *542*, 186–192. [CrossRef]

96. Gholivand, M.B.; Khodadadian, M. Rationally designed molecularly imprinted polymers for selective extraction of methocarbamol from human plasma. *Talanta* **2011**, *85*, 1680–1688. [CrossRef]

97. Zhang, B.; Fan, X.; Zhao, D. Computer-aided design of molecularly imprinted polymers for simultaneous detection of clenbuterol and its metabolites. *Polymers* **2018**, *11*, 17. [CrossRef] [PubMed]

98. Huang, Y.; Zhu, Q. Computational Modeling and Theoretical Calculations on the Interactions between Spermidine and Functional Monomer (Methacrylic Acid) in a Molecularly Imprinted Polymer. *J. Chem.* **2015**, *2015*, 216983. [CrossRef]

99. Wang, L.; Fu, W.; Shen, Y.; Tan, H.; Xu, H.; McPhee, D.J. Molecularly imprinted polymers for selective extraction of Oblongifolin C from garcinia Yunnanensis Hu. *Molecules* **2017**, *22*, 508. [CrossRef]

100. Krishnan, H.; Islam, K.M.S.; Hamzah, Z.; Ahmad, M.N. Rational computational design for the development of andrographolide molecularly imprinted polymer. In *AIP Conference Proceedings*; AIP Publishing LLC: Melville, NY, USA, 2017; Volume 1891, p. 020083. [CrossRef]

101. Estarellas, C.; Escudero, D.; Frontera, A.; Quiñonero, D.; Deyà, P.M. Theoretical ab initio study of the interplay between hydrogen bonding, cation-π and π-π Interactions. *Theor. Chem. Acc.* **2009**, *122*, 325–332. [CrossRef]

102. Dai, Z.; Liu, J.; Tang, S.; Wang, Y.; Wang, Y.; Jin, R. Optimization of enrofloxacin-imprinted polymers by computer-aided design. *J. Mol. Model.* **2015**, *21*, 2–10. [CrossRef] [PubMed]

103. Hasanah, A.N.; Yulianti, A.B.; Rahayu, D. Selective atenolol determination in blood using molecular imprinted polymer with itaconic acid as functional monomer. *Int. J. Appl. Pharm.* **2019**, *11*, 136–143. [CrossRef]

104. Scorrano, S.; Mergola, L.; del Sole, R.; Vasapollo, G. Synthesis of molecularly imprinted polymers for amino acid derivates by using different functional monomers. *Int. J. Mol. Sci.* **2011**, *12*, 1735–1743. [CrossRef]

105. Chen, L.; Xu, S.; Li, J. Recent advances in molecular imprinting technology: Current status, challenges and highlighted applications. *Chem. Soc. Rev.* **2011**, *40*, 2922–2942. [CrossRef]

106. Thordarson, P. Determining association constants from titration experiments in supramolecular chemistry. *Chem. Soc. Rev.* **2011**, *40*, 1305–1323. [CrossRef]

107. Hasanah, A.N.; Rahayu, D.; Pratiwi, R.; Rostinawati, T.; Megantara, S.; Saputri, F.A.; Puspanegara, K.H. Extraction of atenolol from spiked blood serum using a molecularly imprinted polymer sorbent obtained by precipitation polymerization. *Heliyon* **2019**, *5*, e01533. [CrossRef]

108. Hasanah, A.N.; Susanti, I.; Marcellino, M.; Maranata, G.J. Microsphere molecularly imprinted solid—Phase extraction for diazepam analysis using itaconic acid as a monomer in propanol. *Open Chem.* **2021**, *19*, 604–613. [CrossRef]

109. Rui, W.; Yu, Z. Validity and Reliability of Benesi Hildebrand Method. *Acta Phys. Chim. Sin.* **2007**, *23*, 1353–1359.

110. Wang, L.; Zhi, K.; Zhang, Y.; Liu, Y.; Zhang, L.; Yasin, A.; Lin, Q. Molecularly imprinted polymers for gossypol via sol-gel, bulk, and surface layer imprinting-A comparative study. *Polymers* **2019**, *11*, 602. [CrossRef]

111. Reyes, J.; Wyrick, S.D.; Borriero, L.; Benos, D.J. Membrane actions of male contraceptive gossypol tautomers. *BBA Biomembr.* **1986**, *863*, 101–109. [CrossRef]

112. Zhang, W.J.; Xu, Z.R.; Zhao, S.H.; Sun, J.Y.; Yang, X. Development of a microbial fermentation process for detoxification of gossypol in cottonseed meal. *Anim. Feed Sci. Technol.* **2007**, *135*, 176–186. [CrossRef]

113. Fan, M.; Dai, D.; Huang, B. Fourier Transform Infrared Spectroscopy for Natural Fibres. *Fourier Transform.-Mater. Anal.* **2012**, 45–68. [CrossRef]

114. Munajad, A.; Subroto, C.; Suwarno. Fourier transform infrared (FTIR) spectroscopy analysis of transformer paper in mineral oil-paper composite insulation under accelerated thermal aging. *Energies* **2018**, *11*, 364. [CrossRef]

115. Tadi, K.K.; Motghare, R.V. Computational and experimental studies on oxalic acid imprinted polymer. *J. Chem. Sci.* **2013**, *125*, 413–418. [CrossRef]
116. Nagy, P.I. Competing intramolecular vs. Intermolecular hydrogen bonds in solution. *Int. J. Mol. Sci.* **2014**, *15*, 19562–19633. [CrossRef]
117. Andersson, H.S.; Nicholls, I.A. Spectroscopic evaluation of molecular imprinting polymerization systems. *Bioorg. Chem.* **1997**, *25*, 203–211. [CrossRef]
118. Dong, X.; Sun, H.; Lü, X.; Wang, H.; Liu, S.; Wang, N. Separation of ephedrine stereoisomers by molecularly imprinted polymers—Influence of synthetic conditions and mobile phase compositions on the chromatographic performance. *Analyst* **2002**, *127*, 1427–1432. [CrossRef]
119. Zarycz, M.N.C.; Fonseca Guerra, C. NMR 1H-Shielding Constants of Hydrogen-Bond Donor Reflect Manifestation of the Pauli Principle. *J. Phys. Chem. Lett.* **2018**, *9*, 3720–3724. [CrossRef] [PubMed]
120. Quaglia, M.; Chenon, K.; Hall, A.J.; De Lorenzi, E.; Sellergren, B. Target analogue imprinted polymers with affinity for folic acid and related compounds. *J. Am. Chem. Soc.* **2001**, *123*, 2146–2154. [CrossRef]
121. Whitcombe, M.J.; Martin, L.; Vulfson, E.N. Predicting the selectivity of imprinted polymers. *Chromatographia* **1998**, *47*, 457–464. [CrossRef]
122. Scheiner, S. Assessment of the presence and strength of H-bonds by means of corrected NMR. *Molecules* **2016**, *21*, 1426. [CrossRef] [PubMed]
123. Liu, S.; Pan, J.; Zhu, H.; Pan, G.; Qiu, F.; Meng, M.; Yao, J.; Yuan, D. Graphene oxide based molecularly imprinted polymers with double recognition abilities: The combination of covalent boronic acid and traditional non-covalent monomers. *Chem. Eng. J.* **2016**, *290*, 220–231. [CrossRef]
124. Abraham, R.; Fisher, J.; Loftur, P. Introduction to NMR Spectroscopy. In *Magnetic Resonance in Chemistry*; Wiley: New York, NY, USA, 1989; Volume 27, pp. 1004–1005.
125. O'Mahony, J.; Molinelli, A.; Nolan, K.; Smyth, M.R.; Mizaikoff, B. Towards the rational development of molecularly imprinted polymers: 1H NMR studies on hydrophobicity and ion-pair interactions as driving forces for selectivity. *Biosens. Bioelectron.* **2005**, *20*, 1884–1893. [CrossRef]
126. Courtois, J. Monolithic Separation Media Synthesized in Capillaries and Their Applications for Molecularly Imprinted Networks, Department of Chemistry. 2006. Available online: https://www.avhandlingar.se/avhandling/467f6112bc/ (accessed on 2 June 2006).
127. Skogsberg, U.; Meyer, C.; Rehbein, J.; Fischer, G.; Schauff, S.; Welsch, N.; Albert, K.; Hall, A.J.; Sellergren, B. A solid-state and suspended-state magic angle spinning nuclear magnetic resonance spectroscopic investigation of a 9-ethyladenine molecularly imprinted polymer. *Polymer* **2007**, *48*, 229–238. [CrossRef]
128. Kengne-Momo, R.P.; Daniel, P.; Lagarde, F.; Jeyachandran, Y.L.; Pilard, J.F.; Durand-Thouand, M.J.; Thouand, G. Protein Interactions Investigated by the Raman Spectroscopy for Biosensor Applications. *Int. J. Spectrosc.* **2012**, *2012*, 462901. [CrossRef]
129. Mohsin, M.A.; Edwards, H.G.M. Application of FT-Raman spectroscopy for the characterisation of new functionalised macromers. *Spectrochim. Acta A Mol. Biomol. Spectrosc.* **2003**, *59*, 3287–3294. [CrossRef]
130. Singh, A.; Gangopadhyay, D.; Nandi, R.; Sharma, P.; Singh, R.K. Raman signatures of strong and weak hydrogen bonds in binary mixtures of phenol with acetonitrile, benzene and orthodichlorobenzene. *J. Raman Spectrosc.* **2016**, *47*, 712–719. [CrossRef]
131. Xi, W.; Volkert, A.A.; Boller, M.C.; Haes, A.J. Vibrational Frequency Shifts for Monitoring Noncovalent Interactions between Molecular Imprinted Polymers and Analgesics. *J. Phys. Chem. C* **2018**, *122*, 23068–23077. [CrossRef]
132. Huang, W.; Frech, R. Vibrational spectroscopic and electrochemical studies of the low and high temperature phases of $LiCo_{1-x}M_xO_2$ (M = Ni or Ti). *Solid State Ion.* **1996**, *86–88*, 395–400. [CrossRef]
133. Abramczyk, H.; Paradowska-Moszkowska, K. The correlation between the phase transitions and vibrational properties by Raman spectroscopy: Liquid-solid β and solid β-solid α acetonitrile transitions. *Chem. Phys.* **2001**, *265*, 177–191. [CrossRef]
134. Schindler, W.; Sharko, P.T.; Jonas, J. Raman study of pressure effects on frequencies and isotropic line shapes in liquid acetone. *J. Chem. Phys.* **1982**, *76*, 3493–3496. [CrossRef]
135. Torii, H.; Tatsumi, T.; Tasumi, M. Effects of hydration on the structure, vibrational wavenumbers, vibrational force field and resonance raman intensities of N-methylacetamide. *J. Raman Spectrosc.* **1999**, *29*, 537–546. [CrossRef]
136. Kostrewa, S.; Emgenbroich, M.; Klockow, D.; Wulff, G. Surface-enhanced Raman scattering on molecularly imprinted polymers in water. *Macromol. Chem. Phys.* **2003**, *204*, 481–487. [CrossRef]
137. Liu, P.; Liu, R.; Guan, G.; Jiang, C.; Wang, S.; Zhang, Z. Surface-enhanced Raman scattering sensor for theophylline determination by molecular imprinting on silver nanoparticles. *Analyst* **2011**, *136*, 4152–4158. [CrossRef] [PubMed]
138. Hankus, M.E.; Stratis-Cullum, D.N.; Pellegrino, P.M. Surface enhanced Raman scattering (SERS)-based next generation commercially available substrate: Physical characterization and biological application. In *Biosensing and Nanomedicine IV*; International Society for Optics and Photonics: Bellingham, WA, USA, 2011; Volume 8099, p. 80990N. [CrossRef]
139. Shea, K.J.; Sasaki, D.Y. On the control of microenvironment shape of functionalized network polymers prepared by template polymerization. *J. Am. Chem. Soc.* **1989**, *111*, 3442–3444. [CrossRef]
140. Hofstraat, J. Polymer characterization by fluorescence spectroscopy. In *Applied Polymer Science: 21st Century*; Elsevier: Amsterdam, The Netherlands, 2000; pp. 829–850. [CrossRef]

141. Piletsky, S.A.; Andersson, H.S.; Nicholls, I.A. Combined Hydrophobic and Electrostatic Interaction-Based Recognition in Molecularly Imprinted Polymers. *Macromolecules* **1999**, *32*, 633–636. [CrossRef]
142. Jun, M.; Hiroyuki, K.; Takeuchi, T. Molecularly Imprinted Fluorescent-Shift Receptors Prepared with 2-(Trifluoromethyl)acrylic Acid. *Anal. Chem.* **2000**, *72*, 3286–3290. [CrossRef]
143. Gokhale, A.A. To Investigate the Effectiveness of Computer Simulation versus Laboratory Experience, and the Sequencing of Instruction, in Teaching Logic Circuits. Ph.D. Thesis, Iowa State University, Ames, IA, USA, 1989. [CrossRef]
144. Royani, I.; Assaidah; Widayani; Abdullah, M.; Khairurrijal. The effect of atrazine concentration on galvanic cell potential based on molecularly imprinted polymers (MIPS) and aluminium as contact electrode. In *Journal of Physics: Conference Series*; IOP Publishing: Bristol, UK, 2019; Volume 1282. [CrossRef]
145. Reusch, W. *Spectroscopy*; Libretexts: New York, NY, USA, 2021. Available online: https://chem.libretexts.org/Bookshelves/ Physical_and_Theoretical_Chemistry_Textbook_Maps/Supplemental_Modules_(Physical_and_Theoretical_Chemistry) /Spectroscopy (accessed on 16 August 2020).
146. Spivak, D.A. Optimization, evaluation, and characterization of molecularly imprinted polymers. *Adv. Drug Deliv. Rev.* **2005**, *57*, 1779–1794. [CrossRef]
147. Dračínský, M. The Chemical Bond: The Perspective of NMR Spectroscopy. *Annu. Rep. NMR Spectrosc.* **2017**, *90*, 1–40. [CrossRef]
148. Behringer, K.D.; Blümel, J. Suspension NMR Spectroscopy of Phosphines and Carbonylnickel Complexes Immobilized on Silica. *Z. Naturforsch. Sect. B J. Chem. Sci.* **1995**, *50*, 1723–1728. [CrossRef]
149. Wang, F.; Cao, S.; Yan, R.; Wang, Z.; Wang, D.; Yang, H. Selectivity/specificity improvement strategies in surface-enhanced raman spectroscopy analysis. *Sensors* **2017**, *17*, 2689. [CrossRef] [PubMed]
150. Sanz-Ortiz, M.N.; Sentosun, K.; Bals, S.; Liz-Marzán, L.M. Templated Growth of Surface Enhanced Raman Scattering-Active Branched Gold Nanoparticles within Radial Mesoporous Silica Shells. *ACS Nano* **2015**, *9*, 10489–10497. [CrossRef]
151. Lu, G.; Forbes, T.Z.; Haes, A.J. SERS detection of uranyl using functionalized gold nanostars promoted by nanoparticle shape and size. *Analyst* **2016**, *141*, 5137–5143. [CrossRef] [PubMed]
152. Kim, K.; Kim, K.L.; Choi, J.Y.; Shin, D.; Shin, K.S. Effect of volatile organic chemicals on surface-enhanced Raman scattering of 4-aminobenzenethiol on Ag: Comparison with the potential dependence. *Phys. Chem. Chem. Phys.* **2011**, *13*, 15603–15609. [CrossRef]
153. Hu, J.; Zhao, B.; Xu, W.; Li, B.; Fan, Y. Surface-enhanced Raman spectroscopy study on the structure changes of 4-mercaptopyridine adsorbed on silver substrates and silver colloids. *Spectrochim. Acta Part A Mol. Biomol. Spectrosc.* **2002**, *58*, 2827–2834. [CrossRef]
154. Yang, T.; Zhao, B.; Hou, R.; Zhang, Z.; Kinchla, A.J.; Clark, J.M.; He, L. Evaluation of the Penetration of Multiple Classes of Pesticides in Fresh Produce Using Surface-Enhanced Raman Scattering Mapping. *J. Food Sci.* **2016**, *81*, T2891–T2901. [CrossRef]
155. Zheng, J.; He, L. Surface-Enhanced Raman Spectroscopy for the Chemical Analysis of Food. *Compr. Rev. Food Sci. Food Saf.* **2014**, *13*, 317–328. [CrossRef]
156. Lin, X.M.; Cui, Y.; Xu, Y.H.; Ren, B.; Tian, Z.Q. Surface-enhanced raman spectroscopy: Substrate-related issues. *Anal. Bioanal. Chem.* **2009**, *394*, 1729–1745. [CrossRef] [PubMed]
157. Kubo, H.; Yoshioka, N.; Takeuchi, T. Fluorescent imprinted polymers prepared with 2-acrylamidoquinoline as a signaling monomer. *Org. Lett.* **2005**, *7*, 359–362. [CrossRef]
158. Takeuchi, T.; Mukawa, T.; Shinmori, H. Signaling molecularly imprinted polymers: Molecular recognition-based sensing materials. *Chem. Rec.* **2005**, *5*, 263–275. [CrossRef]
159. Deng, Q.; Wu, J.; Zhai, X.; Fang, G.; Wang, S. Highly Selective Fluorescent Sensing of Proteins Based on a Fluorescent Molecularly Imprinted Nanosensor. *Sensors* **2013**, *13*, 12994–13004. [CrossRef] [PubMed]
160. Zhu, H.; Fan, J.; Du, J.; Peng, X. Fluorescent Probes for Sensing and Imaging within Specific Cellular Organelles. *Acc. Chem. Res.* **2016**, *49*, 2115–2126. [CrossRef] [PubMed]
161. Wu, D.; Sedgwick, A.C.; Gunnlaugsson, T.; Akkaya, E.U.; Yoon, J.; James, T.D. Fluorescent chemosensors: The past, present and future. *Chem. Soc. Rev.* **2017**, *46*, 7105–7123. [CrossRef] [PubMed]

Article

Multi-Element Analysis Based on an Automated On-Line Microcolumn Separation/Preconcentration System Using a Novel Sol-Gel Thiocyanatopropyl-Functionalized Silica Sorbent Prior to ICP-AES for Environmental Water Samples

Natalia Manousi [1], Abuzar Kabir [2], Kenneth G. Furton [2], George A. Zachariadis [1] and Aristidis Anthemidis [1,*]

1 Laboratory of Analytical Chemistry, Department of Chemistry, Aristotle University of Thessaloniki, 54124 Thessaloniki, Greece; nmanousi@chem.auth.gr (N.M.); zacharia@chem.auth.gr (G.A.Z.)
2 International Forensic Research Institute, Department of Chemistry and Biochemistry, Florida International University, Miami, FL 33131, USA; akabir@fiu.edu (A.K.); furtonk@fiu.edu (K.G.F.)
* Correspondence: anthemid@chem.auth.gr

Citation: Manousi, N.; Kabir, A.; Furton, K.G.; Zachariadis, G.A.; Anthemidis, A. Multi-Element Analysis Based on an Automated On-Line Microcolumn Separation/Preconcentration System Using a Novel Sol-Gel Thiocyanatopropyl-Functionalized Silica Sorbent Prior to ICP-AES for Environmental Water Samples. *Molecules* **2021**, *26*, 4461. https://doi.org/10.3390/molecules26154461

Academic Editors: Victoria Samanidou, Irene Panderi and Pawel Pohl

Received: 6 July 2021
Accepted: 21 July 2021
Published: 24 July 2021

Abstract: A sol-gel thiocyanatopropyl-functionalized silica sorbent was synthesized and employed for an automated on-line microcolumn preconcentration platform as a front-end to inductively coupled plasma atomic emission spectroscopy (ICP-AES) for the simultaneous determination of Cd(II), Pb(II), Cu(II), Cr(III), Co(II), Ni(II), Zn(II), Mn(II), Hg(II), and V(II). The developed system is based on an easy-to-repack microcolumn construction integrated into a flow injection manifold coupled directly to ICP-AES's nebulizer. After on-line extraction/preconcentration of the target analyte onto the surface of the sorbent, successive elution with 1.0 mol L^{-1} HNO_3 was performed. All main chemical and hydrodynamic factors affecting the effectiveness of the system were thoroughly investigated and optimized. Under optimized experimental conditions, for 60 s preconcentration time, the enhancement factor achieved for the target analytes was between 31 to 53. The limits of detection varied in the range of 0.05 to 0.24 $\mu g\ L^{-1}$, while the limits of quantification ranged from 0.17 to 0.79 $\mu g\ L^{-1}$. The precision of the method was expressed in terms of relative standard deviation (RSD%) and was less than 7.9%. Furthermore, good method accuracy was observed by analyzing three certified reference materials. The proposed method was also successfully employed for the analysis of environmental water samples.

Keywords: automation; flow injection; inductively coupled plasma; sol-gel; solid-phase extraction; metals

1. Introduction

Natural and anthropogenic processes are two major sources for the continuous release of trace metals in the environment [1]. Some elements or elemental species (e.g., lead, cadmium, mercury, etc.) exhibit toxicity even at low concentration levels [2]. Other metals, such as copper and zinc are essential micronutrients, and they are required for the biological processes of living organisms. As a result, the monitoring of toxic and nutritive elements in environmental samples has aroused considerable concern.

Currently, a wide variety of spectroscopic techniques including flame atomic absorption spectroscopy (FAAS) [3], electrothermal atomic absorption spectroscopy (ETAAS) [4], inductively coupled plasma atomic emission spectroscopy (ICP-AES) [5], and inductively coupled plasma mass spectrometry (ICP-MS) [4] are available for the determination of nutritive and toxic elements [6]. Among the different available spectroscopic techniques, ICP-AES is widely employed in trace and ultra-trace analysis, due to its plethora of benefits, including high sensitivity, extended linear working ranges for the target analytes, as well as its ability to rapidly determine multiple elements [7]. However, the direct determination of elements is a challenging endeavor due to their very low concentration levels, as well as the inherent matrix effects. Therefore, a sample separation/preconcentration step is typically required to improve the sensitivity of common atomic spectroscopic techniques [8]. Flow injection (FI) and related techniques have been proved to be appropriate for on-line fluidic manipulation and for automated sample processing [3]. On-line automated systems are highly attractive sample preparation platforms due to the minimization of reagent consumption and reduced laboratory time and operation cost, in combination with the achievement of high extraction efficiency and enhancement factors [7,9].

Solid-phase extraction (SPE) is by far the most attractive approach for sample preparation and preconcentration, since it offers a plethora of benefits including superior performance in terms of straightforward operation, versatility, reliability, and high separation and enrichment capability of the target analytes [10]. Typically, SPE utilizes packed or disk-phase microcolumns, filled with the desired sorbent and placed within the flow network prior to the detection system. As a result, the sorptive phase is considered an integral component of the flow system, and it is repeatedly used for the loading and the elution of the sample solution [9]. However, the applications of on-line automated systems for multi-element separation/preconcentration as a front-end to ICP-AES systems reported in the literature are limited. Chen et al. [11] evaluated the utilization of thiacalix[4]arene tetracarboxylate derivative modified mesoporous TiO_2 for the extraction of vanadium, copper, lead, and chromium. Peng et al. [12] synthesized a multi-wall carbon nanotube chemically modified silica adsorbent for the on-line SPE of Zn(II), Cu(II), Cd(II), Cr(III), V(V), and As(V) from environmental water samples. Chitosan-based materials have also been reported for the development of on-line platforms for multi-element determination [13–15].

Unequivocally, over the past few years, attention has been directed toward the development and evaluation of novel sorptive phases for on-line systems aiming to develop automated methods characterized by high accuracy and sensitivity. In this frame, a variety of sorbents have arisen including carbon nanotubes [16], metal oxides [17], 3D-printed materials [18], and functionalized silicas [19]. Among them, silica-based materials appear to be an excellent choice of support to develop sorptive phases for the extraction of metal ions. Due to their high surface area in combination with the presence of highly reactive silanol groups in its structure, the surface of silicas enables the chemical modification through immobilization of O-, N-, and S-containing organic functional groups [19,20]. Sol-gel technology has been proved to be a significant tool for the preparation of advanced hybrid inorganic–organic polymer coatings. This technology enables the chemical integration of sol-gel sorbent to the substrate in a wide variety of forms (e.g., as particles, fiber, fabric, etc.). Sol-gel materials exhibit various advantages, such as tunable porosity, selectivity, as well as good chemical and thermal stability, and thus, they offer an excellent choice for fabricating automated on-line renewable microcolumn preconcentration platforms for multi-element analysis [3]. Although sol-gel materials have been proved to be powerful sorbents for the microextraction of organic compounds, the applications of sol-gel materials for the development of multi-element analytical techniques are limited, and they are typically applied as in-tube or capillary surface coatings [3,21].

Toward the exploration of sol-gel sorbents for the extraction and preconcentration of metals, Castro et al. [19] prepared a column packed with silica obtained by sol-gel method and functionalized with 2-aminothiazole. The proposed sorbent was employed for the preconcentration of cadmium, copper, and nickel from water samples offering fast kinetics and high adsorption capacity. In 2016, Anthemidis et al. [22] evaluated four different sol-gel sorbents (i.e., sol-gel polytetrahydrofuran, sol-gel polydimethyldiphenylsiloxane, sol-gel triblock copolymers of poly(ethylene oxide) and poly(propylene oxide), and sol-gel graphene) for the flow injection-fabric disk sorptive extraction (FI-FDSE) of lead and cadmium from environmental samples. This approach was an automated alternative to conventional fabric phase extraction (FPSE), which is a novel sample preparation technique introduced by Kabir and Furton [23]. However, the exploration of sol-gel materials in multi-element analytical techniques for the simultaneous determination of toxic and nutrient elements is a new research field that has yet to be explored.

The thiocyanate functional group has long been known for its strong affinity toward metals, and several researchers have reported its use in metal extraction [24–26] as well as in creating ion-imprinted polymers for a specific metal ion [27,28]. Although, some of these studies utilized the sol-gel process to implant a thiocyanate functional group on silica substrate, the processes are complex and time consuming and involve multi-steps. To simplify the synthesis process, herein we propose a two-step reaction scheme: (a) acidic hydrolysis of sol-gel precursors, trimethoxysilane and 3-thiocayanatopropyl triethoxysilane; (b) base-catalyzed polycondensation of the hydrolyzed sol-gel precursors that results in a solid three-dimensional network of 3-thiocyanatopropyl functionalized silica. Recently, our research group designed and utilized a 3-thiocyanatopropyl functionalized sol-gel silica sorbent for the automated SPE of Cd(II), Co(II), Cu(II), and Pb(II) coupled with FAAS. This sorbent showed good performance characteristics, and it enabled the sensitive and accurate determination of those analytes in environmental and biological samples. As such, the 3-thiocyanatopropyl-functionalized sol-gel silica sorbent could be a good choice also for multi-elemental extraction/preconcentration of toxic and nutrient metals [29].

In the present study, an automated on-line sample separation and/or preconcentration platform based on the synthesized 3-thiocyanatopropyl-functionalized silica particles was developed as a front-end to inductively coupled plasma atomic emission spectroscopy. The proposed platform was fabricated by packing the sol-gel thiocyanatopropyl-functionalized silica sorbent in an easy-to-repack microcolumn format, and it was coupled with ICP-AES for multi-element analysis. In this way, we aimed to expand the applications of this sorbent and to develop a multi-element method for the simultaneous determination of a wide range of analytes. The chemical and hydrodynamic parameters affecting the performance of the proposed platform were thoroughly investigated and optimized to achieve the highest sensitivity. The accuracy of the proposed method was determined by analyzing spiked samples as well as a certified reference material. Finally, the herein developed platform was successfully employed for the determination of cadmium, lead, copper, chromium, cobalt, nickel, zinc, manganese, mercury, and vanadium in environmental and drinking water samples.

2. Results and Discussion

2.1. Characterization of the Sol-Gel Thiocyanatopropyl-Functionalized Silica Sorbent

To understand the functional and elemental composition as well as the surface morphology of the particles, thiocyanatopropyl-functionalized sol-gel silica particles were characterized using Fourier transform-infrared spectroscopy (FT-IR) and scanning electron microscopy-energy dispersive spectrometry (SEM-EDS).

2.1.1. Fourier Transform-Infrared Spectroscopy Analysis

The FT-IR spectra of (a) tetramethyl orthosilicate (TMOS), (b) 3-thiocynatopropyl triethoxysilane (3-TCPTES), and (c) 3-thiocyanatopropyl-functionalized sol-gel silica sorbent is depicted in Figure 1. The dominant features of the TMOS FT-IR spectra include bands at 2948 and 2845 cm^{-1} that correspond to the asymmetric and symmetric vibrations of -CH$_2$- groups, respectively. The bands at 1463 and 1076 cm^{-1} correspond to the vibration absorption of Si-O-C and Si-O-Si, respectively. The band at 819 cm^{-1} is attributed to Si-C bonds [30]. Major characteristic features of 3-TCPTES FT-IR spectra include bands at 2972, 2157, and 780 cm^{-1}, which are attributed to the C-H stretching vibration of CH$_2$ group, the C≡N stretching vibration, and the C-S stretching vibration of thiocyanatopropyl functional group. The presence of bands at ~2115, ~1063, and ~796 cm^{-1} in both the 3-TCPTES and 3-thiocyanopropyl-functionalized sol-gel silica FT-IR spectra convincingly suggests the successful integration of 3-thiocyanopropyl functional groups into the 3D sol-gel silica network.

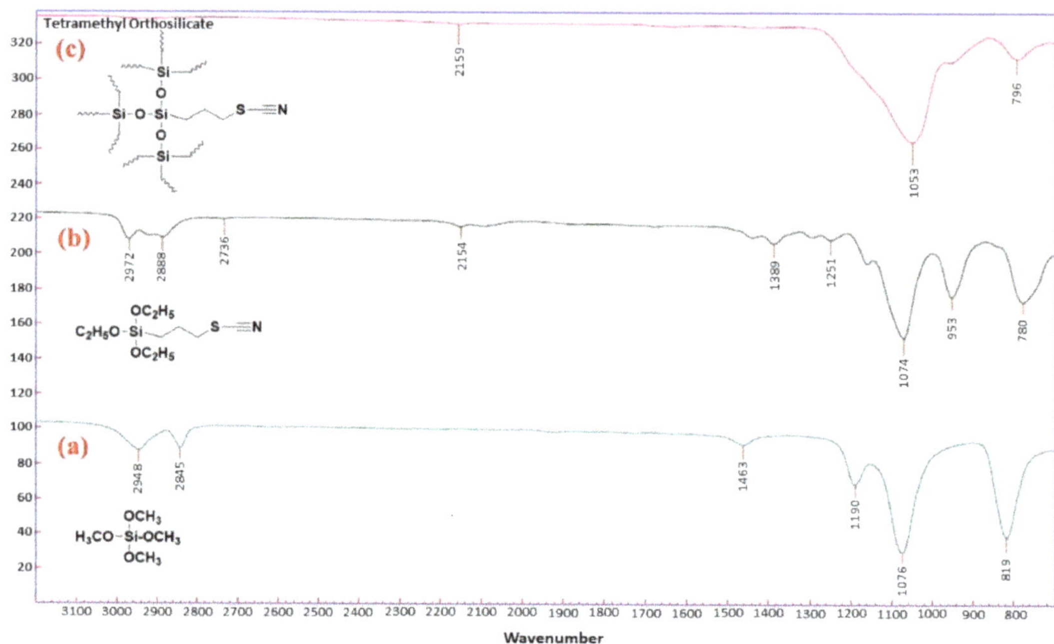

Figure 1. FT-IR spectra of (**a**) tetramethyl orthosilicate (TMOS); (**b**) 3-thiocyanatopropyl triethoxysilane; and (**c**) 3-thiocyanatopropyl-functionalized sol-gel silica sorbent.

2.1.2. Scanning Electron Microscopy-Energy Dispersive Spectrometry (SEM-EDS) Analysis

The scanning electron microscopy image, presented in Supplementary Materials Figure S1, reveals the rough surface morphology of 3-thiocyanopropyl-functionalized sol-gel silica. As the particles were obtained by manual grinding of a monolithic bed using a mortar and pestle, the particles are not homogeneous in size and shape. A mechanical grinder may provide a monodisperse particle size of the functional silica sorbent. The elemental analysis of sol-gel silica sorbent using EDS revealed the composition as 48.09% Si, 47.13% O, 3.11% S, and 1.67% N. The percentage of S and N in the sorbent can be easily enhanced by adding a higher molar ratio of 3-TCPTES in the sol solution or using only 3-TCPTES as a single sol-gel precursor in the sol solution formulation.

2.2. Mechanism of Extraction

The thiocyanate functional group is known to possess strong affinity toward d-block elements. Schematic representation of the sol-gel sorbent with metal ions (M^+) was depicted in Figure 2. It can interact with the metal ions either through the nitrogen atom (known as isothiocyanate binding mode) or through the sulfur atom (known as thiocyanate binding mode) [24]. Matveichuk et al. [24] have conducted a detailed study on a group of d-block elements to understand how an individual element interacts with the thiocyanate moiety and experimentally proved that Zn^{2+} and Co^{2+} interact with thiocyanate via nitrogen, whereas Hg^{2+}, Mn^{2+}, Cd^{2+}, Ni^{2+}, and Fe^{2+} interact with thiocyanate via the sulfur atom. As such, the thiocyanatopropyl functional group appears to possess a universal affinity toward the d-block elements. When the analytical challenge is to indiscriminately isolate and separate all these elements from the aqueous solution as in the case of wastewater treatment plants, water filtration units, and other water treatment processes, 3-thiocyanaopropyl-functionalized sol-gel silica sorbent may be an ideal choice.

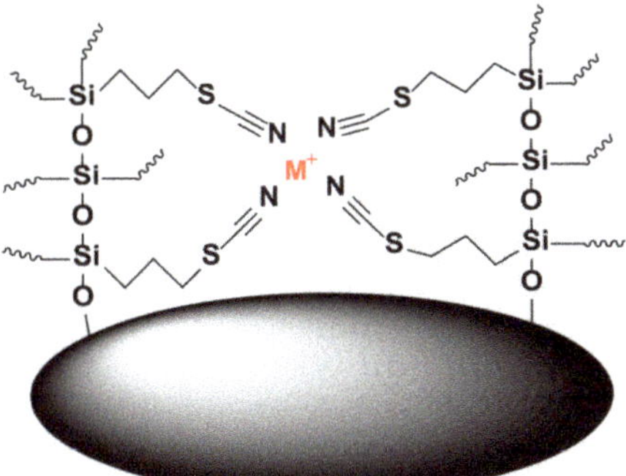

Figure 2. Schematic demonstration of the interaction between 3-thiocyanatopropyl-functionalized sol-gel silica sorbent.

2.3. Optimization of the FI-ICP-AES System

The optimization of the ICP-AES parameters is discussed in the Supplementary Materials. For the optimization of the flow injection ICP-AES (FI/ICP-AES) system, the main chemical and hydrodynamic parameters affecting its performance were thoroughly investigated and optimized using the well-established one-variable-at-a-time (OVAT) approach. As such, each parameter was individually examined within a studied range, while all other factors remained constant. For this purpose, a multi-element standard solution containing all studied analytes at a concentration of 25 μg L^{-1} was used throughout the experiments.

The chemical parameters include the pH value of the sample solution, the type, and the concentration of the eluent the eluent and the concentration of the eluent. The pH of the sample is an important factor that significantly affects the performance on-line column preconcentration procedure. A suitable pH value improves the retention efficiency and, in many cases, reduces the matrix interference. The pH value of the sample was investigated in the range of 2.0 to 7.0. Dilute NaOH and HNO_3 solutions were used to adjust the appropriate pH value. The experimental results are shown in Supplementary Materials, Figure S2 for Cd, Co, Ni, Pb, Zn, Hg and in Figure S3 for Cu, Cr, Mn, and V. As can be observed, the emission intensity was increased by increasing the pH and the extraction efficiency was decreased significant at pH lower than 4 for all analytes. The recorded intensity was the highest from pH 4.0 to pH 6.0 for all studied metal ions. At higher pH, possible hydrolysis phenomena result in gradual signal decrease. As a compromise, a pH value of 5.0 was adopted for further experiments since it was found to be beneficial for most of the target analytes.

The elution procedure of the analytes from the sol-gel thiocyanatopropyl-functionalized silica microcolumn is of a great important to avoid carryover effects that could limit the applicability of the sorbent material. Since the retention efficiency of the analytes was decreased sharply at low pH values, nitric and hydrochloric acid were examined as eluents. However, the utilization of hydrochloric acid did not seem beneficial since it could form salts with some of the target analytes (e.g., mercury and lead), which could sediment in the microcolumn and block the functional groups of the sorbent as well as the frits of the column. Nitric acid solution is an efficient eluent that can elute all retained analytes in a very small segment of the eluent zone. In addition, nitric acid is quite reconcilable with the nebulizer/injector of the ICP atomization system. Diverse nitric acid solutions with concentrations ranged between 0.1 and 2.0 mol L^{-1} were examined. As shown, the intensity increased by increasing the nitric acid concentration up to 1.0 mol L^{-1} and leveled off for concentrations higher than 1.0 mol L^{-1} nitric acid. Thus, 1.0 mol L^{-1} HNO_3 was used as the eluent during the elution step (Supplementary Materials, Table S2).

A key factor that significantly affects the performance of on-line column preconcentration systems is the loading flow rate (LFR). For a specific time (preconcentration time),

LFR determines the volume of sample solution that takes place in the extraction procedure. Although high flow rates are typically desired to achieve high preconcentration factors, adverse phenomena might take place by increasing the flow rate due to column back-pressure, which can negatively impact the sorption efficiency [31]. The effect of LFR on the emission intensity of target ions was studied in the rage 5.0–10.0 mL min^{-1}.

As shown in Supplementary Materials, Figure S4, the recorded signal increased practically linearly by increasing the LFR for Cu, Cr, Mn, and V indicating that the adsorption kinetics of the sol-gel thiocyanatopropyl-functionalized silica sorbent is very effective for the studied ions. The same behavior was observed for all analytes. For further experiments, a sample loading flow rate of 10.0 mL min^{-1} was adopted as a compromise considering the sensitivity, the sample consumption, and the time of analysis of the on-line FI/ICP-AES method.

The flow rate (EFR) of the eluent influences the efficiency of the elution procedure of the adsorbed metal ions and the dispersion of the analytes into the eluent zone. Moreover, EFR contributes to the nebulization and atomization process since it defines the amount of analyte mass injected into ICP. The effect of nitric acid flow rate was examined in the range 1.0 to 3.0 mL min^{-1} for an elution time of 30 s (Supplementary Materials, Table S2). The recorded results for all studied analytes show that the signals were increasing up to 2.6 mL min^{-1} and leveled off for higher flow rate. Hence, a flow rate of 2.6 mL min^{-1} was adopted for further experiments.

The on-line flow injection column preconcentration systems are based on the loading time to increase the sensitivity and the enhancement factor (EF). The influence of the loading time (preconcentration time) on the recovery of the metal ions was examined by loading a standard solution containing all target analytes for preconcentration times between 30 to 180 s. The obtained results (Supplementary Materials, Figure S5) revealed a practical, proportionate increase for Cd, Co, Ni, and Zn by increasing the preconcentration time up to 180 s. The same behavior was observed for all analytes. As a compromise between high sensitivity and low time of analysis and sample consumption, a preconcentration time of 60 s was selected. However, it must be noted that if higher sensitivity is required, the preconcentration time can be prolonged based on the needs of the analysis.

2.4. Figures of Merit

The analytical performance characteristics of the herein developed method for each metal ion, under the optimized operating conditions and a preconcentration time of 60 s, are presented in Table 1. The sampling frequency of the automatic FI/ICP-AES method was 36 h^{-1}. The enhancement factor calculated by the ratio of the slopes of calibration curves with and without preconcentration are given in Table 1.

Table 1. Figures of merit of the proposed FI/ICP-AES method (for 60 s preconcentration time).

	Cd(II)	Co(II)	Cr(III)	Cu(II)	Mn(II)	Ni(II)	Pb(II)	Zn(II)	Hg(II)	V(II)
Enhancement factor	53	35	46	34	36	36	46	34	39	31
Linear range (μg L^{-1})	0.33–100	0.17–80	0.44–80	0.20–100	0.26–50	0.49–80	0.79–100	0.33–80	0.62–80	0.18–80
Correlation coefficient (*r*)	0.9993	0.9981	0.9954	0.9978	0.9984	0.9993	0.9985	0.9988	0.9986	0.9992
Sensitivity (slope), μg^{-1} L	169.0	197.6	3323.9	5206.1	4072.5	38.8	11.4	160.1	199.7	911.3
Detection limit (3 s), μg L^{-1}	0.10	0.05	0.13	0.06	0.08	0.15	0.24	0.10	0.18	0.05
Quantification limit (10 s), μg L^{-1}	0.33	0.17	0.44	0.20	0.26	0.49	0.79	0.33	0.62	0.18
Precision (RSD, *n* = 8), %	3.9	2.6	0.8	1.6	2.0	2.2	7.9	2.5	1.7	2.9
Slope (without preconcentration), μg L^{-1}	3.18	5.68	72.33	155.38	112.93	1.07	0.25	4.76	5.16	29.22

As can be observed, the enhancement factors ranged between 31 and 53 for the target analytes. The linearity of the proposed method was assessed by linear regression analysis through the construction of calibration curves for multi-element standard solutions subjected to the FI/ICP-AES method. As can be observed, the proposed method exhibited good linearity and a wide linear range for all the examined elements. The detection and quantification limit are calculated by 3 and 10 s criteria, respectively, according to IUPAC recommendation, as 3- or 10-times the standard deviation of the blank solution measurements ($n = 10$) divided by the slope of the corresponding calibration equation. For the target analytes, the LOD values ranged between 0.05 and 0.24 μg L^{-1}, while the LOQ values ranged between 0.17 and 0.79 μg L^{-1}. The precision of the method was expressed in terms of relative standard deviation (RSD%) for each examined metal at 10.0 μg L^{-1} ($n = 8$) the concentration level varied from 0.8% up to 7.9%.

The accuracy of the developed FI-/ICP-AES multi-element method was examined by analyzing three different standard reference materials: CRM 1643e, IAEA-433, and Seronorm^{TM} Trace Elements Urine Level 1 (4), and it was expressed in terms of relative error between the nominal and experimentally found concentration. As the certified values of cadmium and mercury in Seronorm^{TM} were below the detection limits, they were not evaluated. Student *t*-test was adopted for statistically significant difference between the certified values and the recorded analytical values' investigation. The calculated t_{exp} values are presented in Supplementary Materials, Table S3.

All the t_{exp} values are lower than the critical value, $t_{crit, 95\%} = 4.3$, meaning that no statistically significant differences were observed for each analyte at a 95% probability level, and the method exhibited satisfactory accuracy.

2.5. Interference Studies

The effect of diverse, usually coexisting, ions on the sensitivity of the proposed FI/ICP-AES method for the determination of the studied metal was studied. Consequently, an aqueous solution containing Cd(II), Co(II), Cr(III), Cu(II), Mn(II), Ni(II), Pb(II), Zn(II), Hg(II), and V(II) at a concentration level of 10.0 μg L^{-1} for each metal and the examined interfering ions was analyzed using the developed analytical method. A variant of the emission intensity greater than $\pm$5% was considered as interference. The tolerance limit for the common matrix ions was Na$^+$, K$^+$ up to 1000 mg L^{-1}; Ca^{2+}, Mg^{2+} up to 500 mg L^{-1}; SO$_4^{2-}$, NO$_3^-$, HCO$_3^-$ up to 2000 mg L^{-1}, and Ag$^+$, Al^{3+}, Ba^{2+}, Fe^{3+} up to 20 mg L^{-1}. Hence, the developed method can be used for multi-element analysis in most cases of environmental water samples without applying any masking agent.

2.6. Applications in Spiked Environmental Water Samples

The proposed method was applied to the analysis of natural water samples collected from the Northern Greece area: Axios river and Volvi lake. The analytical results are presented in Supplementary Materials, Table S4. The calculated recoveries varied within the range 90.0–106.0%, indicating the applicability of the method for trace multi-element determination in environmental water samples.

2.7. Comparison of the Proposed FI/ICP-AES Method with Other Selected On-Line Column Preconcentration ICP-AES Methods

For comparative purposes, the performance characteristics of the proposed method and previously published on-line column preconcentration ICP-AES methods were compared in terms of sampling frequency, detection limit, precision, and enhancement factor, as shown in Table 2.

As can be observed, the precision of the method is similar to or better than the precision of other studies, since similar or better RSD values were obtained. The enhancement factors of the proposed study were better than those reported in most on-line SPE methods [11–13,30,31]. However, for Cd and Co [14,15] and Pb [14,15] higher enrichment factors have been reported. Finally, the sensitivity of the proposed method was satisfactory compared to that of the other on-line column preconcentration ICP-AES methods.

Table 2. Comparison of the performance characteristics of the developed method against selected on-line SPE procedures for multi-element determination with ICP-AES.

Sorbent	Analytical Characteristics	Cd	Co	Cr	Cu	Hg	Mn	Ni	Pb	Zn	V	Ref.
Modified mesoporous TiO$_2$	c_L			0.15	0.23				0.50		0.09	[11]
	s_r (%)	-	-	2.9	3.9	-	-	-	4.6	-	1.7	
	PF			20	20				20		20	
MWCNTs chemically modified silica	c_L	0.11		0.27	0.91					0.45	0.55	[12]
	s_r (%)	3.1	-	3.1	4.0	-	-	-	-	4.1	7.3	
	PF	10		10	10					10	10	
Chitosan-modified ordered mesoporous silica	c_L	0.05			0.30	0.93			0.96		0.33	[13]
	s_r (%)	4.0	-	-	6.7	5.3	-	-	1.8	-	2.8	
	PF	20			20	20			20		20	
N-(2-hydroxyethyl) glycine-type chitosan chelating resin	c_L	0.004	0.023		0.068		0.018	0.13	0.085	0.05	0.17	[14]
	s_r (%)	NA	NA	-	NA	-	NA	NA	NA	NA	NA	
	EF	106	87		32		14	16	25	21	18	
EDTriA-type chitosan chelating resin	c_L	0.002	0.022		0.066		0.018	0.12	0.080	0.048	0.15	[15]
	s_r (%)	<10	<10	-	<10	-	<10	<10	<10	<10	<10	
	EF	116	93		35		14	16	112	25	19	
HyperSep SCX	c_L	0.07	0.06	0.2	0.08		0.1	0.07	0.2	0.08		[31]
	s_r (%)	4.5	3.7	3.9	4.5	-	3.7	4.2	2.5	4.2	-	
	EF	18.4	18.4	18.4	18.4		18.4	18.4	18.4	18.4		
AG50W-X8 Cation exchange	c_L	1.0						4.0	2.0			[32]
	s_r (%)	2.5	-	-	-	-	-	2.6	4.0	-	-	
	PF	10						10	10			
Sol-gel-functionalized silica	c_L	0.10	0.05	0.13	0.06	0.18	0.08	0.15	0.24	0.10	0.05	This work
	s_r (%)	3.9	2.6	0.8	1.6	1.7	2.0	2.2	7.9	2.5	2.9	
	EF	53	35	46	34	39	36	36	46	34	31	

f, sampling frequency; c_L, detection limit (values in μg L^{-1}); s_r, relative standard deviation; EF, enhancement or preconcentration factor based on their availability, NA, not available.

The analytical characteristics of the reported study for Cd(II), Co(II), Cu(II), and Pb(II) were also compared to our previously reported study utilizing a 3-thiocyanatopropyl-functionalized sol-gel silica sorbent as a front-end to FAAS [29]. The LODs of this method were lower (i.e., 0.05–0.24 μg L^{-1}) compared to the method utilizing FAAS as detection system (i.e., 0.15–1.9 μg L^{-1}). On the other hand, higher enrichment factors were obtained for the automated on-line FAAS method (i.e., 73–152), and lower enrichment factors were obtained for the automated on-line ICP-AES method (i.e, 34–53). Moreover, comparable repeatability in terms of RSDs was obtained in both cases for Cd(II), Co(II), and Cu(II) (RSD < 3.9 for the on-line ICP-AES method and RSD < 3.8 for the on-line FAAS method), while the utilization of the on-line FAAs system resulted in better repeatability for Pb(II) (i.e., RSD = 3.8%) compared to that for the on-line ICP-AES system (i.e., RSD = 7.9%). Finally, the utilization of the sol-gel silica sorbent and the ICP-AES detection system enabled the simultaneous determination of Cd(II), Pb(II), Cu(II), Cr(III), Co(II), Ni(II), Zn(II), Mn(II), Hg(II), and V(II) due to the multi-elemental nature of the ICP-AES instrument.

3. Materials and Methods

3.1. Reagents Materials and Samples

Tetramethyl orthosilicate (TMOS), hydrochloric acid, and isopropanol were purchased from Sigma-Aldrich (St. Louis, MO, USA). 3-Thiocyanatopropyl triethoxysilane was purchased from Gelest Inc. (Morrisville, PA, USA). Ultra-pure deionized water (18.2 MΩ) used in the sol solution synthesis was obtained from a Barnstead NANOPure Diamond (Model D11911) deionized water system (APS Water Services Corporation, Lake Barbara, CA, USA).

Nitric acid (HNO$_3$) (65%) and ammonia solution (25%) were supplied by Merck (Darmstadt, Germany). For ultra-pure-quality water, a Milli-Q (Millipore, Bedford, TX, USA) purification system was used during method development, validation, and application.

Single stock standard solutions (1000 mg L^{-1}) of cadmium, chromium, copper, cobalt, nickel, lead, zinc, manganese, mercury, and vanadium were supplied by Merck (Darmstadt, Germany). The stock standard solutions were prepared in HNO_3 0.5 mol L^{-1} for cadmium, chromium, copper, cobalt, nickel, lead, zinc, manganese, and vanadium and in HNO_3 2.0 mol L^{-1} for mercury. Multi-element working standard solutions were prepared daily by appropriate serial dilution from the single-element stock standards. The pH of the working solutions was adjusted accordingly to be used in the extraction/preconcentration step.

In order to avoid contamination, laboratory glassware and storage bottles were rinsed with water and soaked in 10% (v/v) nitric acid overnight. Prior to their use, all sample preparation apparatuses were extensively washed with Milli-Q water.

The following standard reference materials (SRMs) were analyzed for evaluation of the developed method: The NIST (National Institute of Standard and Technology, Gaithersburg, MD, USA) certified reference material (CRM) 1643e, which contains trace elements in water; the International Atomic Energy Agency, IAEA-433, which is a marine sediment; and the Seronorm^{TM} Trace Elements Urine L1. A mass of approximately 0.5 g of sediment SRM was precisely weighed and transferred into Teflon bombs. An appropriate volume of nitric acid perchloric and hydrofluoric acid in a volume ratio of 3/2/1 ($HNO_3/HClO_4/HF$) was added. The certified urine sample was digested using concentrated HNO_3. The digestion procedure was carried out at 130–140 °C in a shielded Teflon beaker stainless-steel pressurized bomb, according to the manufacturer's recommendations. After cooling the system, the digests were properly diluted in ultra-pure water and used for the analysis.

Environmental water samples were collected from sampling sites located in Northern Greece during February 2021: Axios river and Volvi lake. All samples were filtered through 0.45 μm membrane filters, acidified to approximately pH 2 with dilute nitric acid and stored at 4 °C in acid-cleaned polyethylene bottles until the analysis. These solutions were used for the determination of the "dissolved" fraction of the metal in the samples.

3.2. Apparatus

Centrifugation of different sol solutions to obtain particle free gel was carried out in an Eppendorf Centrifuge Model 5415R (Eppendorf North America Inc., Framingham, MA, USA). A 2510 BRANSON Ultrasonic Cleaner (Branson Inc., Brookfield, CT, USA) was used to obtain bubble-free sol solution prior to the gelation process. An Agilent Carry 670 FT-IR Spectrometer equipped with Universal ATR Sampling Accessory (Agilent Technologies, Santa Clara, CA, USA) was used to perform FT-IR characterization of the sol-gel sorbent. A JEOL JSM 5900 LV scanning electron microscope equipped with an EDS-UTW detector (Jeol USA Inc., Peabody, MA, USA) was used to obtain SEM images as well as elemental composition analysis of the sol-gel sorbent.

All experiments were carried out using the Optima 3100XL axial viewing inductively coupled plasma atomic emission spectrometer (ICP-AES) of Perkin-Elmer (Norwalk, CT, USA, https://www.perkinelmer.com/) for the multi-element analysis, under the operating conditions presented in Supplementary Materials, Table S1. Optima 3100XL was equipped with an Echelle polychromator (resolution: 0.006 nm at 200 nm) and a segmented-array charge-coupled detector with 235 sub-arrays. Two different spray-chamber/nebulizer configurations, namely, Cyclonic/Babington and Scott double-pass/Gem tip crossflow, were examined considering their efficiency/sensitivity. The wavelengths of the studied elements were fixed according to the sensitivity order of atomic (I) or ionic (II) lines as they are listed in Supplementary Materials, Table S1.

To determine accurate peak wavelengths on peak algorithm, peak area processing was chosen. Due to the transient concentration of analyte into the segment of the eluent, during the elution/measurement step, a read time of 1 s and 10 replicates were adopted for signal recording. The selection of the emission line was based on the sensitivity of the instrument in combination with the absence of spectral interferences.

The flow injection system for the on-line column preconcentration of the target analytes coupled with the ICP-AES is shown schematically in Figure 3. In brief, it comprises two peristatic pumps (Gilson Minipuls 3, www.gilson.com) for sample (P1) and eluent (P2) solutions delivery, as well as two 6-port 2-position injection valves V_1 and V_2 (Supelco Rheodyne, Sigma-Aldrich Chemie GmbH, Taufkirchen, Germany, https://www.sigmaaldrich.com/). The microcolumn C was fixed at ports 1 and 4 of injection valve V1. All tubing of the flow manifold was made of polytetrafluoroethylene (PTFE). The connecting tube between the V_1 and the nebulizer of ICP-AES was only 50 cm length, 0.5 mm internal diameter to eliminate possible dispersion of the analytes into the segment of the eluent.

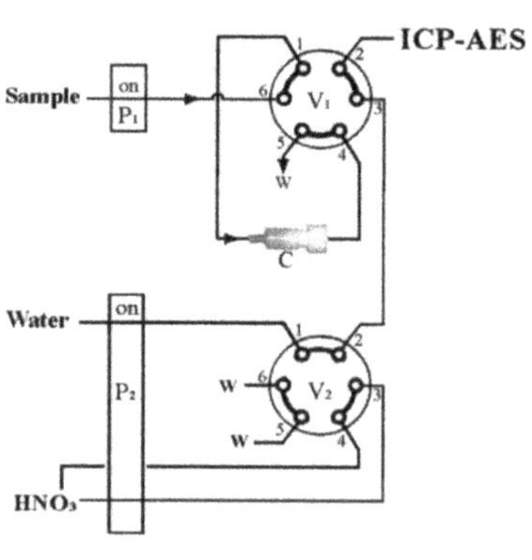

Figure 3. Schematic diagram of the on-line column preconcentration manifold for multi-element analysis by ICP-AES. P_1, P_2—peristaltic pumps; V_1, V_2—injection valves, V_1 in "load" position; V_2 in "A" position; W—waste; C—microcolumn packed with sol-gel thiocyanatopropyl silica sorbent.

A METROHM 654 pH-meter (Metrohm AG, Herisau, Switzerland) was employed for the pH adjusting of sample solutions.

3.3. Synthesis of 3-Thiocyanatopropyl-Functionalized Sol-Gel Silica Sorbent

The sol solution was prepared by sequential addition of tetramethyl orthosilicate (TMOS), 3-thiocyanatopropyl triethoxysilane, and 2-propanol in a 50 mL amber reaction bottle at a molar ratio of 1:0.3:15, respectively, with vortexing for 2 min after adding each ingredient. Subsequently, 0.1 mol L^{-1} HCl was added to the mixture at a molar ratio between TMOS and 0.1 mol L^{-1} HCl 1:5. After thorough mixing of the sol solution, the mixture was subjected to prolonged hydrolysis at 50 °C for 8 h. The solution was then transferred into a wide-mouth glass reaction vessel and a Teflon-coated bar magnet was added to it. In order to initiate polycondensation, 1.0 mol L^{-1} NH_4OH (impregnated with 0.25 mol L^{-1} NH_4F) was added in droplets under constant stirring on a magnetic stir bar. The ratio between TMOS and 1.0 M NH_4OH was maintained at 1:1.2. The sol solution was turned into transparent gel in 30 min. The sol-gel monolithic bed was thermally conditioned and aged for 24 h at 50 °C.

Subsequently, the monolithic bed was crushed and subjected to drying for 24 h at 70 °C. The dried sol-gel sorbent was then crushed and pulverized into fine particles using a mortar and a pestle. To clean the sol-gel sorbent from unreacted precursors, solvent, and reaction by-products, the powder was loaded in a fiber-glass thimble and subjected to Soxhlet extraction for 4 h using methanol:methylene chloride 50:50 (v/v) as the cleaning solution. The sol-gel sorbent was then dried at 70 °C for 24 h. The dried sol-gel sorbent was then ready for upstream processing such as characterization and application. The construction of the repacked microcolumn is described in the Supplementary Materials.

3.4. Automatic On-Line Operational Procedure

The automatic on-line flow injection microcolumn preconcentration analytical procedure for the multi-element determination was operated in four main steps, which are presented in Supplementary Materials, Table S2.

In case a new sample or standard solution was introduced for the first time, a pre-fill step was utilized to fill the tubing of the pump P_1 up to valve V_1.

In loading steps (step 1 and 2, Figure 1), valve V1 was switched to the "load" position, while P_1 was turned on for sample or standard solution delivery through the microcolumn at flow rate of 6.4 mL L^{-1}. The analytes were quantitatively retained in the microcolumn. For the elution (step 3), the injection valve V_1 was in the "elute" position. The eluent 1.0 mol L^{-1} HNO_3 was delivered through the microcolumn to desorb and to deliver the analytes into the ICP nebulizer and plasma torch for atomization.

This step lasts 30 s, which was found to be appropriate for complete washing of the column. It is noteworthy that the eluent flows through the microcolumn in the opposite direction to that of the sample or standard solution, resulting in minimum dispersion of the analytes into the eluent segment. In this way, gradual one-side compaction of the sorbent was also evaded. In all cases, five replicate measurements were performed.

4. Conclusions

In this study, a streamlined approach for synthesizing thiocyanatopropyl-functionalized silica sorbent using sol-gel process was described. The sol-gel based sorbent is deployed in an on-line automated sample preconcentration and separation platform utilizing an easy-to-repack microcolumn for ICP-AES multi-elemental analysis. The novel type of constructed microcolumn reveals some extra advantages such as convenience in preparation and handling on a flow injection manifold. The packing of the column with the sol-gel thiocyanatopropyl-functionalized silica offered very low flow resistance, excellent packing reproducibility, good preconcentration efficiency, as well as satisfactory sensitivity and high reusability calculated as more than 700 loading/elution cycles. It can be concluded that the sol-gel chemistry is an interesting alternative to the development of metal adsorbents, without requiring the addition of chelating agents for complex formation. The proposed method is fast, simple, sensitive, and selective for the multi-element determination of metals in environmental water and urine samples.

Supplementary Materials: The following are available online. Table S1. Instrumental conditions and description of the ICP-AES system. Table S2. Operation sequences of the on-line microcolumn preconcentration system coupled with ICP-AES, Table S3. Determination of trace metals in certified reference materials with the proposed FI/SPE-ICPAES method. Mean value ± standard deviation based on three replicates (n = 3), $t_{exp.}$, experimental value; $t_{crit.}$ = 4.3 (at 95 % probability level). Table S4. Multi-element analysis of trace metals in spiked natural water samples. Figure S1. SEM-EDS image of 3-thiocyanatopropyl-functionalized sol-gel silica particles (at 5.00× magnification). Figure S2. Effect of pH on the emission intensity of Cd, Co, Ni, Zn, Pb, and Hg at the 25 µg L^{-1} concentration level of each metal ion. All other experimental parameters as presented in Table S1. Figure S3. Effect of pH on the emission intensity of Cu, Cr, Mn, and V at the 25 µg L^{-1} concentration level of each metal ion. Figure S4. Effect of loading flow rate on the emission intensity of Cu, Cr, Mn, and V at 25 µg L^{-1} concentration level each metal ion. All other experimental parameters as presented in Table S1. Figure S5. Effect of preconcentration time on the emission intensity for Cd, Co, Ni, and Zn at the 10.0 µg L^{-1} concentration level of each metal ion. All other experimental parameters as presented in Table S1. Figure S6. Construction and the main components of the repacked microcolumn. All other experimental parameters as presented in Table 1. Construction of the repacked microcolumn, examination of the ICP-AES parameters, interference studies.

Author Contributions: Conceptualization, N.M., A.K., G.A.Z., and A.A.; methodology, N.M., A.K., G.A.Z., and A.A.; validation, N.M. and A.A.; formal analysis, N.M., A.K., G.A.Z., and A.A.; investigation, N.M., A.K., G.A.Z., and A.A.; resources, A.K., G.A.Z., and A.A.; writing—original draft preparation, N.M.; writing—review and editing, A.K., K.G.F., G.A.Z., and A.A.; supervision, A.K., K.G.F., G.A.Z., and A.A.; project administration, A.A. All authors have read and agreed to the published version of the manuscript.

Funding: This research received no external funding.

Institutional Review Board Statement: Not applicable.

Informed Consent Statement: Not applicable.

Data Availability Statement: Not applicable.

Conflicts of Interest: The authors declare no conflict of interest.

Sample Availability: Samples of the compounds are not available from the authors.

References

1. Miró, M.; Estela, J.M.; Cerdà, V. Application of flowing stream techniques to water analysis: Part III. Metal ions: Alkaline and alkaline-earth metals, elemental and harmful transition metals, and multielemental analysis. *Talanta* **2004**, *63*, 201–223. [CrossRef]
2. Goldhaber, S.B. Trace element risk assessment: Essentiality vs. toxicity. *Regul. Toxicol. Pharmacol.* **2003**, *38*, 232–242. [CrossRef]
3. Lazaridou, E.; Kabir, A.; Furton, K.G.; Anthemidis, A. A Novel Glass Fiber Coated with Sol-Gel Poly-Diphenylsiloxane Sorbent for the On-Line Determination of Toxic Metals Using Flow Injection Column Preconcentration Platform Coupled with Flame Atomic Absorption Spectrometry. *Molecules* **2020**, *26*, 9. [CrossRef] [PubMed]
4. Wang, J.; Hansen, E.H.; Miró, M. Sequential injection-bead injection-lab-on-valve schemes for on-line solid phase extraction and preconcentration of ultra-trace levels of heavy metals with determination by electrothermal atomic absorption spectrometry and inductively coupled plasma mass. *Anal. Chim. Acta* **2003**, *499*, 139–147. [CrossRef]
5. Ferreira, S.L.C.; De Brito, C.F.; Dantas, A.F.; Lopo De Araújo, N.M.; Costa, A.C.S. Nickel determination in saline matrices by ICP-AES after sorption on Amberlite XAD-2 loaded with PAN. *Talanta* **1999**, *48*, 1173–1177. [CrossRef]
6. Anthemidis, A.N.; Arvanitidis, V.; Stratis, J.A. On-line emulsion formation and multi-element analysis of edible oils by inductively coupled plasma atomic emission spectrometry. *Anal. Chim. Acta* **2005**, *537*, 271–278. [CrossRef]
7. Das, D.; Dutta, M.; Cervera, M.L.; de la Guardia, M. Recent advances in on-line solid-phase pre-concentration for inductively-coupled plasma techniques for determination of mineral elements. *Trends Anal. Chem.* **2012**, *33*, 35–45. [CrossRef]
8. Butler, O.T.; Cairns, W.R.L.; Cook, J.M.; Davidson, C.M.; Mertz-Kraus, R. Atomic spectrometry update-a review of advances in environmental analysis. *J. Anal. At. Spectrom.* **2018**, *33*, 8–56. [CrossRef]
9. Miró, M.; Hansen, E.H. On-line sample processing involving microextraction techniques as a front-end to atomic spectrometric detection for trace metal assays: A review. *Anal. Chim. Acta* **2013**, *782*, 1–11. [CrossRef]
10. Fischer, L.; Hann, S.; Worsfold, P.J.; Miró, M. On-line sample treatment coupled with atomic spectrometric detection for the determination of trace elements in natural waters. *J. Anal. At. Spectrom.* **2020**, *35*, 643–670. [CrossRef]
11. Chen, D.; Hu, B.; He, M.; Huang, C. Micro-column preconcentration/separation using thiacalix[4]arene tetracarboxylate derivative modified mesoporous TiO2 as packing materials on-line coupled to inductively coupled plasma optical emission spectrometry for the determination of trace heavy met. *Microchem. J.* **2010**, *95*, 90–95. [CrossRef]
12. Peng, H.; Zhang, N.; He, M.; Chen, B.; Hu, B. Multi-wall carbon nanotubes chemically modified silica microcolumn preconcentration/separation combined with inductively coupled plasma optical emission spectrometry for the determination of trace elements in environmental waters. *Int. J. Environ. Anal. Chem.* **2016**, *96*, 212–224. [CrossRef]
13. Chen, D.; Hu, B.; Huang, C. Chitosan modified ordered mesoporous silica as micro-column packing materials for on-line flow injection-inductively coupled plasma optical emission spectrometry determination of trace heavy metals in environmental water samples. *Talanta* **2009**, *78*, 491–497. [CrossRef] [PubMed]
14. Katarina, R.K.; Oshima, M.; Motomizu, S. High-capacity chitosan-based chelating resin for on-line collection of transition and rare-earth metals prior to inductively coupled plasma-atomic emission spectrometry measurement. *Talanta* **2009**, *79*, 1252–1259. [CrossRef]
15. Katarina, R.K.; Oshima, M.; Motomizu, S. On-line collection/concentration and determination of transition and rare-earth metals in water samples using Multi-Auto-Pret system coupled with inductively coupled plasma-atomic emission spectrometry. *Talanta* **2009**, *78*, 1043–1050. [CrossRef]
16. Anthemidis, A.N.; Paschalidou, M. Unmodified Multi-Walled Carbon Nanotubes as Sorbent Material in Flow Injection on-Line Sorbent Extraction Preconcentration System for Cadmium Determination by Flame Atomic Absorption Spectrometry. *Anal. Lett.* **2012**, *45*, 1098–1110. [CrossRef]
17. Huang, C.; Jiang, Z.; Hu, B. Mesoporous titanium dioxide as a novel solid-phase extraction material for flow injection micro-column preconcentration on-line coupled with ICP-OES determination of trace metals in environmental samples. *Talanta* **2007**, *73*, 274–281. [CrossRef] [PubMed]

18. Su, C.K.; Lin, J.Y. 3D-Printed Column with Porous Monolithic Packing for Online Solid-Phase Extraction of Multiple Trace Metals in Environmental Water Samples. *Anal. Chem.* **2020**, *92*, 9640–9648. [CrossRef] [PubMed]
19. Castro, G.R.; Cristante, V.M.; Padilha, C.C.F.; Jorge, S.M.A.; Florentino, A.O.; Prado, A.G.S.; Padilha, P.M. Determination of Cd(II), Cu(II) and Ni(II) in aqueous samples by ICP-OES after on-line preconcentration in column packed with silica modified with 2-aminothiazole. *Microchim. Acta* **2008**, *160*, 203–209. [CrossRef]
20. Menezes, M.L.; Moreira, J.C.; Campos, J.T.S. Adsorption of various ions from acetone and ethanol on silica gel modified with 2-, 3-, and 4-aminobenzoate. *J. Colloid Interface Sci.* **1996**, *179*, 207–210. [CrossRef]
21. Kabir, A.; Furton, K.G.; Malik, A. Innovations in sol-gel microextraction phases for solvent-free sample preparation in analytical chemistry. *Trends Anal. Chem.* **2013**, *45*, 197–218. [CrossRef]
22. Anthemidis, A.; Kazantzi, V.; Samanidou, V.; Kabir, A.; Furton, K.G. An automated flow injection system for metal determination by flame atomic absorption spectrometry involving on-line fabric disk sorptive extraction technique. *Talanta* **2016**, *156–157*, 64–70. [CrossRef] [PubMed]
23. Alampanos, V.; Kabir, A.; Furton, K.G.; Samanidou, V.; Papadoyannis, I. Fabric phase sorptive extraction for simultaneous observation of four penicillin antibiotics from human blood serum prior to high performance liquid chromatography and photo-diode array detection. *Microchem. J.* **2019**, *149*, 103964. [CrossRef]
24. Matveichuk, Y.V.; Rakhman'Ko, E.M.; Yasinetskii, V.V. Thiocyanate complexes of d metals: Study of aqueous solutions by UV, visible, and IR spectrometry. *Russ. J. Inorg. Chem.* **2015**, *60*, 100–104. [CrossRef]
25. Wilson, R.E.; Carter, T.J.; Autillo, M.; Stegman, S. Thiocyanate complexes of the lanthanides, Am and Cm. *Chem. Commun.* **2020**, *56*, 2622–2625. [CrossRef] [PubMed]
26. Sui, D.P.; Chen, H.X.; Li, D.W. Sol–gel-derived thiocyanato-functionalized silica gel sorbents for adsorption of Fe(III) ions from aqueous solution: Kinetics, isotherms and thermodynamics. *J. Sol-Gel Sci. Technol.* **2016**, *80*, 504–513. [CrossRef]
27. Wu, J.B.; Zang, S.Y.; Yi, Y.L. Sol-gel derived ion imprinted thiocyanato-functionalized silica gel as selective adsorbent of cadmium(II). *J. Sol-Gel Sci. Technol.* **2013**, *66*, 434–442. [CrossRef]
28. Fan, H.T.; Sun, T. Selective removal of iron from aqueous solution using ion imprinted thiocyanato-functionalized silica gel sorbents. *Korean J. Chem. Eng.* **2012**, *29*, 798–803. [CrossRef]
29. Manousi, N.; Kabir, A.; Furton, K.G.; Zachariadis, G.A.; Anthemidis, A. Automated Solid Phase Extraction of Cd (II), Co (II), Cu (II) and Pb (II) Coupled with Flame Atomic Absorption Spectrometry Utilizing a New Sol-Gel Functionalized Silica Sorbent. *Separations* **2021**, *8*, 100. [CrossRef]
30. Saini, S.S.; Kabir, A.; Rao, A.L.J.; Malik, A.K.; Furton, K.G. A novel protocol to monitor trace levels of selected polycyclic aromatic hydrocarbons in environmental water using fabric phase sorptive extraction followed by high performance liquid chromatography-fluorescence detection. *Separations* **2017**, *4*, 22. [CrossRef]
31. Anthemidis, A.N.; Giakisikli, G.; Zachariadis, G.A. The HyperSep SCX micro-cartridge for on-line flow injection inductively coupled plasma atomic emission spectrometric determination of trace elements in biological and environmental samples. *Anal. Methods* **2011**, *3*, 2108–2114. [CrossRef]
32. Miranda, C.E.S.; Reis, B.F.; Baccan, N.; Packer, A.P.; Giné, M.F. Automated flow analysis system based on multicommutation for Cd, Ni and Pb on-line pre-concentration in a cationic exchange resin with determination by inductively coupled plasma atomic emission spectrometry. *Anal. Chim. Acta* **2002**, *453*, 301–310. [CrossRef]

Article

Microwave-Assisted Extraction Coupled to HPLC-UV Combined with Chemometrics for the Determination of Bioactive Compounds in Pistachio Nuts and the Guarantee of Quality and Authenticity

Natasa P. Kalogiouri [1], Petros D. Mitsikaris [2], Athanasios N. Papadopoulos [2] and Victoria F. Samanidou [1,*]

[1] Laboratory of Analytical Chemistry, Department of Chemistry, Aristotle University of Thessaloniki, 54124 Thessaloniki, Greece; kalogiourin@chem.auth.gr

[2] Laboratory of Chemical Biology, Department of Nutritional Sciences and Dietetics, International Hellenic University, Sindos, 57400 Thessaloniki, Greece; petrosmitsikaris@gmail.com (P.D.M.); papadnas@ihu.gr (A.N.P.)

* Correspondence: samanidu@chem.auth.gr

Abstract: Two novel microwave-assisted extraction (MAE) methods were developed for the isolation of phenols and tocopherols from pistachio nuts. The extracts were analyzed by reversed-phase high-pressure liquid chromatography coupled with a UV detector (RP-HPLC-UV). In total, eighteen pistachio samples, originating from Greece and Turkey, were analyzed and thirteen phenolic compounds, as well as α-tocopherol, $(\beta + \gamma)$-tocopherol, and δ-tocopherol, were identified. The analytical methods were validated and presented good linearity ($r^2 > 0.990$) and a high recovery rate over the range of 82.4 to 95.3% for phenols, and 93.1 to 96.4% for tocopherols. Repeatablility was calculated over the range 1.8–5.8%RSD for intra-day experiments, and reproducibility over the range 3.2–9.4%RSD for inter-day experiments, respectively. Principal component analysis (PCA) was employed to analyze the differences between the concentrations of the bioactive compounds with respect to geographical origin, while agglomerative hierarchical clustering (AHC) was used to cluster the samples based on their similarity and according to the geographical origin.

Keywords: microwave-assisted extraction; tocopherols; phenolics; flavonoids; authenticity; HPLC-UV

Citation: Kalogiouri, N.P.; Mitsikaris, P.D.; Papadopoulos, A.N.; Samanidou, V.F. Microwave-Assisted Extraction Coupled to HPLC-UV Combined with Chemometrics for the Determination of Bioactive Compounds in Pistachio Nuts and the Guarantee of Quality and Authenticity. *Molecules* **2022**, *27*, 1435. https://doi.org/10.3390/molecules27041435

Academic Editor: Gregory Chatel

Received: 3 February 2022
Accepted: 17 February 2022
Published: 21 February 2022

Publisher's Note: MDPI stays neutral with regard to jurisdictional claims in published maps and institutional affiliations.

1. Introduction

Nuts are appreciated for their distinctive taste and beneficial health properties. Various clinical and epidemiological studies have demonstrated a direct correlation between the consumption of nuts and numerous improvements in different health markers, such as cholesterol levels [1], glycemic control [2], and waist circumference [3]. The pistachio nut (*Pistacia vera* L.) is a prominent member of the nut family and is valued globally for both its sensory attributes and nutritional value.

Several studies have already displayed the cardioprotective, antioxidant, and anti-inflammatory properties of pistachios [4–8]. The health properties of pistachios are related to their favorable macro- and micronutrient profile. Pistachios have a rich phytochemical content; in particular, they are rich in phenolic compounds and tocopherols, which are two characteristic classes of chemical compounds with proven antioxidant effects [9,10]. According to the literature, the content of both phenols and tocopherols is affected by numerous factors, including plant genetics, variety, geographical origin, pre- and post-harvest factors, and climate conditions [11,12]. All of the aforementioned factors could positively or negatively affect the amounts of these bioactive constituents, hence directly affecting the quality of the nut. The analysis of pistachios could provide useful information in terms of evaluating and differentiating between different cultivars and geographical origins. For this reason, developing and optimizing rapid, easy-to-apply, and widely applicable analytical methods could ensure that nut products hold to a high standard of quality and also protect against adulteration incidents [13].

The isolation and determination of bioactive constituents in food products contains challenges in several distinct steps of the analytical procedure. Sample preparation is pivotal, since it is of the utmost importance to ensure that as little as possible of the targeted substances are lost during the process. There are a handful of protocols that have been introduced to the literature, though most of them propose the use of large volumes of toxic organic solvents [14]. The recent trends in sample preparation necessitate the development of extraction protocols that are compatible with sustainable green chemistry processes [15–17]. Ultrasound-assisted extraction has been successfully applied in the extraction of phenolics and tocopherols using methanol–water mixtures as extraction solvents [18,19]. Microwave-assisted extraction (MAE) is an automated green extraction process that enables the isolation of the target analytes with short times, dramatically decreases solvent consumption, and improves sample throughput [20]. MAE has been widely used for the isolation of functional compounds from plant matrices [19,21,22]; however, only a few works have investigated the use of green solvents such as ethanol–water mixtures, proposing them to be effective solvents that allow for safe extraction operations [17]. This gap in the literature has to be filled. The determination of phenols and tocopherols is commonly achieved through high-pressure liquid chromatography (HPLC) coupled to UV, photodiode arrays (DAD), mass spectrometric (MS) detectors, or a combination of the above [23–30].

The further coupling of the analytical methods with exploratory and discriminatory chemometric techniques allows for an in-depth interpretation of the results in the fields of food authenticity and traceability. Authenticity issues are multivariate, covering different aspects such as characterization, mislabeling, and adulteration. The development of chemometric models enables the extraction of useful information from food authenticity studies, enabling the discrimination of the samples according to their variety, geographical origin, and type of cultivar, among others [31,32].

This study reports, for the first time, the effective use of MAE in the isolation of phenolic compounds and tocopherols from pistachio nuts and pistachio oils, respectively, and their further determination by HPLC-UV. The phenolic and tocopherol content of *Pistacia vera* L. originating from Greece and Turkey was examined for the first time, and principal component analysis (PCA) and agglomerative hierarchical clustering (AHC) were used in order to explore similarities between the samples originating from different geographical regions.

2. Results

2.1. Method Validation

2.1.1. Validation Results of the MAE-HPLC-UV Method for the Determination of Phenolic Compounds

The analytical performance of the MAE-HPLC-UV method was evaluated after measuring its trueness and precision, as well as calculating its linearity, limits of detection (LODs), and limits of quantification (LOQs). The validation results are presented in the Supplementary Materials in Table S1, and as it is observed that the coefficients of determination ranged between 0.991 and 0.997, showing a good linearity for all the analytes. The LODs and LOQs ranged between 0.10 and 0.50 µg/g and 0.20 and 1.80 µg/g, respectively. The RSD% of the within-day (n = 6) and between-day (n = 3 × 3) assays was lower than 5.8 and 8.8, respectively, presenting an adequate precision. The trueness was evaluated by means of relative percentage of recovery (%R) at the lowest, medium, and highest concentration level (0.5, 5, 10 µg/g) and ranged between 83.2 and 95.3% (for the within-day assay (n = 6), as presented in Table S2) and between 82.4 and 94.8% (for the between-day assay (n = 3 × 3), as presented in Table S3).

2.1.2. Validation Results of the MAE-HPLC-UV Method for the Determination of Tocopherols

Table S4 presents the validation results of the MAE-HPLC-UV method for the determination of tocopherols. The calibration curves were linear in the entire working range (5–50 µg/g). The LODs and LOQs ranged from 0.10 to 0.30 and 0.30 to 0.90, respectively. The precision was good, as the RSD% of the within-day (n = 6) and between-day assays (n = 3 × 3) was lower than 5.3 and 8.7, respectively. Trueness was assessed by means of the relative percentage of recovery at the lowest, medium, and highest concentration levels (0.5, 5, 10 µg/g) and ranged between 93.1 and 96.8% for the within-day assay (n = 6), as shown in Table S5, and between 93.3 and 96.4% for the between-day assay (n = 3 × 3), as shown in Table S6.

2.2. Pistachio Nuts Analysis

2.2.1. Identification and Quantification Results of Phenolics

The developed methods were applied in the analysis of phenols and tocopherols in eighteen pistachio samples produced in Turkey and Greece. In total, thirteen phenolic compounds were determined. Catechin, diosmin, epicatechin, epigallocatechin, gallic acid, luteolin, rosmarinic acid, sinapic acid, syringaldehyde, syringic acid, trans-cinnamic acid, vanillic acid, and vanillin were all determined. Figure 1 shows a characteristic chromatogram of a pistachio sample spiked at 5 µg/g with the target analytes. The concentration ranges of all of the determined analytes, as well as the mean values for each sample, are presented in Table 1. All samples were analyzed in triplicate (n = 3, µg/g ±SD).

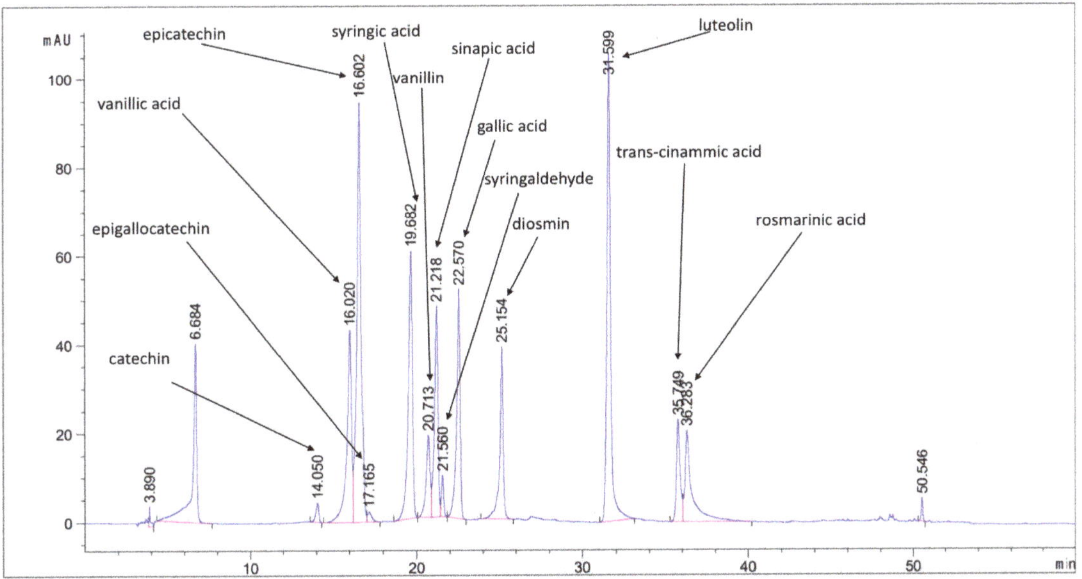

Figure 1. Characteristic chromatogram of a pistachio sample spiked at 5 µg/g and monitored at 280 nm.

The concentration ranges of the determined phenolic compounds are in accordance with the current literature [33,34]. According to Table 2, diosmin, epigallocatechin, and rosmarinic acid were not detected in the pistachio nuts originating from Turkey. On the contrary, the average concentration of diosmin, epigallocatechin, and rosmarinic acid was equal to 24.73, 6.51, and 8.55 µg/g, respectively, in the pistachio nuts produced in Greece. Gallic acid exhibited the highest concentration in all of the samples. The determined phenolic compounds have been associated with numerous health-promoting effects, such as antioxidant, anti-inflammatory, antimicrobial, anti-diabetic, anti-mutagenic, and cytoprotective effects, which corroborate the beneficial health effects associated with pistachios' consumption.

Table 1. Concentration ranges and mean values of the phenolic analytes determined in pistachios originating from Greece and Turkey.

	Greek Pistachios		Turkish Pistachios	
Compound	**Concentration Range (μg/g)**	**Mean Value ($\pm$SD, μg/g)**	**Concentration Range (μg/g)**	**Mean Value ($\pm$SD, μg/g)**
catechin	25.21–46.80	37.08 $\pm$ 5.42	5.96–22.00	13.02 $\pm$ 4.00
diosmin	22.60–29.48	24.73 $\pm$ 2.04	ND	ND
epicatechin	78.20–124.58	90.12 $\pm$ 13.28	ND–5.64	3.21 $\pm$ 1.95
epigallocatechin	ND–12.60	6.51 $\pm$ 4.20	ND	ND
gallic acid	225.77–274.00	249.83 $\pm$ 14.25	122.00–188.00	151.23 $\pm$ 21.74
luteolin	12.97–29.74	21.20 $\pm$ 4.82	ND–5.86	4.27 $\pm$ 1.63
rosmarinic acid	4.32–14.60	8.55 $\pm$ 3.32	ND	ND
sinapic acid	39.14–66.40	55.64 $\pm$ 7.86	ND–2.21	0.54 $\pm$ 0.80
syringaldehyde	15.12–23.79	20.40 $\pm$ 3.03	1.85–12.40	7.13 $\pm$ 3.02
syringic acid	12.60–15.60	14.07 $\pm$ 1.05	ND–2.45	1.32 $\pm$ 1.01
trans-cinammic acid	ND–0.88	0.18 $\pm$ 0.35	1.05–2.24	1.67 $\pm$ 0.34
vanillic acid	ND	ND	3.21–5.32	4.42 $\pm$ 0.61
vanillin	3.22–8.32	6.27 $\pm$ 1.49	1.06–3.33	1.86 $\pm$ 0.71

Table 2. Concentration ranges and mean values of the tocopherols determined in pistachios originating from Greece and Turkey.

Tocopherols	**Greek Pistachios**		**Turkish Pistachios**	
	Concentration Range (μg/g)	**Mean Value ($\pm$SD, μg/g)**	**Concentration Range (μg/g)**	**Mean Value ($\pm$SD, μg/g)**
α-tocopherol	36.00–78.00	57.77 $\pm$ 11.99	13.00–25.00	16.56 $\pm$ 3.50
($\beta + \gamma$)-tocopherol	78.00–152.00	115.44 $\pm$ 23.97	105.09–156.00	129.12 $\pm$ 17.56
δ-tocopherol	10.99–18.90	15.35 $\pm$ 2.46	20.80–27.40	24.12 $\pm$ 2.36

2.2.2. Determination of Tocopherols

The separation of tocopherols was achieved within 12 min. The gradient elution program enabled the separation of δ-tocopherol (Rt = 8.9 min) and α-tocopherol (Rt = 11.1 min). β- and γ-tocopherol co-eluted (Rt = 9.8 min) and were analyzed together, hereafter referred to as ($\beta + \gamma$)-tocopherol [35]. A representative chromatogram of a 5 μg/g standard solution mixture is shown in Figure 2. Each pistachio oil was analyzed in triplicate. The quantification ranges and the mean values ($\pm$SD) are presented in Table 2. The determined quantification ranges were in accordance with previously reported data [36]. According to the results, α-tocopherol, ($\beta + \gamma$)-tocopherol, and δ-tocopherol existed within the samples in relatively high concentrations. The highest average concentrations were observed in the mixture of ($\beta + \gamma$)-tocopherol, equal to 115.44 μg/g in the Aegina type pistachios from Greece, and 129.12 μg/g in the Antep type pistachios from Turkey. Lower α-tocopherol concentration ranges were determined in pistachios produced in Turkey (13.00–25.00 μg/g) as compared to the pistachios produced in Greece (36.00–78.00 μg/g). On the other hand, the average concentration of δ-tocopherol was higher in the pistachios originating from Turkey (24.12 μg/g) as compared to those produced in Greece (10.99 μg/g).

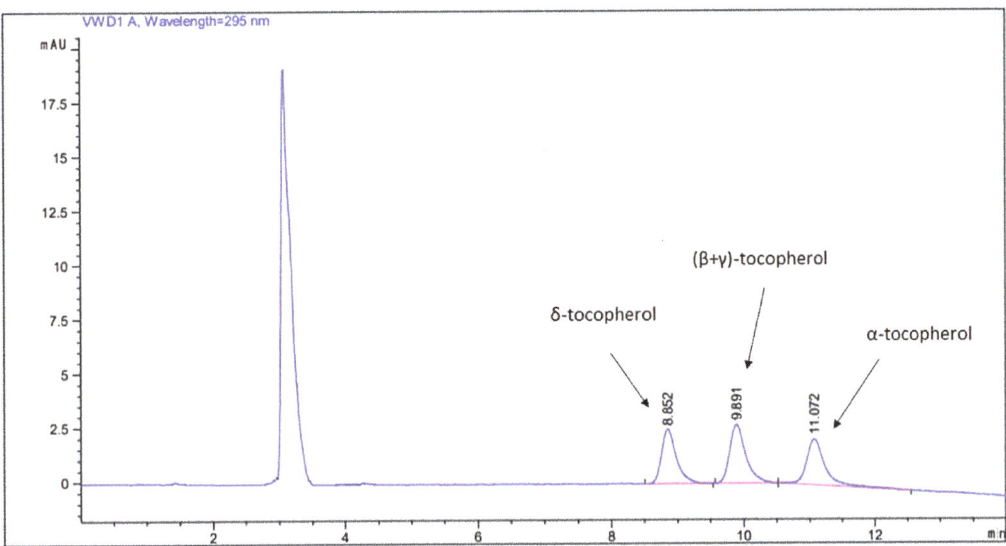

Figure 2. Characteristic chromatogram of a pistachio sample spiked at 5 μg/g and monitored at 295 nm.

2.3. Chemometrics

2.3.1. Principal Component Analysis

PCA was employed to explore distribution of the samples and formation of groups on the basis of the concentrations of the determined bioactive compounds. The data matrix consisted of eighteen pistachio samples originating from Greece and Turkey and sixteen features (the concentration results of phenolics and tocopherols). The MetaboAnalyst package was used for PCA, and the data matrix was auto-scaled [37]. Figure 3 presents the score plot and it is observed that the samples clustered into two individual groups according to the country of production. The pistachio nuts originating from Greece were grouped into the green ellipse, and those produced in Turkey were grouped into the red ellipse. The first two principal components (PCs) explained 55% of the total variance, establishing two individual groups of samples according to the country of production. The biplot presented in Figure S1 graphically shows, using the loadings, the effects of the variables in each PC.

2.3.2. Agglomerative Hierarchical Clustering

AHC was employed as a cluster analysis method to build a hierarchy of clusters in a tree diagram (also known as a dendrogram) to identify samples that present similarities and to group all of the objects that share similar characteristics into a large cluster [37]. The dendrogram presented in Figure 4 shows that the observations clustered into two major groups, one which was comprised of all of the pistachio samples produced in Greece (G1–G9), and a second one which comprised all of the samples originating from Turkey (T1–T9).

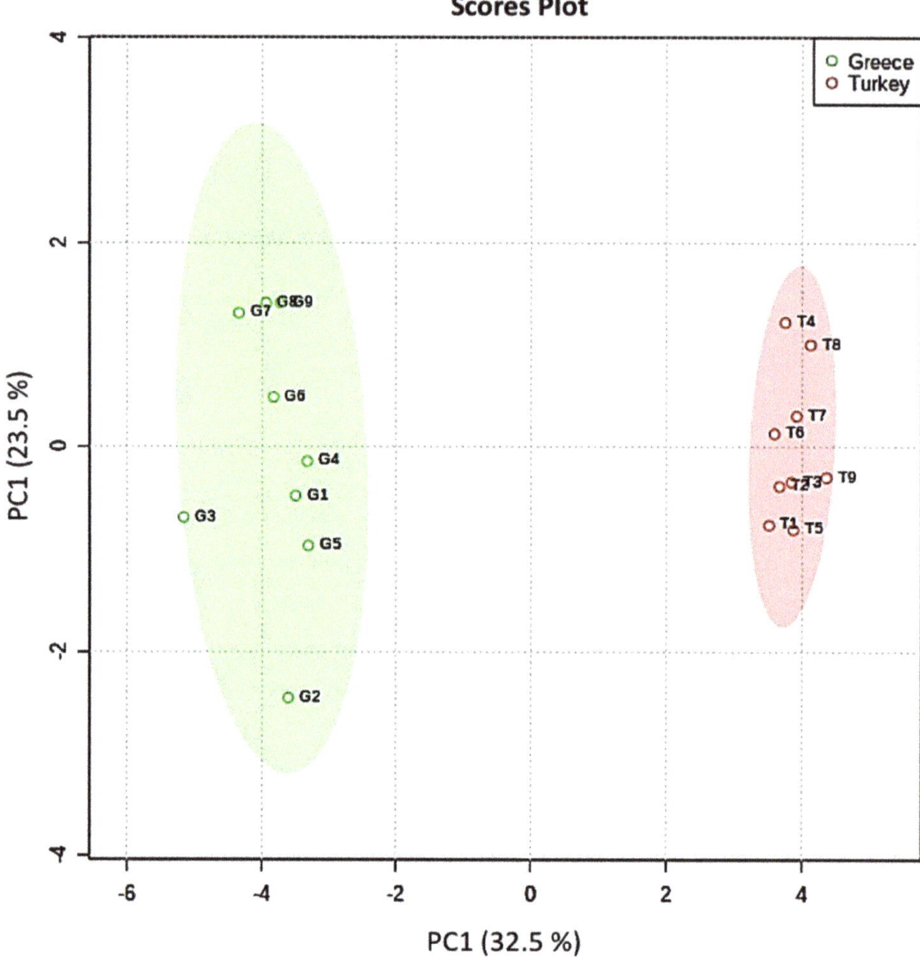

Figure 3. PCA scores plot illustrating the clustering between pistachio nuts originating from Greece and Turkey.

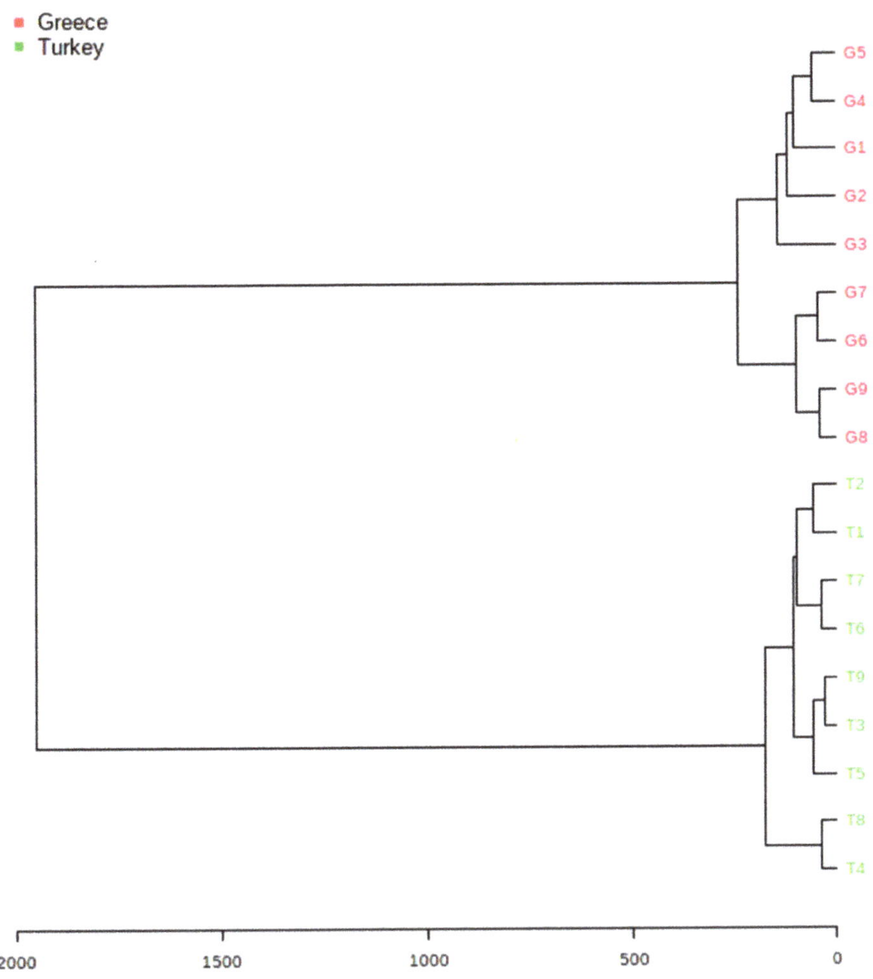

Figure 4. AHCdendrogram of pistachios originating from Greece (G1–G9) and Turkey (T1–T9) clustered into two major groups.

3. Materials and Methods

3.1. Chemicals and Reagents

HPLC-grade Methanol (MeOH) and HPLC-grade acetonitrile (ACN) were purchased from Carl Roth (Carlsruhe, Germany). Isopropanol (IPA) was purchased from Panreac-AppliChem (Darmstadt, Germany). Acetic acid (99%) was purchased from Sigma-Aldrich (Steinheim, Germany). To obtain ultrapure water, a Milli-Q purification system (Millipore, Bedford, MA, USA) was utilized. Analytical standards of catechin (98%), diosmin (97%), epicatechin (97%), epigallocatechin (98%), gallic acid (98%), trans-cinnamic acid (97%), syringaldehyde (98%), rosmarinic acid (98%), sinapic acid (95%), syringic acid (95%), vanillic acid (97%), luteolin (98%), vanillin (98%), α-tocopherol (96%), β-tocopherol (96%), γ-tocopherol (96%), and δ-tocopherol (96%) were acquired from Sigma-Aldrich (Steinheim, Germany). Stock standard solutions at 1000 μg/g were prepared for each analyte in methanol and stored at –20 °C.

3.2. Sampling and Pre-Treatment

Eighteen pistachio samples (approximately 300 g each) originating from Greece (Aegina type) and Turkey (Antep type) were collected from local producers and importers. These samples were available in the Greek market in 2020. The pistachio seeds were separated from the endocarps, and each sample of 300 g was homogenized and then crushed using a mortar and pestle. The samples were lyophilized for 24 h. The lyophilized samples were stored at –20 °C until analysis.

3.3. Instrumentation

A 1220 Infinity HPLC-UV system from Agilent Technologies (Santa Clara, CA, USA) was used for the chromatographic analysis of the phenolic compounds and tocopherols. The HPLC system consisted of a degasser, a column oven, a manual injector, and a UV detector. For monitoring the analysis, the OpenLAB software (Agilent Technologies, Santa Clara, CA, USA) was used with the Method and Run Control package. For peak identification and data processing, the Data Analysis software package (Agilent Technologies, Santa Clara, CA, USA) was used. QMax RR syringe filters (0.22 μm Nylon; and 0.22 PTFE) were acquired from Frisenette ApS (Knebel, Denmark) to filter the samples prior to analysis. For agitation, a vortex mixer from VELP Scientifica (Usmate Velate, Italy) was used. Centrifugation was carried out using a 3-16PK centrifuge system from Sigma (Osterode am Harz, Germany). For the green extractions, a MARS X Model 1000 microwave oven (CEM, Matthews, NC, USA) equipped with a 14-position holder, PTFE vessels, and a stirring mechanism was used.

3.4. Microwave-Assisted Extraction

For the extraction of the phenolics, a modified MAE protocol previously reported by Gallo et al. [38] was used. Extractions were performed at 200 W and 50 °C, and a magnetic stirring rod was added into each vessel. Approximately 1 g of each lyophilized sample was extracted with a mixture of ethanol and water (50:50 *v/v*, 5 mL), at 200 W using magnetic stirring at 50% of nominal power and a temperature of 50 °C for 15 min. The extract was collected and centrifuged for 10 min at 12,000 rpm. Then, 0.5 mL of the supernatant was collected, dried under nitrogen air, and reconstituted in 0.5 mL of MeOH–mobile phase A (10:90, *v/v*) and filtered through 0.22 μm nylon syringe filters. Finally, 20 μL was injected into the chromatographic system.

Prior to the isolation of the tocopherols, a MAE protocol was employed for oil extraction [39]. Approximately 1 g of lyophilized sample was extracted using a mixture of acetone (2:1 *v/v*, 5 mL), at a microwave power of 420 W and temperature of 80 °C for 30 min. After extraction, the solvent was distilled with a rotary evaporator at 60 °C. The oil was collected. For the extraction of tocopherols, a modified version of a previously reported protocol [40] was applied using 20 mg of oil dissolved in 1 mL of ethanol in a 2 mL Eppendorf tube. The mixture was vortexed for 1 min at 3000 rpm and then centrifuged for 10 min at 12,000 rpm. The supernatant was collected and filtered through 0.22 μm PTFE syringe filters and 20 μL was injected into the HPLC system.

3.5. Chromatographic Analysis

The analysis of the polyphenols was conducted at 280 nm in a Macherey-Nagel (Düren, Nordrhein-Westfalen, Germany) Nucleosil RP-18 analytical column (250 mm × 4.6 mm, 5 μm particle size). The mobile phase comprised of 1% *v/v* formic acid in water (A) and ACN 0.5% (B), at a 1 mL/min flow rate. The column was thermostatically controlled at 30 °C. The flow rate was set at 1 mL/min and each chromatographic run had a duration of 60 min. The chromatographic program started with 5% B, gradually increased to 20% during the first 15 min, then to 50% until 40 min, then from 50% to 90% over the following 5 min, and then kept stable until 50 min. Then, the organic phase decreased to 5% between 50 and 55 min, and then remained constant for the next 5 min.

For the analysis of the tocopherols, a Kromasil RP-18 analytical column (125 mm × 4.6 mm, 5 μm particle size) was acquired from Macherey-Nagel. The UV wavelength was set to 295 nm. The mobile phase consisted of MeOH (A) and ACN (B). The column was thermostatically controlled at 30 °C, and the flow rate was set to 1 mL/min. Each chromatographic run lasted for 15 min. The elution program started with 50% B and remained stable for 7 min, then decreased to 0% over the following 5 min and remained stable for 3 min. For the identification of the compounds, the RTs of the peaks in the extracts were compared with the RTs of their respective standards. Spiked extracts of different concentrations (e.g., 0.5, 5, 10, and 20 μg/g), depending on the peak intensities of the real samples, were also injected into the chromatographic system and the RTs were matched with the corresponding neat extracts to verify their presence.

3.6. Method Validation

The developed methods were validated in terms of linearity, trueness, precision, limits of detection (LODs), and limits of quantification (LOQs). To assess linearity, calibration curves were constructed over the range 0.5–20 μg/g for phenolic analytes and over the range 5–50 μg/g for tocopherols using 8 concentration points. The LODs were equal to 3.3 multiplied by the signal-to-noise ratio (S/N), while LOQs were equal to 10 multiplied by the S/N ratio. Trueness and precision were evaluated using a real spike at three concentration levels: 0.5, 10, and 20 μg/g for the phenolic compounds, and 5, 25, and 50 μg/g for the tocopherols [40]. Within-day precision (repeatability) was assessed using six replicate spiked samples (n = 6), while between-day precision (reproducibility) was estimated by analyzing spiked samples in triplicate over three days (n = 3 × 3).

3.7. Chemometric Analysis

PCA is an unsupervised technique used for exploratory chemometric data analysis. PCA is a dimensionality reduction method that computes the PCs and uses them make changes in a dataset. AHC was used to cluster the samples according to their resemblance. PCA and AHC were executed in R using the MetaboAnalyst 5.0 package [41].

4. Conclusions

Two MAE-HPLC-UV methods were proposed as innovative green approaches for the analysis of phenolic compounds and tocopherols in pistachio nuts. Eighteen pistachio samples produced in Turkey and Greece were analyzed, and the phenolic compounds catechin, diosmin, epicatechin, epigallocatechin, gallic acid, luteolin, rosmarinic acid, sinapic acid, syringaldehyde, syringic acid, trans-cinammic acid, vanillic acid, and vanillin were determined. In pistachio oils, α-tocopherol, δ-tocopherol, and the sum of β- and γ-tocopherols were determined. The quantification results were analyzed using PCA, and the samples were distributed into two individual groups according to the geographical origin. The first two PCs explained 56% of the total variance. Furthermore, an AHCdendrogram was also created, which also clustered the samples, on the basis of their similarities, into two major groups. Overall, two green extraction methods are suggested for the isolation of phenolics and tocopherols, supporting the idea that these compounds could be used as markers for the authentication of pistachios.

Supplementary Materials: The following supporting information can be downloaded online. Table S1: Analytical parameters of the MAE-HPLC-UV method for the determination of phenolics; Table S2: Intra-day recoveries and repeatability results of the MAE-HPLC-UV for the determination of phenolics; Table S3: Inter-day recoveries (%R) and reproducibility results of the MAE-HPLC-UV for the determination of phenolics; Table S4: Analytical parameters of the MAE-HPLC-UV for the determination of tocopherols; Table S5: Inter-day recoveries and repeatability results of the MAE-HPLC-UV method for the determination of tocopherols; Table S6: Inter-day recoveries (%R) and reproducibility results of the MAE-HPLC-UV for the determination of tocopherols; Figure S1: PCA biplot.

Author Contributions: Conceptualization, N.P.K.; methodology, N.P.K.; validation, N.P.K.; formal analysis, N.P.K. and P.D.M.; investigation, N.P.K.; resources, A.N.P.; data curation, N.P.K. and P.D.M.; writing—original draft preparation, N.P.K. and P.D.M.; writing—review and editing, N.P.K., A.N.P. and V.F.S.; visualization, N.P.K.; supervision, V.F.S.; project administration, N.P.K. and V.F.S. All authors have read and agreed to the published version of the manuscript.

Funding: This research was co-financed by Greece and the European Union (European Social Fund, ESF) through the Operational Program "Human Resources Development, Education and Lifelong Learning" in the context of the project "Reinforcement of Postdoctoral Researchers—2nd Cycle" (MIS-5033021), implemented by the State Scholarships Foundation (IKY) with grant number 2019-050-0503-17749.

Institutional Review Board Statement: Not applicable.

Informed Consent Statement: Not applicable.

Data Availability Statement: Not applicable.

Conflicts of Interest: The authors declare no conflict of interest.

Sample Availability: Samples of the compounds are not available from the authors.

References

1. Del Gobbo, L.C.; Falk, M.C.; Feldman, R.; Lewis, K.; Mozaffarian, D. Effects of tree nuts on blood lipids, apolipoproteins, and blood pressure: Systematic review, meta-analysis, and dose-response of 61 controlled intervention trials. *Am. J. Clin. Nutr.* **2015**, *102*, 1347–1356. [CrossRef] [PubMed]
2. Tindall, A.M.; Johnston, E.A.; Kris-Etherton, P.M.; Petersen, K.S. The effect of nuts on markers of glycemic control: A systematic review and meta-Analysis of randomized controlled trials. *Am. J. Clin. Nutr.* **2019**, *109*, 297–314. [CrossRef] [PubMed]
3. Li, H.; Li, X.; Yuan, S.; Jin, Y.; Lu, J. Nut consumption and risk of metabolic syndrome and overweight/obesity: A meta-analysis of prospective cohort studies and randomized trials. *Nutr. Metab.* **2018**, *15*, 1–10. [CrossRef] [PubMed]
4. Sari, I.; Baltaci, Y.; Bagci, C.; Davutoglu, V.; Erel, O.; Celik, H.; Ozer, O.; Aksoy, N.; Aksoy, M. Effect of pistachio diet on lipid parameters, endothelial function, inflammation, and oxidative status: A prospective study. *Nutrition* **2010**, *26*, 399–404. [CrossRef]
5. Gebauer, S.K.; West, S.G.; Kay, C.D.; Alaupovic, P.; Bagshaw, D.; Kris-Etherton, P.M. Effects of pistachios on cardiovascular disease risk factors and potential mechanisms of action: A dose-response study. *Am. J. Clin. Nutr.* **2008**, *88*, 651–659. [CrossRef]
6. Edwards, K.; Kwaw, I.; Matud, J.; Kurtz, I. Effect of Pistachio Nuts on Serum Lipid Levels in Patients with Moderate Hypercholesterolemia. *J. Am. Coll. Nutr.* **1999**, *18*, 229–232. [CrossRef]
7. Kocyigit, A.; Koylu, A.A.; Keles, H. Effects of pistachio nuts consumption on plasma lipid profile and oxidative status in healthy volunteers. *Nutr. Metab. Cardiovasc. Dis.* **2006**, *16*, 202–209. [CrossRef]
8. Saito, M.; Iwabuchi, H.; Yang, P.; Tang, G.; King, M.D.; Sekiguchi, M. Ice particle morphology andmicrophysical properties of cirrus clouds inferred from combined CALIOP-IIR measurements. *J. Geophys. Res.* **2017**, *122*, 4440–4462. [CrossRef]
9. Liao, X.; Brock, A.A.; Jackson, B.T.; Greenspan, P.; Pegg, R.B. The cellular antioxidant and anti-glycation capacities of phenolics from Georgia peaches. *Food Chem.* **2020**, *316*, 126234. [CrossRef]
10. Karmowski, J.; Hintze, V.; Kschonsek, J.; Killenberg, M.; Böhm, V. Antioxidant activities of tocopherols/tocotrienols and lipophilic antioxidant capacity of wheat, vegetable oils, milk and milk cream by using photochemiluminescence. *Food Chem.* **2015**, *175*, 593–600. [CrossRef]
11. Ghisoni, S.; Lucini, L.; Rocchetti, G.; Chiodelli, G.; Farinelli, D.; Tombesi, S.; Trevisan, M. Untargeted metabolomics with multivariate analysis to discriminate hazelnut (*Corylus avellana* L.) cultivars and their geographical origin. *J. Sci. Food Agric.* **2020**, *100*, 500–508. [CrossRef] [PubMed]
12. Bolling, B.W.; Chen, C.Y.O.; McKay, D.L.; Blumberg, J.B. Tree nut phytochemicals: Composition, antioxidant capacity, bioactivity, impact factors. A systematic review of almonds, Brazils, cashews, hazelnuts, macadamias, pecans, pine nuts, pistachios and walnuts. *Nutr. Res. Rev.* **2011**, *24*, 244–275. [CrossRef] [PubMed]
13. Kalogiouri, N.P.; Samanidou, V.F. HPLC Fingerprints for the Characterization of Walnuts and the Detection of Fraudulent Incidents. *Foods* **2021**, *10*, 2145. [CrossRef] [PubMed]
14. de Albuquerque Mendes, M.K.; dos Santos Oliveira, C.B.; Veras, M.D.A.; Araújo, B.Q.; Dantas, C.; Chaves, M.H.; Lopes Júnior, C.A.; Vieira, E.C. Application of multivariate optimization for the selective extraction of phenolic compounds in cashew nuts (*Anacardium occidentale* L.). *Talanta* **2019**, *205*, 120100. [CrossRef]
15. Bodoira, R.; Maestri, D. Phenolic Compounds from Nuts: Extraction, Chemical Profiles, and Bioactivity. *J. Agric. Food Chem.* **2020**, *68*, 927–942. [CrossRef]
16. Gałuszka, A.; Migaszewski, Z.; Namieśnik, J. The 12 principles of green analytical chemistry and the SIGNIFICANCE mnemonic of green analytical practices. *TrAC Trends Anal. Chem.* **2013**, *50*, 78–84. [CrossRef]

17. Kalogiouri, N.P.; Manousi, N.; Rosenberg, E.; Zachariadis, G.A.; Samanidou, V.F. Advances in the Chromatographic Separation and Determination of Bioactive Compounds for Assessing the Nutrient Profile of Nuts. *Curr. Anal. Chem.* **2020**, *16*, 1–17. [CrossRef]
18. Bodoira, R.; Cittadini, M.C.; Velez, A.; Rossi, Y.; Martínez, M.; Maestri, D. An overview on extraction, composition, bioactivity and food applications of peanut phenolics. *Food Chem.* **2022**, 132250. [CrossRef]
19. Vinatoru, M.; Mason, T.J.; Calinescu, I. Ultrasonically assisted extraction (UAE) and microwave assisted extraction (MAE) of functional compounds from plant materials. *TrAC Trends Anal. Chem.* **2017**, *97*, 159–178. [CrossRef]
20. Llompart, M.; Celeiro, M.; Dagnac, T. Microwave-assisted extraction of pharmaceuticals, personal care products and industrial contaminants in the environment. *TrAC Trends Anal. Chem.* **2019**, *116*, 136–150. [CrossRef]
21. Eskilsson, C.S.; Bjorklund, E. Analytical-scale microwave-assisted extraction. *J. Chromatogr. A* **2000**, *902*, 227–250. [CrossRef]
22. Barreca, D.; Laganà, G.; Leuzzi, U.; Smeriglio, A.; Trombetta, D.; Bellocco, E. Evaluation of the nutraceutical, antioxidant and cytoprotective properties of ripe pistachio (*Pistacia vera* L., variety Bronte) hulls. *Food Chem.* **2016**, *196*, 493–502. [CrossRef] [PubMed]
23. Noguera-Artiaga, L.; Salvador, M.D.; Fregapane, G.; Collado-González, J.; Wojdyło, A.; López-Lluch, D.; Carbonell-Barrachina, Á.A. Functional and sensory properties of pistachio nuts as affected by cultivar. *J. Sci. Food Agric.* **2019**, *99*, 6696–6705. [CrossRef] [PubMed]
24. Noguera-Artiaga, L.; Pérez-López, D.; Burgos-Hernández, A.; Wojdyło, A.; Carbonell-Barrachina, Á.A. Phenolic and triterpenoid composition and inhibition of α-amylase of pistachio kernels (*Pistacia vera* L.) as affected by rootstock and irrigation treatment. *Food Chem.* **2018**, *261*, 240–245. [CrossRef]
25. Ballistreri, G.; Arena, E.; Fallico, B. Influence of ripeness and drying process on the polyphenols and tocopherols of *Pistacia vera* L. *Molecules* **2009**, *14*, 4358–4369. [CrossRef]
26. Schlörmann, W.; Birringer, M.; Böhm, V.; Löber, K.; Jahreis, G.; Lorkowski, S.; Müller, A.K.; Schöne, F.; Glei, M. Influence of roasting conditions on health-related compounds in different nuts. *Food Chem.* **2015**, *180*, 77–85. [CrossRef]
27. Delgado-Zamarreño, M.M.; Fernández-Prieto, C.; Bustamante-Rangel, M.; Pérez-Martín, L. Determination of tocopherols and sitosterols in seeds and nuts by QuEChERS-liquid chromatography. *Food Chem.* **2016**, *192*, 825–830. [CrossRef]
28. Ryan, E.; Galvin, K.; O'Connor, T.P.; Maguire, A.R.; O'Brien, N.M. Fatty acid profile, tocopherol, squalene and phytosterol content of brazil, pecan, pine, pistachio and cashew nuts. *Int. J. Food Sci. Nutr.* **2006**, *57*, 219–228. [CrossRef]
29. Stuetz, W.; Schlörmann, W.; Glei, M. B-vitamins, carotenoids and α-/γ-tocopherol in raw and roasted nuts. *Food Chem.* **2017**, *221*, 222–227. [CrossRef]
30. Wang, Y.-P.; Zou, Y.-R.; Shi, J.-T.; Shi, J. Review of the chemometrics application in oil-oil and oil-source rock correlations. *J. Nat. Gas Geosci.* **2018**, *3*, 217–232. [CrossRef]
31. Jurado-Campos, N.; García-Nicolás, M.; Pastor-Belda, M.; Bußmann, T.; Arroyo-Manzanares, N.; Jiménez, B.; Viñas, P.; Arce, L. Exploration of the potential of different analytical techniques to authenticate organic vs. conventional olives and olive oils from two varieties using untargeted fingerprinting approaches. *Food Control* **2021**, *124*, 107828. [CrossRef]
32. Ojeda-Amador, R.M.; Salvador, M.D.; Fregapane, G.; Gómez-Alonso, S. Comprehensive Study of the Phenolic Compound Profile and Antioxidant Activity of Eight Pistachio Cultivars and Their Residual Cakes and Virgin Oils. *J. Agric. Food Chem.* **2019**, *67*, 3583–3594. [CrossRef] [PubMed]
33. Arjeh, E.; Akhavan, H.R.; Barzegar, M.; Carbonell-Barrachina, Á.A. Bio-active compounds and functional properties of pistachio hull: A review. *Trends Food Sci. Technol.* **2020**, *97*, 55–64. [CrossRef]
34. Gliszczynska-Swiglo, A.; Sikorska, E.; Khmelinskii, I.; Sikorski, M. Tocopherol content in edible plant oils. *Polish J. Food Nutr. Sci.* **2007**, *57*, 157–161.
35. Özrenk, K.; Javidipour, I.; Yarilgac, T.; Balta, F.; Gündoğdu, M. Fatty acids, tocopherols, selenium and total carotene of pistachios (*P. vera* L.) from Diyarbakir (Southestern Turkey) and walnuts (*J. regia* L.) from Erzincan (Eastern Turkey). *Food Sci. Technol. Int.* **2012**, *18*, 55–62. [CrossRef] [PubMed]
36. Granato, D.; Santos, J.S.; Escher, G.B.; Ferreira, B.L.; Maggio, R.M. Use of principal component analysis (PCA) and hierarchical cluster analysis (HCA) for multivariate association between bioactive compounds and functional properties in foods: A critical perspective. *Trends Food Sci. Technol.* **2018**, *72*, 83–90. [CrossRef]
37. Gallo, M.; Ferracane, R.; Graziani, G.; Ritieni, A.; Fogliano, V. Microwave assisted extraction of phenolic compounds from four different spices. *Molecules* **2010**, *15*, 6365–6374. [CrossRef]
38. Hu, B.; Zhou, K.; Liu, Y.; Liu, A.; Zhang, Q.; Han, G.; Liu, S.; Yang, Y.; Zhu, Y.; Zhu, D. Optimization of microwave-assisted extraction of oil from tiger nut (*Cyperus esculentus* L.) and its quality evaluation. *Ind. Crops Prod.* **2018**, *115*, 290–297. [CrossRef]
39. Kalogiouri, N.P.; Mitsikaris, P.D.; Klaoudatos, D.; Papadopoulos, A.N.; Samanidou, V.F. A rapid HPLIC-UV protocol coupled to chemometric analysis for the determination of the major phenolic constituents and tocopherol content in almonds and the discrimination of the geographical origin. *Molecules* **2021**, *26*, 5433. [CrossRef]
40. European Commission. SANTE/12682/2019 Guidance document on analytical quality control and method validation for pesticide residues analysis in food and feed. *Saf. Food Chain Pestic. Biocides Eur. Comm.* **2019**, 1–48.
41. Pang, Z.; Chong, J.; Zhou, G.; De Lima Morais, D.A.; Chang, L.; Barrette, M.; Gauthier, C.; Jacques, P.É.; Li, S.; Xia, J. MetaboAnalyst 5.0: Narrowing the gap between raw spectra and functional insights. *Nucleic Acids Res.* **2021**, *49*, W388–W396. [CrossRef] [PubMed]

Article

Application of Bar Adsorptive Microextraction for the Determination of Levels of Tricyclic Antidepressants in Urine Samples

Mariana N. Oliveira [1], Oriana C. Gonçalves [1], Samir M. Ahmad [1,2,3], Jaderson K. Schneider [4], Laiza C. Krause [4], Nuno R. Neng [1,5,*], Elina B. Caramão [4,6] and José M. F. Nogueira [1,5,*]

1   Centro de Química Estrutural, Faculdade de Ciências, Universidade de Lisboa, Campo Grande, 1749-016 Lisboa, Portugal; mariananetoliveira@hotmail.com (M.N.O.); ocgp98@gmail.com (O.C.G.); samir.marcos.ahmad@gmail.com (S.M.A.)
2   Molecular Pathology and Forensic Biochemistry Laboratory, Centro de Investigação Interdisciplinar Egas Moniz (CiiEM), Instituto Universitário Egas Moniz (IUEM), Campus Universitário—Quinta da Granja, Monte da Caparica, 2829-511 Caparica, Portugal
3   Forensic and Psychological Sciences Laboratory Egas Moniz, Campus Universitário—Quinta da Granja, Monte da Caparica, 2829-511 Caparica, Portugal
4   Instituto de Química, Universidade Federal do Rio Grande do Sul, 91509-900 Porto Alegre, Brazil; jadersonqmc@gmail.com (J.K.S.); laiza_canielas@hotmail.com (L.C.K.); elina@ufrgs.br (E.B.C.)
5   Departamento de Química e Bioquímica, Faculdade de Ciências, Universidade de Lisboa, Campo Grande, 1749-016 Lisboa, Portugal
6   Programa de Pós-Graduação em Biotecnologia Industrial, Universidade Tiradentes, 49032-490 Aracaju, Brazil
*   Correspondence: ndneng@fc.ul.pt (N.R.N.); nogueira@fc.ul.pt (J.M.F.N.)

Citation: Oliveira, M.N.; Gonçalves, O.C.; Ahmad, S.M.; Schneider, J.K.; Krause, L.C.; Neng, N.R.; Caramão, E.B.; Nogueira, J.M.F. Application of Bar Adsorptive Microextraction for the Determination of Levels of Tricyclic Antidepressants in Urine Samples. *Molecules* **2021**, *26*, 3101. https://doi.org/10.3390/molecules26113101

Academic Editors: Victoria Samanidou and Irene Panderi

Received: 15 April 2021
Accepted: 18 May 2021
Published: 22 May 2021

Publisher's Note: MDPI stays neutral with regard to jurisdictional claims in published maps and institutional affiliations.

Abstract: This work entailed the development, optimization, validation, and application of a novel analytical approach, using the bar adsorptive microextraction technique (BAμE), for the determination of the six most common tricyclic antidepressants (TCAs; amitriptyline, mianserin, trimipramine, imipramine, mirtazapine and dosulepin) in urine matrices. To achieve this goal, we employed, for the first time, new generation microextraction devices coated with convenient sorbent phases, polymers and novel activated carbons prepared from biomaterial waste, in combination with large-volume-injection gas chromatography-mass spectrometry operating in selected-ion monitoring mode (LVI-GC-MS(SIM)). Preliminary assays on sorbent coatings, showed that the polymeric phases present a much more effective performance, as the tested biosorbents exhibited low efficiency for application in microextraction techniques. By using BAμE coated with C_{18} polymer, under optimized experimental conditions, the detection limits achieved for the six TCAs ranged from 0.2 to 1.6 μg L^{-1} and, weighted linear regressions resulted in remarkable linearity ($r^2 > 0.9960$) between 10.0 and 1000.0 μg L^{-1}. The developed analytical methodology (BAμE(C18)/LVI-GC-MS(SIM)) provided suitable matrix effects (90.2–112.9%, RSD ≤ 13.9%), high recovery yields (92.3–111.5%, RSD ≤ 12.3%) and a remarkable overall process efficiency (ranging from 84.9% to 124.3%, RSD ≤ 13.9%). The developed and validated methodology was successfully applied for screening the six TCAs in real urine matrices. The proposed analytical methodology proved to be an eco-user-friendly approach to monitor trace levels of TCAs in complex urine matrices and an outstanding analytical alternative in comparison with other microextraction-based techniques.

Keywords: tricyclic antidepressants; urine samples; bar adsorptive microextraction (BAμE); novel sorbent phases; biomaterials waste; flotation sampling technology; GC-MS

1. Introduction

According to the World Health Organization, depression is a common mental disorder, being one of the biggest causes of incapacity around the world. Depression can be characterized by a vast number of symptoms, including but not limited to sadness, low self-esteem, difficulty to sleep, loss of appetite, fatigue, low concentration, and poor decision making. In the most severe cases it can even lead to suicide. Antidepressants are an effective way of treatment, which allows the patients to live a normal life. The first commercially available antidepressant, introduced in 1955, was imipramine, a tricyclic antidepressant (TCA) still widely used today [1–3]. However, these compounds have narrow therapeutic range (between 50 and 300 μg L^{-1} in plasma). When concentration exceeds 500 μg L^{-1}, toxic effects, such as high body temperature, sleepiness, confusion, cardiac arrest, among others, can happen and, when it rises to 1000 μg L^{-1}, death may also occur. Therefore, it is essential to develop sensible, accurate and simple analytical methods together with suitable sample preparation approaches for the determination of these pharmaceutical compounds in forensic matrices [4–6]. Nevertheless, when dealing with very complex matrices like biological matrices, a sample preparation step is always a must. In the past decades there has been a concern in making sample preparation more eco-friendly, leading to new strategies that include several innovative concepts, namely, miniaturization, simplification, much higher selectivity and sensitivity, the elimination of toxic organic solvents, and reduction the sample amount. Apart from other microextraction approaches such as solid-phase microextraction (SPME) or stir bar sorptive extraction (SBSE) [7–13], bar adsorptive microextraction (BAμE), which was introduced in the last decade, presents several advantages and has shown great simplicity and versatility by allowing the choice of sorbent coating for each particular type of application. Materials such as activated carbons (ACs) prepared from several sources, alumina, silica, cork, polystyrene divinylbenzene (PS-DVB), modified pyrrolidine, silica-based polymers, and nanomaterials, such as carbon nanotubes, and multi-walled carbon nanotubes have been used, demonstrating great performance as sorbent phases for BAμE technique [14–19]. Nevertheless, the preparation and application of new materials with specific sorption characteristics is still very important for particular applications, especially those from ecological sources, such as biomaterial waste. In addition, the BAμE devices can be lab-made, the experimental procedure is easy to implement and cost-effective, requires low sample volume and negligible amounts (100 μL) of organic solvent during the back-extraction step, presents remarkable performance, high enrichment factors and analytical limits at the trace level. Recently, new generation BAμE devices were introduced aiming to improve the overall procedure, promoting a better interfacing with the instrumental systems, as well as an alternative option for the routine work. The novel devices are smaller and more flexible, prepared with cylindrical nylon supports coated with suitable adhesive films where the sorbents are fixed [10,14,20–23].

In the present work, a new analytical strategy is proposed for trace determination of amitriptyline (AMT), mianserin (MIA), trimipramine (TRI), imipramine (IMP), mirtazapine (MIR) and dothiepin (DOT) (Figure 1) in urine matrices, using state-of-the-art BAμE devices, coated with several phases, in combination with large-volume-injection gas chromatography-mass spectrometry operating in the selected-ion monitoring mode (LVI-GC-MS(SIM)). It is also our goal to test, compare and discuss the selectivity and performance of several novel sorbents, having particular characteristics, prepared from biomaterials waste. The development, optimization, validation, and application in urine samples is also addressed.

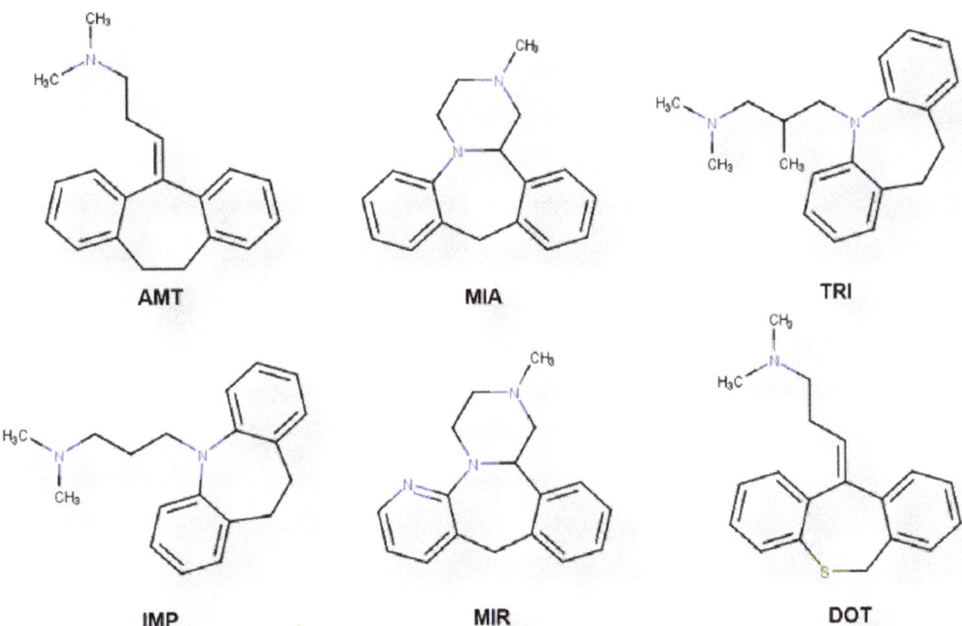

Figure 1. Chemical structures of the six TCAs studied in the present work.

2. Results and Discussion

2.1. LVI-GC-MS(SIM) Optimization

The initial step of this work was establishing the instrumental conditions that better fit the compounds under study. A mix solution of six TCAs was analyzed by GC-MS operating in the full-scan mode acquisition, which allowed to obtain the target ions (base peaks) and quantifier ions, also in accordance with the literature (Table 1) [24–27]. The obtained chromatogram by LVI-GC-MS(SIM) showed symmetrical peak shapes in less than 20 min of running time. The instrumental analytical thresholds were evaluated through the LODs and LOQs, corresponding to S/N of 3:1 and 10:1, in which 0.50 μg L^{-1} and 1.65 μg L^{-1} were achieved, respectively. The instrumental linearity was also assessed with eleven concentration levels ranging from 2.4 to 2500.0 μg L^{-1}. Linear regressions showed plots with good linearity with determination coefficients (r^2) higher than 0.9958.

Table 1. Target ions (base peaks in bold) and quantifier ions for each TCA studied by LVI-GC-MS(SIM), under optimized instrumental conditions.

TCAs	Ions (*m/z*)	Retention Time (min)
AMT	**58**/202/215	12.91
MIA	**193**/220/264	13.13
TRI	**58**/249/294	13.22
IMP	**58**/**234**/280	13.28
MIR	**195**/208/265	13.77
DOT	**58**/202/295	15.96

2.2. Optimization of the BAμE-μLD Efficiency

The first task in the BAμE optimization process is the selection of the most suitable sorbent phase for the target analytes involved. Afterwards, several experimental parameters were studied using one-variable-at-a-time (OVAT) strategy, both for microextraction and back-extraction stages, including the desorption solvent and time, stirring rate, matrix pH, organic modifier, ionic strength, equilibrium time, and sample dilution effect, according to previous reports [8,23,28–36]. Although OVAT does not allow the identification of possible interactions between variables and requires many assays, this strategy is easy to implement in this type of optimizations since the number of variables is low and no relevant interactions are expected in compliance with our expertise [8,37].

2.2.1. Selection of Sorbent Coatings and Back-Extraction Conditions

As usual in the BAμE technique, we started by selecting the best sorbent for the six TCAs and, for the present study, we proposed to test new and conventional sorbents. Thus, four new ACs (AC1, AC2, AC3 and AC4) prepared from biomaterials waste, as well as six commercial polymers (SX, HLB, C18, SDVB, SCN and DVBM) were assayed under the following experimental conditions; extraction stage: 5 mL of ultrapure water spiked with 100 μL of TCAs mix solution (500.0 μg L^{-1}), pH 5.5, 3 h (990 rpm); back-extraction stage: 90 μL of MeOH, 30 min under ultrasonic treatment (42 ± 2.5 kHz, 100 W).

In a first approach, it was evaluated the adsorptive properties of the four new ACs prepared from coconut fiber (AC1), coffee residue (AC2), sugarcane chaff (AC3) and sugarcane bagasse (AC4) wastes, for the six target analytes. These biomaterials have already been successfully tested to remove metal ions from aqueous solutions [38] and proven to be ideal sorbents in environmental applications, namely for water cleaning treatment. Preliminary assays from these new AC phases showed that the recoveries yields were maximized at 12% for the six target TCAs (data not shown). Although these biomaterials have shown remarkable performance for adsorption of metal ions from aqueous media, based on the exploratory data obtained for organic compounds, we postulated that the observed lack of efficiency can be attributed to the back-extraction stage, despite the highly probable efficiency of the previous microextraction stage. Therefore, to evaluate this hypothesis, additional experiments were carried out to verify the ability of the ACs to extract the six TCAs from the aqueous medium. For this purpose, we used similar conditions for microextraction, although assaying a solution of ultrapure water (5 mL) plus 25 μL of the TCAs mix solution (100.0 mg L^{-1}) and 16 h of equilibration time. Subsequently, a portion of the aqueous sample resulted from the microextraction stage was removed and extracted with dichloromethane (50:50, *v:v*), followed by ultrasonic treatment (42 ± 2.5 kHz, 100 W, 5 min) and analysis of the organic phase by LVI-GC-MS(SIM). The results obtained and depicted in Figure 2 shows, apart from the AC3 sorbent, that all the remaining carbon-based coatings (AC1, AC2 and AC4) presented remarkable microextraction efficiencies, since the resulting aqueous media do not present levels of TCAs higher than 15%. In this sense, we found that the proposed back-extraction stage conditions would not be able to efficiently remove the adsorbed TCAs, due to the very strong interactions established with the surface of these novel biosorbents, leading to low overall recovery yields. It should be emphasized that the analytical process of the BAμE technique is always characterized by a two-stage balancing process (adsorption-desorption equilibrium), in which the adsorption phenomena must be sufficiently effective during the microextraction stage, but not too strong that could compromise the effectiveness of the subsequent back-extraction stage. Like this, from the preliminary data achieved, we can deduce that the four biosorbents tested herein seem very effective from the adsorption point of view, ideal for removal processes (e.g., water decontamination, etc.), yet presenting great limitations as coating phases for microextraction-based techniques.

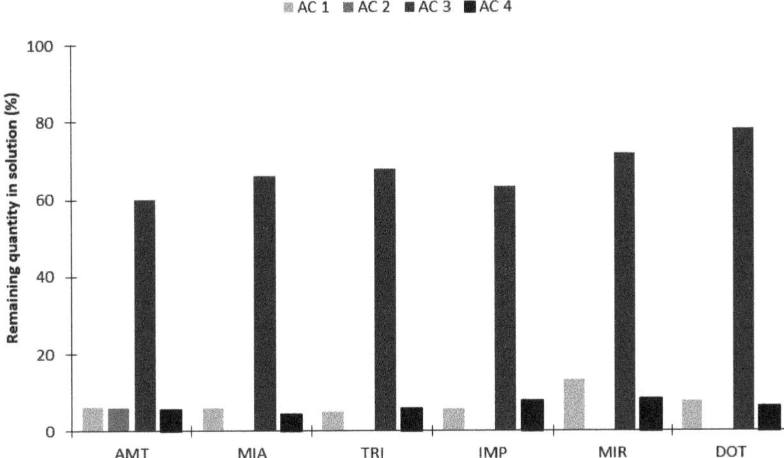

Figure 2. Remaining TCAs present in the aqueous matrix after the microextraction stage using the four different ACs prepared from biomaterials waste.

This finding may also be associated due to the large surface area and small pore size (<20 Å) presented by the four ACs under study, which may promote strong interactions with the target compounds and may hinder the back-extraction stage. For this reason, we decided to discard these biomaterial sorbents and assayed the six commercial polymeric phases (SX, HLB, C18, SDVB, SCN and DVBM), since they seem to be more suitable for the chemical structures of the target molecules involved. Figure 3 present assays performed by using these six polymer-based coatings, showing the efficiency achieved for the six TCAs, under similar experimental conditions, where the best selectivity is obtained for C18 and SCN phases. Although these polymers are structurally very different, the results obtained were expected, once the C_{18} polymer has a long aliphatic chain promoting strong hydrophobic interactions with the aliphatic chains present in the molecular structures of the target TCAs. On the other hand, the SCN polymer can promote dipole-dipole interactions through the cyano group and the electronegative elements (N and/or S) present in the chemical moieties of the six TCAs. Even so, we decided to evaluate the influence of the stripping solvent involved during the desorption step for all six polymer-based coatings. In general, the most common solvents used for the back-extraction stage in BAμE technique are MeOH (Figure 3a), ACN and mixtures of both (Figure 3b) [8,18]. In this sense, the data achieved showed that the use of an equivalent volume of MeOH and ACN presented the best performance, which allowed the selection of three polymeric phases, i.e., SX, C18 and SCN, the former promoting interactions through the N-vinylpyrrolidone group, demonstrating the great influence of the solvent involved during the back-extraction stage.

Even so, to speed up the back-extraction process, ultrasonic treatment was also implemented by using 15, 30, and 45 min of sonification time. The results obtained (data not shown) lead us to quit using the SCN phase once it showed the worst efficiency among the other two sorbent coatings. Therefore, the SX and C18 sorbents were selected for further assays and it has been found that at least 15 and 30 min are respectively needed to fully desorb the six TCAs from the microextraction devices.

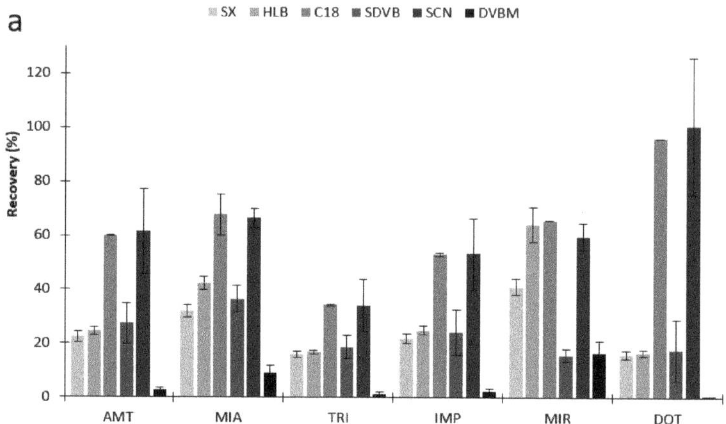

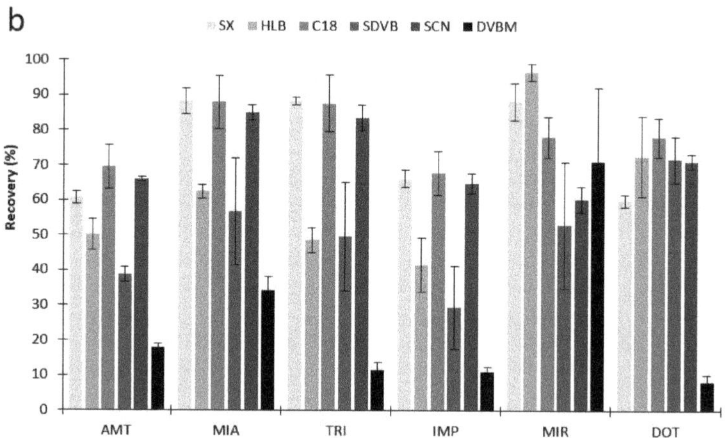

Figure 3. Effect of MeOH (**a**) and MeOH:ACN mix (**b**) stripping solvents on the back-extraction of the six TCAs from the different polymer phases by BAμE-μLD/LVI-GC-MS(SIM). The error bars represent the standard deviation of three replicates.

2.2.2. Microextraction Parameters

In this section, the first parameter optimized was the effect of the agitation stirring, in which three different speeds were tested, i.e., 750, 990, and 1250 rpm. From the results achieved (data not shown), it was observed that the best results occurred when 990 rpm are used, in line with previous reports [29–36] and having been selected for further assays.

Afterwards, the solution pH was also adjusted in order to promote partially or completely non-ionized TCAs, which may condition the the recovery yields through reverse-phase interactions [8,29,39]. The pK_a values for the six TCAs under study vary between 6 and 10, noticing that at pH 12, all of them are in the non-ionized form. Figure 4a,b show the results obtained from assays performed at several pH values (2.0, 5.5, 8.0 and 12.0), where the best data are attained by using the C_{18} coating at pH 12.0, once the six target TCAs become non-ionized [40]. However, when using the SX phase, the recovery yields did not vary significantly with pH variation. This observation can be attributed to additional chemical interactions promoted by the latter sorbent phase, other than reverse-phase type, e.g., dipole-dipole. For this reason, the optimization was pursued using only C_{18} polymer at pH 12.

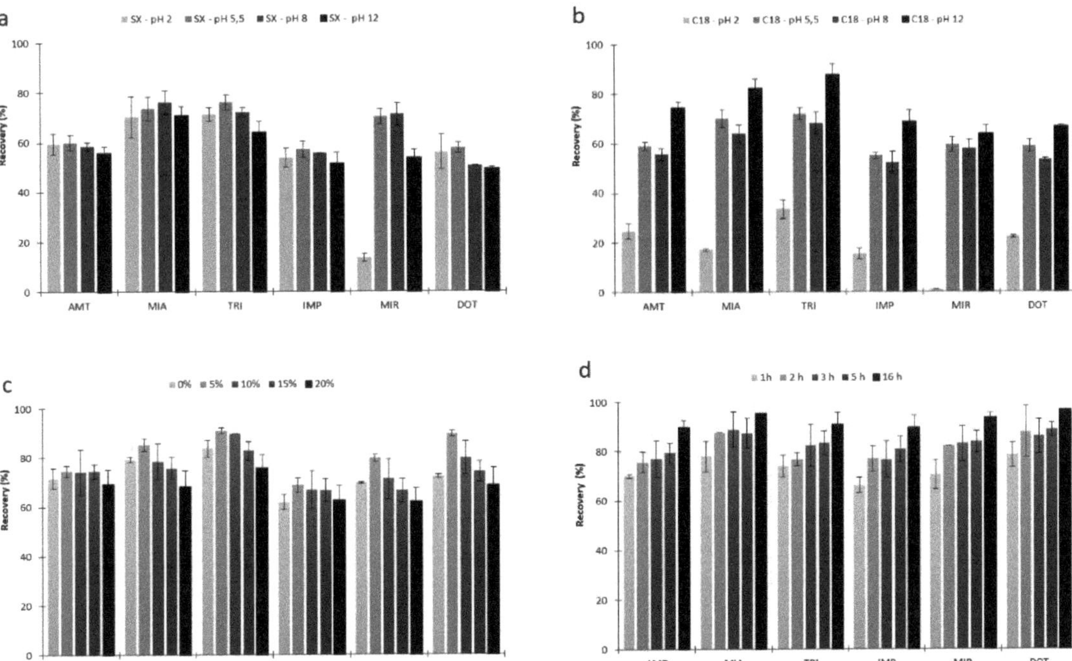

Figure 4. Effect of matrix pH using SX (**a**) and C_{18} (**b**) sorbent phases, percentage of NaCl (**c**) and equilibrium time (**d**) on the microextraction efficiency of six TCAs in aqueous media by BAμE-μLD/LVI-GC-MS(SIM). The error bars represent the standard deviation of three replicates.

The next optimization step was to study the effect of the medium polarity on the recovery yields for the six TCAs, varying the content (0 to 20%) of an organic modifier (MeOH). It is usually employed to minimize the possible adsorption of the analytes to the glass walls of the sampling flask ("wall-effect"), which generally occurs for compounds with log $K_{O/W}$ higher than 4. However, it can also lead to a solubility increment of the analytes in the aqueous media, hindering the microextraction process [8,28]. As expected, and once the six TCAs present log $K_{O/W}$ lower than 5, the best results were attained without MeOH addition.

Afterwards, the matrix ionic strength was evaluated by increasing amounts of NaCl (0 to 20%). An inert salt is usually added to the aqueous matrix to promote the "salting-out" effect, reducing the target compounds solubility in water. Commonly, for analytes presenting a log $K_{O/W}$ higher than 3.5, the addition of an inert salt does not improve the recovery yields through the BAμE approach. This can be caused by several reasons; it can caused by the "oil-effect" phenomena, promoting the migration of non-polar compounds to the surface of the aqueous media, minimizing the contact between the analytes and the microextraction device; it can enhance the viscosity by decreasing the equilibrium kinetics; and the polymeric surface area can be blocked through the salt ions. Nevertheless, discrepancies are sometimes observed, and therefore, this parameter should be always evaluated [8,28]. Figure 4c depicts the obtained results, where the highest recovery yields were obtained by using 5% of NaCl. A possible explanation to this finding is that there is a maximum efficiency until that value, where the "salting-out" effect takes place, i.e., the available water to dissolve the analytes is reduced with the consequent increment on the recovery yields. However, above 5% of NaCl, the "salting-in" effect can also occur, where the analytes can interact with the salt ions through electrostatic interactions and decreasing the recovery, in agreement with previous reports [41].

The equilibrium time was the next parameter to be optimized, since in equilibrium conditions the maximum sensitivity and precision are achieved [8]. In Figure 4d, it is possible to observe that the best results are achieved when the equilibrium time lasts for 16 h, in agreement with previous reports, where a comparable behavior was observed [23]. Even so, despite the substantial time involved, the analytical process can be performed overnight without any special requirement. It is noteworthy that 16 h are needed to achieve maximum recovery yields, although we can greatly reduce the equilibrium time and still obtain efficiencies of about 65% by using only 1 h.

Finally, the effect of the sample dilution was also studied and, as expected, by keeping the amount of the sorbent phase fixed and decreasing the sample volume, the recovery yields increase due to the increase in the concentration factor. Thus, with a larger sample volume, an instrumental response should increase due to the greater enrichment of the extracted analytes, with the consequence of an increase in the equilibrium time [8,42–44]. In this sense, the variation of the sample volume was evaluated for five different levels (1.5, 5.0, 10.0, 25.0 and 40.0 mL), with the results (data not shown) showing that the best response was obtained with the use of only 5 mL of aqueous solution.

From the beginning, the BAµE devices used in this work were prepared with PP subtracts. Nevertheless, to make them more compatible with the instrumental systems, as well as to improve the routine work, new generation devices were already proposed using nylon subtracts [23]. They are much smaller, more flexible, user-friendly, allows the back-extraction stage in an only single step and compatible with the injection operation of the conventional instrumental systems. So far, these devices had only been combined with HPLC systems. In this work, we decided to combine, for the first time, the new generation BAµE devices with a GC-MS system. For this purpose we decided to compare the performance of the conventional devices with the new generation ones, as depicted in Figure 5. From the data obtained, very similar performances are achieved between both devices. Therefore, given the advantages of the nylon subtracts, these were adopted for the validation and application sections.

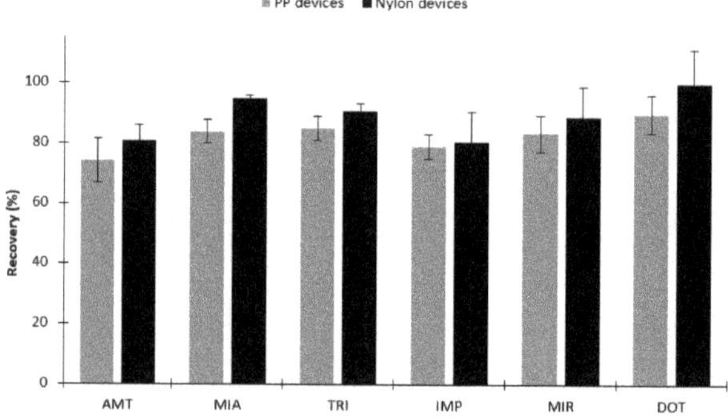

Figure 5. Effect of substrate type on the microextraction efficiency for the six TCAs in aqueous media by BAµE(C18)-µLD/LVI-GC-MS(SIM). The error bars represent the standard deviation of three replicates.

At the end of the optimization procedure, the attained recovery yields were between 80.5 and 99.6% for the six target TCAs (RSD < 12.1%), under the following experimental conditions; extraction stage: 5 mL of ultrapure water spiked with 100 µL of TCAs mix solution (500.0 µg L^{-1}), pH 12, 5% NaCl, 3 h (990 rpm); back-extraction stage: 90 µL MeOH/ACN (50:50), 30 min under ultrasonic treatment (42 ± 2.5 kHz, 100 W).

2.3. Validation of the Proposed Methodology

After the optimization section, we proceeded to the validation of the proposed methodology (BAμE(C18)-μLD/LVI-GC-MS(SIM)) using blank urine matrices. In a first approach, the analytical thresholds were determined through the LODs where values in between 0.20 and 1.56 μg L^{-1}, were achieved. The linearity range was assessed between 10.0 and 1000.0 μg L^{-1} (nine concentration levels). This range was chosen once it included the therapeutic, toxic and lethal concentration values for TCAs [45]. For the calibration plots, we started by plotting a conventional linear regression, which showed good linearity ($r^2 \geq 0.9914$), and according to the lack-of-fit test (at the confidence level 95%), it presented a good fit for the present study, i.e., the F_{calc} was always below the F_{tab}. Nevertheless, by studying the data dispersion, the relative residues were too high (>15%), and it was clearly visible a tendency in the residue plots, which showed heterogeneous profiles. As a result, heteroscedasticity was observed for all TCA plots, which might be caused by the large range of the concentrations considered. For this reason, ordinary least-squares linear regression spiking method was not fit for the attained data, since it results in large errors, in particular for the lower concentrations. Given the evidence of heteroscedasticity, a weighted linear regression method was adopted, once it is the simplest and most effective way to compensate the data dispersion observed, especially for the lower concentration levels. The quality of the fit of weighted regressions can be evaluated by calculating the sum of percentage relative error (%RE). The appropriate weighting factor can be calculated from the inverse of the variance. Nevertheless, this is impractical, once it requires several determinations for each calibration point and a new calibration plot must be done every time that the methodology is applied. Therefore, an empirical weight based on concentrations (variable x) and responses (variable y) were used. Six weighting factors were tested, $1/y^{1/2}$, $1/y$, $1/y^2$, $1/x^{1/2}$, $1/x$ and $1/x^2$. The data achieved shows that a weighting factor of $1/y$ resulted in the lowest %RE across the whole range. The weighting factor should be used in the calculation of the regression equation parameters as described in previous works [45–48].

As stated before, the LLOQ was defined as the lowest concentration value in which it is possible to quantify any analyte with precision and accuracy, i.e., it is the concentration in which the RSD and %RE are lower than 20%. Table 2 summarizes the validation parameters, namely the LODs, LLOQs, calibration equations and the r^2 values achieved.

Table 2. LODs, LLOQs, calibration equations and r^2 achieved for the six TCAs through BAμE(C18)-μLD/LVI-GC-MS(SIM) methodology, under optimized experimental conditions.

TCAs	LODs (μg L^{-1})	LLOQs (μg L^{-1})	Calibration Equations	r^2
AMT			y = 34.4780 x − 0.0174	0.9974
MIA			y = 7.3701 x − 0.0026	0.9974
TRI	0.20		y = 13.0250 x − 0.0033	0.9988
IMP		10.00	y = 3.5938 x − 0.0027	0.9960
MIR	0.39		y = 11.2264 x − 0.0026	0.9982
DOT	1.56		y = 29.1992 x − 0.0094	0.9978

As previously mentioned, accuracy and precision were calculated for inter and intraday at four spiking levels, as summarized in Table 3. The intraday accuracy and precision values were between −12.0 and 14.2%, and ranging from 0.4 to 15.8%, respectively. The interday accuracy and precision ranged from −8.2 to 20.0%, and between 2.1 and 19.9%, respectively. These data show that the proposed analytical approach presented suitable levels of accuracy and precision for trace level analysis of TCAs in urine matrices.

Table 3. Inter and intraday accuracy and precision levels obtained for the six TCAs at four different concentrations by BAμE(C18)-μLD/LVI-GC-MS(SIM) methodology, under optimized experimental conditions.

TCAs	Spiking Level (μg L^{-1})	Intraday			Interday		
		Accuracy (%) $\pm$ Precision (%)			Accuracy (%) $\pm$ Precision (%)		
AMT	10.0	5.1	$\pm$	8.2	20.0	$\pm$	15.0
	50.0	-5.5	$\pm$	4.6	-5.0	$\pm$	8.9
	500.0	1.7	$\pm$	4.8	2.9	$\pm$	12.2
	1000.0	4.8	$\pm$	8.8	-4.4	$\pm$	11.8
MIA	10.0	-6.6	$\pm$	4.3	-2.8	$\pm$	16.7
	50.0	-0.5	$\pm$	11.1	0.1	$\pm$	13.2
	500.0	4.8	$\pm$	8.6	1.9	$\pm$	11.8
	1000.0	-0.1	$\pm$	8.9	2.5	$\pm$	13.2
TRI	10.0	0.9	$\pm$	6.8	13.1	$\pm$	18.1
	50.0	-1.7	$\pm$	6.1	11.0	$\pm$	13.8
	500.0	5.7	$\pm$	4.8	13.0	$\pm$	9.2
	1000.0	-4.1	$\pm$	13.7	5.8	$\pm$	11.7
IMP	10.0	-8.4	$\pm$	9.0	14.8	$\pm$	17.6
	50.0	6.7	$\pm$	12.8	2.4	$\pm$	14.5
	500.0	-1.2	$\pm$	6.3	14.2	$\pm$	14.9
	1000.0	-0.5	$\pm$	11.2	-4.3	$\pm$	10.1
MIR	10.0	1.3	$\pm$	15.8	15.9	$\pm$	19.9
	50.0	13.3	$\pm$	0.9	14.5	$\pm$	2.1
	500.0	6.5	$\pm$	6.0	13.0	$\pm$	12.1
	1000.0	-0.7	$\pm$	7.3	11.1	$\pm$	13.8
DOT	10.0	8.1	$\pm$	0.4	14.1	$\pm$	9.4
	50.0	14.2	$\pm$	2.5	1.0	$\pm$	13.5
	500.0	-8.2	$\pm$	10.3	-4.5	$\pm$	14.5
	1000.0	-12.0	$\pm$	7.2	-8.2	$\pm$	10.1

Matrix effects, average recovery yields and process efficiency were also performed at two different concentration levels (25 and 750 μg L^{-1}), as presented in Table 4.

Table 4. Matrix effects, recovery yields and process efficiency obtained for the six TCAs at two different concentrations by BAμE(C18)-μLD/LVI-GC-MS(SIM) methodology, under optimized experimental conditions.

TCAs	Spiking Level (μg L^{-1})	Matrix Effects (%) $\pm$ RSD (%)			Recovery Yields (%) $\pm$ RSD (%)			Process Efficiency (%) $\pm$ RSD (%)		
AMT	25.0	90.2	$\pm$	6.6	95.3	$\pm$	9.6	86.0	$\pm$	7.5
	750.0	93.7	$\pm$	12.5	107.9	$\pm$	6.9	101.1	$\pm$	12.9
MIA	25.0	100.1	$\pm$	9.8	95.3	$\pm$	7.2	95.4	$\pm$	8.6
	750.0	111.5	$\pm$	11.2	111.5	$\pm$	7.7	124.3	$\pm$	11.7
TRI	25.0	102.1	$\pm$	8.4	103.3	$\pm$	9.3	105.5	$\pm$	12.0
	750.0	99.3	$\pm$	8.2	109.0	$\pm$	5.0	108.2	$\pm$	7.9
IMP	25.0	91.9	$\pm$	4.1	92.3	$\pm$	11.1	84.9	$\pm$	10.4
	750.0	109.3	$\pm$	13.9	106.5	$\pm$	10.0	116.5	$\pm$	12.6
MIR	25.0	95.6	$\pm$	10.9	108.5	$\pm$	12.3	103.6	$\pm$	12.0
	750.0	112.9	$\pm$	14.4	99.0	$\pm$	9.9	111.7	$\pm$	13.9
DOT	25.0	91.3	$\pm$	4.4	99.6	$\pm$	8.9	90.9	$\pm$	8.4
	750.0	99.4	$\pm$	11.6	103.7	$\pm$	7.5	103.1	$\pm$	13.9

Matrix effects were found to be in between 90.2 and 112.9% (RSD < 14.4%). Results above 100% show ionic enrichment and below, ionic suppression; 100% is the ideal, although it is impossible to eliminate matrix effects [49–51]. The average recovery yields achieved were between 92.3 and 111.5% (RSD < 12.4%). Finally, the process efficiency ranged from 84.9 to 124.3% (RSD < 13.9%). These data proved that the developed methodology is well suited to determine trace levels of TCAs in urine matrices.

2.4. Application to Real Urine Samples

To test the applicability of the present methodology to real matrices, fifty-two urine samples from anonymous donors were analyzed, using GC-MS operating in the full-scan mode acquisition, to ensure possible positive identifications. Figure 6 depicts a chromatogram from a urine sample spiked at the 100.0 µg L^{-1} for the six TCAs (a) and a positive sample (b) for AMT with an amount of 158.87 ± 1.93 µg L^{-1}, which is well above the therapeutic levels [52]. The figures of merit presented show that good selectivity and sensitivity are achieved by the proposed methodology.

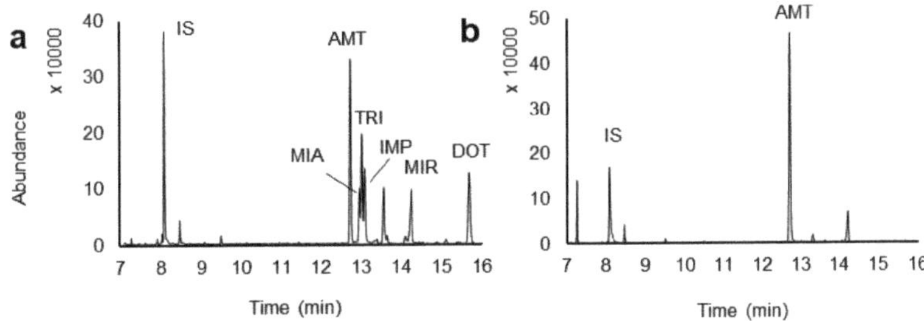

Figure 6. Chromatograms obtained from urine samples spiked at the 100.0 µg L^{-1} for six TCAs (**a**) and a positive anonymous donor without spiking (**b**) analyzed through BAµE(C18)-µLD/LVI-GC-MS(SIM) methodology, under optimized experimental conditions.

2.5. Performance Comparison with Other Microextraction Techniques

In the present contribution, we decided to compare the developed methodology with other microextraction techniques dedicated for the analysis of TCAs in urine samples already reported in the literature [6,27,52–55], as summarized in Table 5.

Table 5. Comparison between the present study and others analytical approaches already reported in the literature for each six TCAs determination in urine and water matrices.

Analytical Method	TCAs	Recovery Yields (%)	RSD (%)	LOD (µg L^{-1})	LOQ (µg L^{-1})	Linear Range (µg L^{-1})	r^2	Ref.
BAµE-µLD/LVI-GC-MS(SIM)	AMT, MIA, TRI, IMP, MIR, DOT	~100.0	<9.6	0.20	10.0 (LLOQ)	10.0–1000.0	0.9974	This study
MSPE/HPLC-UV	AMT, IMP	98.5–99.5 (RR)						[6]
SPE/GC-MS	AMT, MIA, TRI, MIR, DOT	64.4–99.8	6.0–20.6	1.0–2.5	-	1.0–320.0	0.9963–0.9996	[27]
SI-HLLE-DSPE-DLLME-SFO/HPLC-UV	AMT, IMP	69.0–84.0	3.0–4.0	0.2–0.3	0.7–1.1 (LOQ)	0.7–1000.0	0.9960–0.9970	[52]
DLLME/HPLC-UV	TRI	112.0	6.1	0.6	-	2.0–100.0	0.9946	[53]
HF-LPME/HPLC-UV	AMT, IMP,	-	6.8	-	-	-	[54]
DLLME/GC-MS	AMT, IMP	88.2–103.6	7.4–7.9	0.5	2.0	2.0–100.0	0.9990	[55]

DLLME: Dispersive liquid-liquid microextraction GC: Gas chromatography; HF-LPME: Hollow fiber-liquid phase microextraction; HPLC: High performance liquid chromatography; MS: Mass spectrometry; MSPE: Magnetic solid-phase extraction; RR: Relative recovery; SI-HLLE-DSPE-DLLME-SFO: Salt induced-homogenous liquid-liquid extraction, dispersive solid phase extraction, and dispersive liquid–liquid microextraction based on the solidification of floating organic droplet; SPE: Solid-phase extraction; Ultraviolet.

First, the obtained recovery yields and precision levels are much better or alike other analytical approaches, both in water and urine matrices. The attained analytical thresholds (LODs and LLOQs) compares favorably with the other proposed analytical methods. Unlike other methodologies already reported, the present work has the great advantage of encompassing therapeutic, toxic and lethal concentrations in its linear range. In short, the methodology proposed herein can be considered an alternative for the determination of the six TCAs involved in urine samples.

3. Materials and Methods

3.1. Chemicals and Standards

Methanol (MeOH, 99.9%), acetonitrile (ACN, 99.9%), formic acid (99%), acetic acid (99.5%) and dichloromethane (99.9%) were purchased from Carlo Erba (Barcelona, Spain). Sodium hydroxide (NaOH, 98.0%) was obtained from AnalaR BDH Chemicals (Leicestershire, UK). Phosphoric acid (85.0%) and disodium phosphate (99%) were purchased from Panreac (Barcelona, Spain). Diphenylamine (DIF, 98.0%) used as internal standard (IS) was purchased from Sigma-Aldrich (Saint Louis, MO, USA). Hydrochloric acid (HCl, 37%) was purchased from Sigma-Aldrich (Vienna, Austria). Ultra-pure water was obtained from Milli-Q water purification systems from Merck Millipore (Burlington, MA, USA). Six pharmaceutical tablets commercially available in the Portuguese market were used to prepare the standard solutions: ADT produced by Generis (Amadora, Portugal) containing 10 mg of amitriptyline hydrochloride (AMT); Tolvon produced by Merck Sharp & Dohme (Paço de Arcos, Portugal) containing 30 mg of mianserin hydrochloride (MIA); Surmontil produced by Laboratórios Vitória (Amadora, Portugal) containing 35 mg of trimipramine (TRI); Tofranil produced by Amdipharm (Dublin, Ireland) containing 10 mg of imipramine hydrochloride (IMP); Mirtazapina Alter produced by Alter (Alter do Chão, Portugal) containing 15 mg of mirtazapine (MIR); Protiadene produced by Teofarma (Pavia, Italy) containing 75 mg of dothiepin hydrochloride (DOT). Stock solution were prepared according to previously described protocols [56,57], where each tablet was individually meshed and dissolved in 10 mL of MeOH giving different concentrations according to the pill initial dosage. Then, the solutions were sonicated (42 $\pm$ 2.5 kHz, 100 W, Branson 3510, Switzerland) for 15 min, centrifuged at 3000 rpm for 10 min. Diphenylamine stock solutions were prepared by dissolving in MeOH to give a final concentration of 1000 mg L^{-1}, stored at -20 °C in glass flasks and renewed every month.

3.2. Sorbent Phases

The sorbent phases used for coating the BAμE devices were from several ACs and polymers. The novel four ACs were prepared from biomaterials waste, characterized, and provided by the Industrial Biotechnology Laboratory from Tiradentes University (Aracaju, Sergipe, Brazil). AC1 was obtained from coconut fiber (pH$_{PZC}$ 6.3, 1130 m^2 g^{-1} (BET) surface area, <20 Å pore size); AC2 was obtained from coffee residue (pH$_{PZC}$ 7.3, 1308 m^2 g^{-1} (BET) surface area, <20 Å pore size); AC3 was obtained from sugarcane chaff (pH$_{PZC}$ 5.9, 1185 m^2 g^{-1} (BET) surface area, <20 Å pore size); and AC4 was obtained from sugarcane bagasse (pH$_{PZC}$ 6.9, 791 m^2 g^{-1} (BET) surface area, <20 Å pore size). The polymeric sorbents used were Strata-X (SX) (polymer containing N-vinylpyrrolidone; 33 μm particle size, 85 Å pore size, 800 m^2 g^{-1} surface area), Strata-CN (SCN) (reversed phase polymer containing ciano; particle size 55 μm, 70 Å pore size, surface area 500 m^2 g^{-1}) and Strata SDB-L (SDVB) (reversed phase styrene-divinylbenzene polymer; particle size 100 μm, 260 Å pore size, 500 m^2 g^{-1} surface area) from Phenomenex (Torrance, CA, USA); Oasis HLB (HLB) (reversed phase N-vinylpyrrolidone-divinylbenzene co-polymer; 30 μm particle size, 80 Å pore size, 830 m^2 g^{-1} surface area and pH stability at 0–14) from Waters (Milford, MA, USA); ENVI-18 (C18) (reversed phase octadecyl silica polymer; 45 μm particle size, 60 Å pore size, 475 m^2 g^{-1} surface area) from Supelco (Darmstadt, Germany); LiChrolut EN (DVBM) (reversed phase ethylvinylbenzene-divinylbenzene co-polymer;

particle size 40–120 µm, 60 Å pore size, surface area 1200 m^2 g^{-1}) from Merck Millipore (Darmstadt, Germany).

3.3. Urine Matrices

Urine samples were obtained from the Joaquim Chaves Saúde clinic (Algés, Portugal) and provided in total anonymity without any information from the donors. Additional urine control samples were provided for the validation process from healthy volunteers who guaranteed not to have consumed any of the TCAs under study. Upon arrival at the laboratory, the samples were frozen ($-20\ °$C) until use.

3.4. Experimental Set-Up

3.4.1. Preparation of the BAµE Devices

The BAµE devices were lab-made prepared. For the optimization assays, the devices were made with polypropylene (PP) cylindrical subtracts, produced in similar way to previous works [30,31]. PP subtracts with 10×3 mm were coated with a suitable adhesive film, and then coated with powdered sorbents. To evaluate the similarity between the conventional used PP devices and the new generation of BAµE devices, nylon cylindrical subtracts were used. The nylon subtracts having 10×1 mm were coated with a suitable adhesive film, and then coated with powdered sorbents [23]. Before being used, and to remove potential impurities, the devices were cleaned with ultrapure water under magnetic stirring and dried in a clean Kimwipe [23,30,31].

3.4.2. Optimization Assays

For the optimization assays, 5 mL of ultrapure water (pH 5.5) were added to a sampling glass flask and spiked with 100 µL of a mix solution of all six TCAs (500.0 µg L^{-1}), resulting in a final concentration of 10.0 µg L^{-1}. Afterwards, a conventional Teflon magnetic stirring bar and a BAµE device were introduced in the same flask. The microextraction process was performed through floating sampling technology using a multi-point agitation plate (Variomag HþP Labortechnik AG Multipoint 15, Oberschleissheim, Germany) at room temperature (25 °C). After microextraction, the BAµE devices were removed from the glass flasks with clean tweezers and a clean Kimwipe. For the back-extraction stage, the BAµE devices were placed into glass vial inserts having 90 µL of an organic solvent following by ultrasonic treatment (42 ± 2.5 kHz, 100 W) at room temperature (25 °C). Afterwards, 10 µL of a DIF solution (IS; 10,000.0 µg L^{-1}) were introduced to the glass vial inserts, resulting in a final concentration of 1000.0 µg L^{-1}. In the case of the PP devices and after the back-extraction stage, these are removed from the vial before being sealed and proceeding to the instrumental analysis. In the case of the nylon devices this step is not required. Blank assays without spiking were also performed. Unless specified, each assay was done in triplicate. Several parameters were studied to optimize the extraction efficiency following an OVAT strategy. In this approach all variables are fixed, except one, and the extraction efficiency is studied at several levels of this parameter. The selected parameters for the optimization process were desorption solvent (MeOH, ACN, and MeOH/ACN (50/50%, v/v) and time (15, 30, and 45 min), stirring speed (750, 990, and 1250 rpm), pH (2.0, 5.5, 8.0, and 12.0), organic modifier (MeOH: 0, 5, 10, 15, and 20%), matrix ionic strength (NaCl: 0, 5, 10, 15, and 20%), equilibrium time (1, 2, 3, 5, and 16 h) and sample dilution effect using different volumes (1.5, 5.0, 10.0, 25.0, and 40.0 mL), in accordance with previous reports [23,29–36].

3.4.3. Pre-Treatment of Biological Samples

After thawing at room temperature (25 °C), the samples were subjected to an alkaline hydrolysis step to eliminate most of the interfering compounds from the urine samples, e.g., carbamide, uric acid, or calcium salts, etc. This process is similar to a previous described hydrolysis method [53]. In this case, 200 µL of NaOH 10 mol L^{-1} was added to 1 mL of urine and the solution was placed in an automatic evaporator (Laborota 4000 Efficient,

Heidolph, Schwabach, Germany) at 60 °C for 10 min. Afterwards, the samples were centrifuged for 10 min at 2000 rpm, and the supernatant transferred to new clean vials and subjected to ultrasonic treatment (42 ± 2.5 kHz, 100 W) for 5 min. Lastly the samples were filtered (0.45 µm nylon filters, Laborspirit, Lisbon, Portugal).

3.4.4. Validation and Real Sample Assays

Assays performed with urine samples were done after the optimal conditions were found. In this sense, for each assay, 1.2 mL of urine was added to 3.8 mL of buffer solution (pH 12.0, adjusted with HCl 5 mol L^{-1}) with 5% of NaCl. For validation assays, 100 µL of TCAs mix solution was added, except when performing blank assays. For the validation process several parameters were evaluated, such as linearity, analytical thresholds, selectivity, accuracy, precision, recovery, matrix effects and the process efficiency, according to previous works [50,51]. To assess the developed methods selectivity, the optimized BAµE-µLD/LVI-GC-MS method was applied to urine control samples and the absence of interfering compounds at the studied TCAs retention time was verified. Calibration standards were prepared between 10.0 and 1000.0 µg L^{-1} to assess the linearity (estimated with the lack-of-fit test), the coefficients of determination (r^2) and residuals dispersion. In order to determine intra and inter-day accuracy and precision, several assays (n = 6) were performed using different spiking concentrations of TCAs mix solution, including 10.0, 50.0, 500.0, and 1000.0 µg L^{-1}, corresponding to the lower limit of quantification (LLOQ), low, medium, and high concentrations, respectively. Interday assays were performed in three consecutive days and intraday assays were performed in the same day. The acceptance criteria were that the relative residuals and relative standard deviations (RSDs) ≤ 15%, except for LLOQ values were ≤20% values were accepted. The analytical thresholds were assessed by the limit of detection (LOD) and LLOQ. The LOD corresponds to a signal-to-noise (S/N) ratio of 3/1, while LLOQ is the lowest concentration where it is possible to quantify according to accuracy and precision parameters, being the first point in calibration curve, and also it corresponds to a S/N < 10. Lastly, to obtain matrix effects, recovery yields, and process efficiency, three sets of samples, each at two different concentrations (25.0 and 750.0 µg L^{-1}), were prepared. Set A samples consisted of a mix solution with all the TCAs at the previously stated concentrations. Set B samples were fortified after the microextraction and before liquid desorption. Set C samples were spiked before the microextraction. The ration between absolute peak areas of sets A and B allowed for the calculation of matrix effects, between B and C for the recovery yields calculation, and between A and C for the calculation of the process efficiency.

3.5. Instrumental Set-Up

GC-MS analyses were performed with an Agilent Technologies system (Santa Clara, CA, USA) constituted by an Agilent 6890 series gas chromatograph, equipped with an Agilent 7683 automating liquid sampler and a programmed temperature vaporization injector, coupled to an Agilent 5973N mass selective detector. All data recorded and instrumental control was performed in the MS ChemStation software (G1701; version E.02.02.1431). Injection was made in solvent vent mode (vent time: 0.3 min, 50.0 mL/min, pressure 0.0 psi, purge flux: 60.0 mL/min, time 2.0 min) with programmed temperature starting at 80 °C (0.45 min) to 280 °C (600 °C/min, 3 min isothermal). Large-volume-injections (LVI) of 10 µL at 100 µL/min were performed. A capillary column was used Zebron ZB-5 (30.0 m × 0.25 mm × 0.25 µm; 5% phenyl, 95% dimethylpolysiloxane) (Phenomenex) for the GC analysis, along with helium as the carrier gas, in constant pressure mode (9.82 psi). Oven temperature was programmed starting at 80 °C (held for 1 min), to 240 °C (20 °C/min, for 5 min), then to 245 °C (1 °C/min), and finally 300 °C (20 °C/min), achieving a total run time of approximately 22 min. The transfer line temperature was 280 °C, the quadrupole analyzer temperature was 150 °C, and the ion source temperature was 230 °C. A solvent delay of 7 min was selected. Electron ionization was used (70 eV) in a range of masses between 35 and 550 Da, in the full-scan mode, with an ionization current of 34.6 µA and a multiplier voltage of 1200 V. In the selected ion monitoring (SIM) mode several groups of ions were monitored in a defined time frame according to their retention time, maintaining a dwell time of 100 ms^{-1}. All the instrumental data were performed in triplicate and the calculations of each assay were performed by comparing the average peak areas of the extracted compounds to the IS peak area.

4. Conclusions

The analytical methodology proposed in the present work was fully developed, optimized, validated and applied for the determination of trace levels of selected six TCAs in urine matrices. The results show that the proposed approach compares favorably with other analytical strategies already reported in the literature, especially regarding the recovery yields and linear range. The proposed method is simple and user-friendly, suitable for trace analysis of TCAs in urine samples, in compliance with green analytical chemistry principles and can be used as a tool for routine work.

Author Contributions: Conceptualization, M.N.O. and S.M.A.; methodology, M.N.O., O.C.G. and J.K.S.; validation, N.R.N. and L.C.K.; investigation, M.N.O., O.C.G. and J.K.S.; writing-original draft preparation, M.N.O. and S.M.A.; writing-review and editing, S.M.A., N.R.N. and J.M.F.N.; supervision, N.R.N., E.B.C. and J.M.F.N. All authors have read and agreed to the published version of the manuscript.

Funding: The authors thank Fundação para a Ciência e a Tecnologia (Portugal) for financial support (projects: UIDB/00100/2020 and UIDP/00100/2020), the PhD grant (SFRH/BD/107892/2015) and the contract established from DL 57/2016, as well as Conselho Nacional de Desenvolvimento Científico e Tecnológico Coordenação (CNPq, Brazil; project: n° 314526/2014-5).

Institutional Review Board Statement: The study was conducted according to the guidelines of the Declaration of Helsinki, and approved by the Institutional Review Board of the Faculdy of Sciences of The University of Lisbon ('Comissão de Ética para Recolha e Proteção de Dados') nr 01/2019.

Informed Consent Statement: Informed consent was obtained from all subjects involved in the study.

Data Availability Statement: The data supporting reported results can be found in the laboratory databases of the Faculty of Sciences of the University of Lisbon.

Acknowledgments: The authors also wish to thank Carlos Cardoso from Joaquim Chaves Saúde clinic (Algés, Portugal) for providing the urine samples.

Conflicts of Interest: The authors declare no conflict of interest. The funders had no role in the design of the study; in the collection, analyses, or interpretation of data; in the writing of the manuscript, or in the decision to publish the results.

Sample Availability: Samples of the compounds studied in the present work are available from the authors at reasonable request.

References

1. World Health Organization. Mental Disorders. Available online: https://www.who.int/news-room/fact-sheets/detail/mental-disorders (accessed on 28 November 2019).
2. Lovatt, P. Tricyclic antidepressants: Pharmacological profile. *Nurse Prescr.* **2014**, *9*, 38–41. [CrossRef]
3. Hecht, A. *Understanding Drugs: Antidepressants and Antianxiety Drugs*, 1st ed.; Chelsea House Publications: New York, NY, USA, 2011; ISBN 9781604135329.
4. Santos, M.G.; Tavares, I.M.C.; Barbosa, A.F.; Bettini, J.; Figueiredo, E.C. Analysis of tricyclic antidepressants in human plasma using online-restricted access molecularly imprinted solid phase extraction followed by direct mass spectrometry identification/quantification. *Talanta* **2017**, *163*, 8–16. [CrossRef] [PubMed]
5. Mohebbi, A.; Yaripour, S.; Farajzadeh, M.A.; Afshar Mogaddam, M.R. Combination of dispersive solid phase extraction and deep eutectic solvent–based air–assisted liquid–liquid microextraction followed by gas chromatography–mass spectrometry as an efficient analytical method for the quantification of some tricyclic antidep. *J. Chromatogr. A* **2018**, *1571*, 84–93. [CrossRef] [PubMed]
6. Safari, M.; Shahlaei, M.; Yamini, Y.; Shakorian, M.; Arkan, E. Magnetic framework composite as sorbent for magnetic solid phase extraction coupled with high performance liquid chromatography for simultaneous extraction and determination of tricyclic antidepressants. *Anal. Chim. Acta* **2018**, *1034*, 204–213. [CrossRef]

7. Nogueira, J.M.F. Stir-bar sorptive extraction: 15 years making sample preparation more environment-friendly. *TrAC Trends Anal. Chem.* **2015**, *71*, 214–223. [CrossRef]
8. Prieto, A.; Basauri, O.; Rodil, R.; Usobiaga, A.; Fernández, L.A.; Etxebarria, N.; Zuloaga, O. Stir-bar sorptive extraction: A view on method optimisation, novel applications, limitations and potential solutions. *J. Chromatogr. A* **2010**, *1217*, 2642–2666. [CrossRef]
9. Neng, N.R.; Silva, A.R.M.; Nogueira, J.M.F. Adsorptive micro-extraction techniques-novel analytical tools for trace levels of polar solutes in aqueous media. *J. Chromatogr. A* **2010**, *1217*, 7303–7310. [CrossRef] [PubMed]
10. Nogueira, J.M.F. Novel sorption-based methodologies for static microextraction analysis: A review on SBSE and related techniques. *Anal. Chim. Acta* **2012**, *757*, 1–10. [CrossRef] [PubMed]
11. Kabir, A.; Locatelli, M.; Ulusoy, H. Recent Trends in Microextraction Techniques Employed in Analytical and Bioanalytical Sample Preparation. *Separations* **2017**, *4*, 36. [CrossRef]
12. Schmidt, K.; Podmore, I. Solid Phase Microextraction (SPME) Method Development in Analysis of Volatile Organic Compounds (VOCS) as Potential Biomarkers of Cancer. *J. Mol. Biomark. Diagn.* **2015**, *6*. [CrossRef]
13. Nogueira, J.M.F. Extração Sortiva em Barra de Agitação (SBSE): Uma metodologia inovadora para microextração estática. *Sci. Chromatogr.* **2012**, *4*, 259–269. [CrossRef]
14. Abujaber, F.; Ahmad, S.M.; Neng, N.R.; Rodríguez Martín-Doimeadios, R.C.; Guzmán Bernardo, F.J.; Nogueira, J.M.F. Bar adsorptive microextraction coated with multi-walled carbon nanotube phases-Application for trace analysis of pharmaceuticals in environmental waters. *J. Chromatogr. A* **2019**, *1600*, 17–22. [CrossRef]
15. Ahmad, S.M.; Mestre, A.S.; Neng, N.R.; Ania, C.O.; Carvalho, A.P.; Nogueira, J.M.F. Carbon-Based Sorbent Coatings for the Determination of 2 Pharmaceutical Compounds by Bar Adsorptive Microextraction. *ACS Appl. Bio Mater.* **2020**, *3*, 2078–2091. [CrossRef]
16. Mafra, G.; Oenning, A.L.; Dias, A.N.; Merib, J.; Budziak, D.; da Silveira, C.B.; Carasek, E. Low-cost approach to increase the analysis throughput of bar adsorptive microextraction (BAµE) combined with environmentally-friendly renewable sorbent phase of recycled diatomaceous earth. *Talanta* **2018**, *178*, 886–893. [CrossRef] [PubMed]
17. De Souza, M.P.; Rizzetti, T.M.; Francesquett, J.Z.; Prestes, O.D.; Zanella, R. Bar adsorptive microextraction (BAµE) with a polymeric sorbent for the determination of emerging contaminants in water samples by ultra-high performance liquid chromatography with tandem mass spectrometry. *Anal. Methods* **2018**, *10*, 697–705. [CrossRef]
18. Ahmad, S.M.; Nogueira, J.M.F. High throughput bar adsorptive microextraction: A simple and effective analytical approach for the determination of nicotine and cotinine in urine samples. *J. Chromatogr. A* **2019**, *1615*, 460750. [CrossRef] [PubMed]
19. Ahmad, S.M.; Gomes, M.I.; Ide, A.H.; Neng, N.R.; Nogueira, J.M.F. Monitoring traces of organochlorine pesticides in herbal matrices by bar adsorptive microextraction–Application to black tea and tobacco. *Int. J. Environ. Anal. Chem.* **2019**, 1–15. [CrossRef]
20. Oenning, A.L.; Morés, L.; Dias, A.N.; Carasek, E. A new configuration for bar adsorptive microextraction (BAµE) for the quantification of biomarkers (hexanal and heptanal) in human urine by HPLC providing an alternative for early lung cancer diagnosis. *Anal. Chim. Acta* **2017**, *965*, 54–62. [CrossRef]
21. Dias, A.N.; da Silva, A.C.; Simão, V.; Merib, J.; Carasek, E. A novel approach to bar adsorptive microextraction: Cork as extractor phase for determination of benzophenone, triclocarban and parabens in aqueous samples. *Anal. Chim. Acta* **2015**, *888*, 59–66. [CrossRef]
22. Almeida, C.; Nogueira, J.M.F. Determination of steroid sex hormones in real matrices by bar adsorptive microextraction (BAµE). *Talanta* **2015**, *136*, 145–154. [CrossRef] [PubMed]
23. Ide, A.H.; Nogueira, J.M.F. New-generation bar adsorptive microextraction (BAµE) devices for a better eco-user-friendly analytical approach–Application for the determination of antidepressant pharmaceuticals in biological fluids. *J. Pharm. Biomed. Anal.* **2018**, *153*, 126–134. [CrossRef] [PubMed]
24. Ferrari Júnior, E.; Caldas, E.D. Simultaneous determination of drugs and pesticides in postmortem blood using dispersive solid-phase extraction and large volume injection-programmed temperature vaporization-gas chromatography–mass spectrometry. *Forensic Sci. Int.* **2018**, *290*, 318–326. [CrossRef]
25. Wille, S.M.R.; Van Hee, P.; Neels, H.M.; Van Peteghem, C.H.; Lambert, W.E. Comparison of electron and chemical ionization modes by validation of a quantitative gas chromatographic-mass spectrometric assay of new generation antidepressants and their active metabolites in plasma. *J. Chromatogr. A* **2007**, *1176*, 236–245. [CrossRef]
26. Gunnar, T.; Mykkänen, S.; Ariniemi, K.; Lillsunde, P. Validated semiquantitative/quantitative screening of 51 drugs in whole blood as silylated derivatives by gas chromatography-selected ion monitoring mass spectrometry and gas chromatography electron capture detection. *J. Chromatogr. B* **2004**, *806*, 205–219. [CrossRef]
27. Rosado, T.; Gonçalves, A.; Martinho, A.; Alves, G.; Duarte, A.P.; Domingues, F.; Silvestre, S.; Breitenfeld, L.; Víctor, G.; Carlos, O.; et al. Simultaneous Quantification of Antidepressants and Metabolites in Urine and Plasma Samples by GC–MS for Therapeutic Drug Monitoring. *Chromatographia* **2017**, *80*, 301–328. [CrossRef]
28. Camino-Sánchez, F.J.; Rodríguez-Gómez, R.; Zafra-Gómez, A.; Santos-Fandila, A.; Vílchez, J.L. Stir bar sorptive extraction: Recent applications, limitations and future trends. *Talanta* **2014**, *130*, 388–399. [CrossRef]
29. Almeida, C.; Ahmad, S.M.; Nogueira, J.M.F. Bar adsorptive microextraction technique-application for the determination of pharmaceuticals in real matrices. *Anal. Bioanal. Chem.* **2017**, *409*, 2093–2106. [CrossRef]

30. Ahmad, S.M.; Almeida, C.; Neng, N.R.; Nogueira, J.M.F. Bar adsorptive microextraction (BAµE) coated with mixed sorbent phases-Enhanced selectivity for the determination of non-steroidal anti-inflammatory drugs in real matrices in combination with capillary electrophoresis. *J. Chromatogr. B Anal. Technol. Biomed. Life Sci.* **2016**, *1008*, 115–124. [CrossRef] [PubMed]

31. Ahmad, S.M.; Almeida, C.; Neng, N.R.; Nogueira, J.M.F. Application of bar adsorptive microextraction (BAµE) for anti-doping control screening of anabolic steroids in urine matrices. *J. Chromatogr. B Anal. Technol. Biomed. Life Sci.* **2014**, *969*. [CrossRef]

32. Neng, N.R.; Ahmad, S.M.; Gaspar, H.; Nogueira, J.M.F. Determination of mitragynine in urine matrices by bar adsorptive microextraction and HPLC analysis. *Talanta* **2015**, *144*, 105–109. [CrossRef]

33. Ide, A.H.; Ahmad, S.M.; Neng, N.R.; Nogueira, J.M.F. Enhancement for trace analysis of sulfonamide antibiotics in water matrices using bar adsorptive microextraction (BAµE). *J. Pharm. Biomed. Anal.* **2016**, *129*, 593–599. [CrossRef]

34. Ahmad, S.M.; Ide, A.H.; Neng, N.R.; Nogueira, J.M.F. Application of bar adsorptive microextraction to determine trace organic micro-pollutants in environmental water matrices. *Int. J. Environ. Anal. Chem.* **2017**, *97*, 484–498. [CrossRef]

35. Neng, N.R.; Nogueira, J.M.F. Monitoring trace levels of hydroxy aromatic compounds in urine matrices by bar adsorptive microextraction (BAµE). *Anal. Methods* **2017**, *9*, 5260–5265. [CrossRef]

36. Ide, A.H.; Nogueira, J.M.F. Hollow fiber microextraction: A new hybrid microextraction technique for trace analysis. *Anal. Bioanal. Chem.* **2018**, *410*, 2911–2920. [CrossRef]

37. Quintana, J.B.; Rodil, R.; Muniategui-Lorenzo, S.; López-Mahía, P.; Prada-Rodríguez, D. Multiresidue analysis of acidic and polar organic contaminants in water samples by stir-bar sorptive extraction–liquid desorption–gas chromatography–mass spectrometry. *J. Chromatogr. A* **2007**, *1174*, 27–39. [CrossRef] [PubMed]

38. Schneider, J.K.; Krause, L.C.; Carvalho, A.P.; Nogueira, J.M.F.; Caramão, E.B. Adsorption Properties of Activated Biochars Produced from Agro-industrial Residual Biomass. *Int. J. Adv. Eng. Res. Sci.* **2020**, *7*, 121–129. [CrossRef]

39. Almeida, C.; Strzelczyk, R.; Nogueira, J.M.F. Improvements on bar adsorptive microextraction (BAµE) technique-Application for the determination of insecticide repellents in environmental water matrices. *Talanta* **2014**, *120*, 126–134. [CrossRef]

40. MarvinSketch 20.10, version 20.10.0, ChemAxon Marvin 6.2.2 2014.

41. Jafari, M.T.; Saraji, M.; Sherafatmand, H. Electrospray ionization-ion mobility spectrometry as a detection system for three-phase hollow fiber microextraction technique and simultaneous determination of trimipramine and desipramine in urine and plasma samples. *Anal. Bioanal. Chem.* **2011**, *399*, 3555–3564. [CrossRef]

42. Erik, B.; Pat, S.; Frank, D.; Carel, C. Stir bar sorptive extraction (SBSE), a novel extraction technique for aqueous samples: Theory and principles. *J. Microcolumn Sep.* **1999**, *11*, 737–747. [CrossRef]

43. Baltussen, E.; Cramers, C.A.; Sandra, P.J.F. Sorptive sample preparation—A review. *Anal Bioanal Chem* **2002**, *373*, 3–22. [CrossRef]

44. David, F.; Sandra, P. Stir bar sorptive extraction for trace analysis. *J. Chromatogr. A* **2007**, *1152*, 54–69. [CrossRef]

45. Dos Santos, M.F.; Ferri, C.C.; Seulin, S.C.; Leyton, V.; Pasqualucci, C.A.G.; Muñoz, D.R.; Yonamine, M. Determination of antidepressants in whole blood using hollow-fiber liquid-phase microextraction and gas chromatography-mass spectrometry. *Forensic Toxicol.* **2014**, *32*, 214–224. [CrossRef]

46. Mansilha, C.; Melo, A.; Rebelo, H.; Ferreira, I.M.P.L.V.O.; Pinho, O.; Domingues, V.; Pinho, C.; Gameiro, P. Quantification of endocrine disruptors and pesticides in water by gas chromatography-tandem mass spectrometry. Method validation using weighted linear regression schemes. *J. Chromatogr. A* **2010**, *1217*, 6681–6691. [CrossRef] [PubMed]

47. Martins, F.I.C.C.; Barbosa, P.G.A.; Zocolo, G.J.; do Nascimento, R.F. Method Validation Using Normal and Weighted Linear Regression Models for Quantification of Pesticides in Mango (*Mangifera indica* L.) Samples. *Chromatographia* **2018**, *81*, 677–688. [CrossRef]

48. Da Silva, C.P.; Emídio, E.S.; De Marchi, M.R.R. Method validation using weighted linear regression models for quantification of UV filters in water samples. *Talanta* **2015**, *131*, 221–227. [CrossRef] [PubMed]

49. Rutkowska, E.; Lozowicka, B.; Kaczynski, P. Three approaches to minimize matrix effects in residue analysis of multiclass pesticides in dried complex matrices using gas chromatography tandem mass spectrometry. *Food Chem.* **2019**, *279*, 20–29. [CrossRef]

50. Montenarh, D.; Wernet, M.P.; Hopf, M.; Maurer, H.H.; Schmidt, P.H.; Ewald, A.H. Quantification of 33 antidepressants by LC-MS/MS-Comparative validation in whole blood, plasma, and serum. *Anal. Bioanal. Chem.* **2014**, *406*, 5939–5953. [CrossRef] [PubMed]

51. Amaratunga, P.; Thomas, C.; Lemberg, B.L.; Lemberg, D. Quantitative Measurement of XLR11 and UR-144 in Oral Fluid by LC–MS-MS. *J. Anal. Toxicol.* **2014**, *38*, 315–321. [CrossRef] [PubMed]

52. Mohebbi, A.; Farajzadeh, M.A.; Yaripour, S.; Afshar, M.R. Determination of Tricyclic Antidepressants in Human Urine Samples by the Three–Step Sample Pretreatment Followed by HPLC–UV Analysis: An Efficent Analytical Method for Further Pharmcokinetic and Forensic Studies. *EXCLI J.* **2018**, *17*, 952–963.

53. Shamsipur, M.; Mirmohammadi, M. High performance liquid chromatographic determination of ultra-traces of two tricyclic antidepressant drugs imipramine and trimipramine in urine samples after their dispersive liquid-liquid microextraction coupled with response surface optimization. *J. Pharm. Biomed. Anal.* **2014**, *100*, 271–278. [CrossRef]

54. Esrafili, A.; Yamini, Y.; Shariati, S. Hollow fiber-based liquid phase microextraction combined with high-performance liquid chromatography for extraction and determination of some antidepressant drugs in biological fluids. *Anal. Bioanal. Chem.* **2007**, *604*, 127–133. [CrossRef]

55. Ito, R.; Ushiro, M.; Takahashi, Y.; Saito, K.; Ookubo, T.; Iwasaki, Y. Improvement and validation the method using dispersive liquid–liquid microextraction with in situ derivatization followed by gas chromatography–mass spectrometry for determination of tricyclic antidepressants in human urine samples. *J. Chromatogr. B.* **2011**, *879*, 3714–3720. [CrossRef] [PubMed]
56. Uddin, M.N.; Samanidou, V.F.; Papadoyannis, I.N. Development and Validation of an HPLC Method for the Determination of Six 1,4-Benzodiazepines in Pharmaceuticals and Human Biological Fluids. *J. Liq. Chromatogr. Relat. Technol.* **2008**, *31*, 1258–1282. [CrossRef]
57. Ahmad, S.M.; Nogueira, J.M.F. High throughput bar adsorptive microextraction: A novel cost-effective tool for monitoring benzodiazepines in large number of biological samples. *Talanta* **2019**, *199*, 195–202. [CrossRef] [PubMed]

Article

Micro Salting-Out Assisted Matrix Solid-Phase Dispersion: A Simple and Fast Sample Preparation Method for the Analysis of Bisphenol Contaminants in Bee Pollen

Jianing Zhang [1], Fengjie Yu [1], Yunmin Tao [1], Chunping Du [2,3], Wenchao Yang [1,2,3], Wenbin Chen [2,3,*] and Xijuan Tu [2,3,*]

1   College of Food Science, Fujian Agriculture and Forestry University, Fuzhou 350002, China; zjnlo1225@163.com (J.Z.); yfj918@126.com (F.Y.); july766@126.com (Y.T.); beesyang@gmail.com (W.Y.)

2   College of Bee Science, Fujian Agriculture and Forestry University, Fuzhou 350002, China; du_cp39@163.com

3   College of Animal Sciences, Fujian Agriculture and Forestry University, Fuzhou 350002, China

*   Correspondence: wbchen@fafu.edu.cn (W.C.); xjtu@fafu.edu.cn (X.T.)

Abstract: In the present work, a novel sample preparation method, micro salting-out assisted matrix solid-phase dispersion (μ-SOA-MSPD), was developed for the determination of bisphenol A (BPA) and bisphenol B (BPB) contaminants in bee pollen. The proposed method was designed to combine two classical sample preparation methodologies, matrix solid-phase dispersion (MSPD) and homogenous liquid-liquid extraction (HLLE), to simplify and speed-up the preparation process. Parameters of μ-SOA-MSPD were systematically investigated, and results indicated the significant effect of salt and ACN-H_2O extractant on the signal response of analytes. In addition, excellent clean-up ability in removing matrix components was observed when primary secondary amine (PSA) sorbent was introduced into the blending operation. The developed method was fully validated, and the limits of detection for BPA and BPB were 20 μg/kg and 30 μg/kg, respectively. Average recoveries and precisions were ranged from 83.03% to 94.64% and 1.76% to 5.45%, respectively. This is the first report on the analysis of bisphenol contaminants in bee pollen sample, and also on the combination of MSPD and HLLE. The present method might provide a new strategy for simple and fast sample preparation of solid and semi-solid samples.

Keywords: sample preparation; matrix solid-phase dispersion; salting-out; homogenous liquid-liquid extraction; bisphenol; bee pollen

Citation: Zhang, J.; Yu, F.; Tao, Y.; Du, C.; Yang, W.; Chen, W.; Tu, X. Micro Salting-Out Assisted Matrix Solid-Phase Dispersion: A Simple and Fast Sample Preparation Method for the Analysis of Bisphenol Contaminants in Bee Pollen. *Molecules* **2021**, *26*, 2350. https://doi.org/10.3390/molecules26082350

Academic Editors: Victoria Samanidou and Irene Panderi

Received: 19 March 2021
Accepted: 16 April 2021
Published: 18 April 2021

Publisher's Note: MDPI stays neutral with regard to jurisdictional claims in published maps and institutional affiliations.

1. Introduction

In the past decade, matrix solid-phase dispersion (MSPD) has achieved great progress in the sample preparation of complex samples [1–4]. The major merit of MSPD is the accomplishing of extraction and clean-up procedures in one step especially for solid and semi-solid samples [5,6]. Due to its simplicity and flexibility, MSPD has been widely applied in the analysis of food, environmental, and biological matrices [3]. In recent years, modifications of the classical MSPD have been reported to refine the original method. In the ultrasonic-assisted MSPD, ultrasonication was performed on the MSPD column for speeding the process and improving the extraction yield [7,8]. Additionally, vortex and homogenization have been reported to replace the column elution and blending in the classical MSPD, respectively, to simplify the procedure of preparation [9–14]. Another method for the replacement of column was the magnetically assisted MSPD, in which magnetic ion liquid [15] or particles [16] were used as dispersants. Thus, analytes could be simply extracted by magnetic isolation. Additionally, micro/mini-MSPD, which reduced the sample amount in the protocol, was a good choice for reducing the consumption of sample and material [17–24] and improving the greenness of MSPD [25]. These improved methods have promoted the development of sample preparation technology for solid and semi-solid sample.

Homogenous liquid-liquid extraction (HLLE) is an alternative method to the traditional liquid-liquid extraction. In HLLE, a mixture of water and water-miscible solvent is applied for the liquid extraction, which is triggered to form into two separate phases after the introduction of a phase separation agent or condition [26–29]. Acetonitrile (ACN)-water-based protocol is in the hot area of HLLE studies, particularly the use of salt as a phase separation agent combined with dispersive solid-phase extraction (d-SPE), has been developed into the popular QuEChERS method [30,31]. Compared with traditional liquid-liquid extraction, HLLE shows the advantage of extraction over a wider polarity range. Furthermore, ACN is compatible with chromatography systems; this means that the obtained extract could be directly injected without additional operation of solvent exchange. These enable HLLE to be widely applied in the analysis of multiple analytes in complex matrices [26–28,31].

In the present work, we combined the principles of MSPD and HLLE to develop a novel sample preparation method named micro salting-out assisted MSPD (μ-SOA-MSPD). Impacts of its parameters were systematically investigated, and the proposed method was demonstrated to be simple, rapid, and effective with combining advantages of both MSPD and HLLE. Bisphenol compounds, a group of widespread environmental contaminants that can potentially pollute honeybee products [32,33], were successfully determined in bee pollen matrix by using the μ-SOA-MSPD and HPLC-fluorescence detection. To the best of our knowledge, this is the first report of HLLE-modified MSPD, and also the first report on the analysis of bisphenols contaminant in bee pollen sample.

2. Results and Discussion

2.1. Micro Salting-Out Assisted MSPD

The schematic procedure of the proposed μ-SOA-MSPD method is shown in Figure 1. Firstly, the sample is blended with salt and sorbent by way of the classical MSPD methodology. According to the principle of MSPD, this step is designed for the disruption of solid sample by using the shearing and grinding force in the blending operation [5]. Afterwards, the blended materials are transferred into a tube and vortexed with solution of ACN-H_2O mixture. Under vortexing, salt is dissolved into the extractant, triggering the phase separation of ACN from its aqueous solution. After short-time centrifugation to make the phase separation clear, analytes are partitioned into the upper ACN phase with high efficiency. Finally, aliquot of the ACN phase is collected and analyzed by HPLC system.

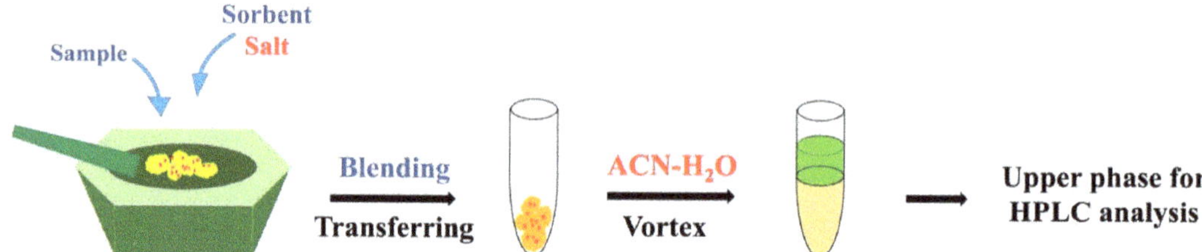

Figure 1. Schematic presentation of micro salting-out assisted matrix solid-phase dispersion (μ-SOA-MSPD).

The unique aspect of this μ-SOA-MSPD method is the introduction of salting-out induced phase separation into the procedure, which makes the HLLE simultaneously accomplished in the MSPD process. As a result, analytes are directly partitioned into the upper ACN phase in this modified MSPD method, which enables the concentration of analytes and provides a substantial clean-up effect as the majority of the matrix is isolated into the lower H_2O phase. Furthermore, the sorbent material used in the blending step could provide an additional clean-up effect similar to the dispersive solid-phase extraction (d-SPE) [30]. Based on these clean-up effects resulting from the phase partition and the d-SPE behavior of sorbent, the cartridge generally used in the classical MSPD is eliminated in the proposed method. This means that the column wash and elution, the most solvent-

and time-consuming steps, are also omitted. Thus, the proposed method possesses the merit of a much simpler procedure, as well as less consumption of time, labor, and organic solvents. These would make the modified MSPD procedure greener than the classical one according to the idea of green analytical chemistry [34].

2.2. Salting-Out Parameters

Salt is designed as both the dispersant for the disruption of sample and the phase separation agent in the following partition performance. $MgSO_4$ and $NaCl$ were investigated, due to their high efficiency for the phase separation of ACN-H_2O mixture [29,35]. Application of $MgSO_4$ in ACN-H_2O-based HLLE has shown high extraction yields for compounds with a wide polarity range due to the large volume of separated ACN phase. As regards the $NaCl$, it provides a relatively small volume of ACN phase, and thus a higher signal response for the target compounds. The effects of salts and ACN-H_2O mixture on the calculated recovery and the signal response of the analytes are compared in Figure 2.

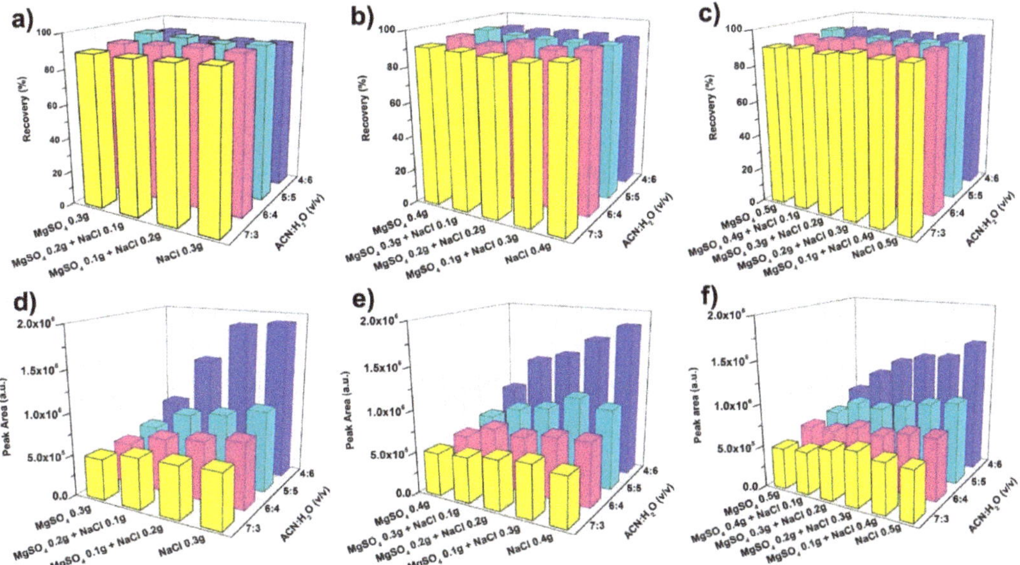

Figure 2. Effects of salts and ACN-H_2O mixture on the calculated recovery (**a**–**c**) and signal response (**d**–**f**) of bisphenol A (BPA). The total mass of salts were 0.3 g (**a,d**), 0.4 g (**b,e**), and 0.5 g (**c,f**). Mean values of triplicate experiments are presented.

High recovery values for both bisphenol A (BPA) and bisphenol B (BPB) were observed in extensive salting-out conditions. As shown in Figure 2a–c, the calculated recoveries of BPA were in the range between 88.08% and 93.89%. Under the same recipe of salts, recovery of BPA was slightly increased as the concentration of ACN in the ACN-H_2O mixture increased from 4:6 to 7:3 (v/v). Meanwhile, with the same ACN-H_2O solution, the recipe of salts did not show significant effect on the recovery values of BPA. For instance, under the concentration of 4:6 (v/v), as the recipe of salts changed from 0.5 g of $MgSO_4$ to 0.5 g of $NaCl$, recovery of BPA varied slightly from 90.43% to 90.71%.

Different from the results on recovery, the signal response of bisphenols was significantly affected by the salts and the ACN-H_2O mixture (Figure 2d–f). Under the same recipe of salts, the volume of upper phase decreased as the concentration of ACN in the ACN-H_2O mixture was reduced. As a result, concentration of analytes in the ACN phase could be significantly increased. For example, with the presence of 0.3 g of $NaCl$, as the concentration of ACN was reduced from 7:3 to 4:6 (v/v), the signal response of BPA increased to about triple. On the other hand, since $NaCl$ resulted into smaller volume of ACN phase than $MgSO_4$ [35,36], signal response of BPA became much higher when the recipe of salts changed from total $MgSO_4$ to $MgSO_4$-NaCl mixture, and then to total NaCl. Typically, in the ACN-H_2O mixture of 4:6 (*v/v*), the signal response of BPA obtained by 0.3 g of $MgSO_4$ was only about 40% of that using the same mass of NaCl. It is important to notice that increasing the mass of salts also led to the increase of the volume of ACN phase, and in consequence reducing the signal. When the concentration of ACN was 4:6 (*v/v*), as the mass of NaCl increased from 0.3 g to 0.5 g (Figure 2d–f), signal response of BPA decreased about 20%. It should be pointed out that the effects of salts and ACN-H_2O mixture on BPB (Figure S1 in Supplementary Materials) were similar to those on BPA. Based on the above results, 0.3 g of NaCl and ACN-H_2O mixture with concentration of 4:6 (*v/v*) were selected as the optimal salting-out conditions, as they provided the highest signal response.

2.3. Clean-Up Performance

In this µ-SOA-MSPD method, sorbent could be introduced into the blending step to improve clean-up performance. Bisphenol analytes were separated in reversed-phase HPLC and detected by fluorescence detector (FLD) (Figure 3). To better illustrate the clean-up effect, UV-Vis signal was recorded by diode array detector (DAD). As shown in Figure 4a, introduction of PSA in the blending exhibited remarkable reduction of matrix peaks. It was noticed that the clean-up behavior was happened in the lower retention time, which implied that relative polar compounds in the matrix might be removed by PSA. This was consistent with the reported results using PSA-based solid-phase extraction, that the PSA pipette column showed an excellent retention of phenolic compounds in bee pollen [37]. Therefore, the HPLC condition for the separation of phenolic compounds [29,38] was implemented to further demonstrate the clean-up ability. As indicated in Figure 4b, intensive matrix peaks were observed in the chromatogram of extract prepared in the absence of sorbent. Interestingly, these peaks were dramatically reduced with the addition of PSA in the blending. As the mass of PSA increased to 0.4 g, peaks area located from RT 50 min to 70 min were significantly reduced to only 4% of that obtained without addition of PSA. These DAD results demonstrated the excellent clean-up ability when the sorbent was added in the blending process. Meanwhile, for the fluorescence detection of bisphenols, the HPLC chromatogram was selective and clear enough, and no additional improvement in FLD was observed when the PSA was introduced (Figure S2). Therefore, in the case of bisphenols determination in bee pollen using HPLC-FLD, µ-SOA-MSPD could be performed by simply blended with salt without the presence of PSA.

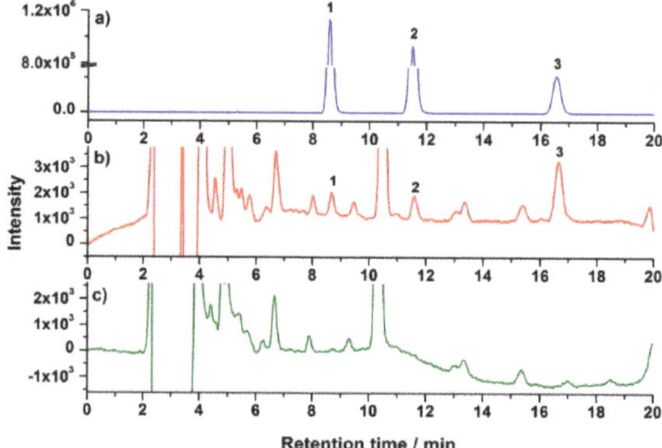

Figure 3. Representative HPLC-fluorescence detector (FLD) chromatograms of (**a**) bisphenol standards, (**b**) spiked bee pollen sample, and (**c**) blank bee pollen sample. Peak 1: bisphenol A (BPA); 2: bisphenol B (BPB); 3: internal standard (IS). The spiked concentrations in (**b**) were 60 µg/kg and 80 µg/kg for BPA and BPB, respectively.

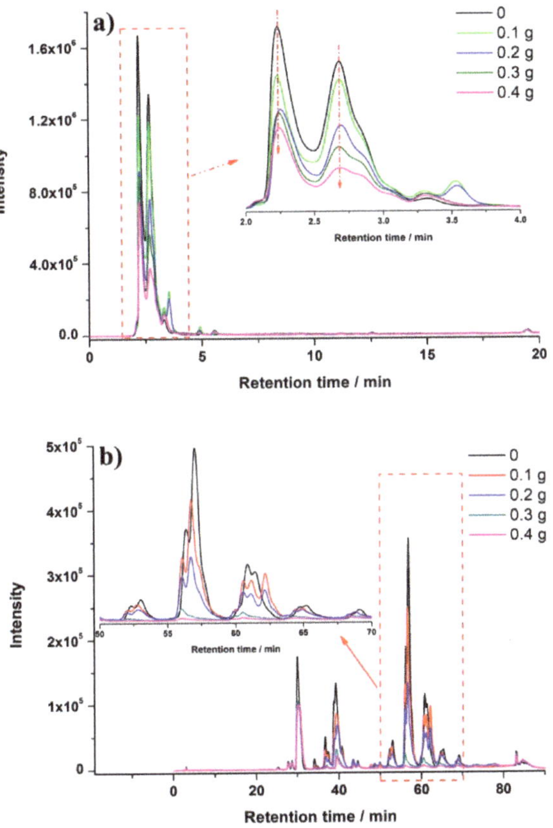

Figure 4. Representative HPLC-DAD chromatograms of extract under different masses of PSA. (**a**) Separation was performed for the analysis of bisphenols (λ = 210 nm), (**b**) separation was performed for the analysis of phenolic compounds (λ = 280 nm).

2.4. Method Validation

Seven levels of calibration curves ranging from 0.002 μg/mL to 0.4 μg/mL were applied for the quantification. Good linearity with correlation coefficients of >0.9995 were achieved. The limits of detection (LOD, S/N = 3) for BPA and BPB in bee pollen sample were 20 μg/kg and 30 μg/kg, respectively; and the limits of quantification (LOQ, S/N = 10) were 60 μg/kg and 80 μg/kg, respectively. With spiked level of LOQ, concentration of analytes in the final extractant was about 0.008 μg/mL. The accuracy and precision were investigated in blank bee pollen sample spiked at three levels (1×LOQ, 5×LOQ, and 10×LOQ). Results of the calculated recoveries and RSDs are shown in Table 1. The average recoveries for BPA and BPB were between 87.70% and 94.64%, and between 83.03% and 89.59%, respectively. The precisions were in the range of 1.76% to 5.21%, and 2.07% to 5.45% for BPA and BPB, respectively. All the results were in the acceptable range according to the AOAC [39]. In addition, eleven commercial rape bee pollen samples collected from local markets were analyzed using the proposed method. The results showed that no bisphenols were detected in these samples. More samples using different species of bee pollen should be investigated to reveal the contamination level of bisphenols. In addition, improving the enriching effect of the sample preparation method would be helpful for detecting ultra-low level of bisphenols in bee pollen. We observed lower volumes of extractant phase when a lower mass of salts and lower concentration of ACN in the ACN-H_2O mixture were used in the μ-SOA-MSPD. This would be valuable for the further development of micro-extraction methodology. These studies are under way.

Table 1. Accuracy and precision of the proposed method at three spiked levels.

Analytes	Spiked Levels (μg/kg)	Intra-Day						Inter-Day	
		Day 1		Day 2		Day 3			
		Recovery (Mean $\pm$ SD, %, $n = 6$)	RSD (%, $n = 6$)	Recovery (Mean $\pm$ SD, %, $n = 6$)	RSD (%, $n = 6$)	Recovery (Mean $\pm$ SD, %, $n = 6$)	RSD (%, $n = 6$)	Recovery (Mean $\pm$ SD, %, $n = 18$)	RSD (%, $n = 18$)
BPA	60	92.70 $\pm$ 4.13	4.46	90.19 $\pm$ 4.41	4.89	94.64 $\pm$ 3.59	3.79	92.51 $\pm$ 4.24	4.59
	300	89.13 $\pm$ 1.57	1.76	88.44 $\pm$ 2.75	3.11	87.70 $\pm$ 4.57	5.21	88.43 $\pm$ 3.07	3.47
	600	89.04 $\pm$ 3.08	3.46	90.16 $\pm$ 2.92	3.24	88.30 $\pm$ 2.25	2.55	89.16 $\pm$ 2.81	3.15
BPB	80	85.14 $\pm$ 2.31	2.71	89.59 $\pm$ 4.46	4.98	85.25 $\pm$ 4.65	5.45	86.66 $\pm$ 4.28	4.94
	400	84.29 $\pm$ 2.26	2.68	83.18 $\pm$ 2.20	2.64	83.46 $\pm$ 2.50	3.00	83.64 $\pm$ 2.24	2.68
	800	83.03 $\pm$ 1.72	2.07	83.26 $\pm$ 1.69	2.03	83.68 $\pm$ 2.61	3.12	83.32 $\pm$ 1.99	2.39

3. Materials and Methods

3.1. Materials

ACN (HPLC grade) was obtained from Merck (Darmstadt, Germany). Standards of BPA, BPB, and 4, 4′-cyclohexylidenebisphenol (internal standard, IS) were purchased from Aladdin (Shanghai, China). $MgSO_4$ was from Sinopharm Chemical Reagent Co., Ltd. (Shanghai, China) and NaCl was from Xilong Science Co., Ltd. (Guangdong, China). The PSA was obtained from Sepax (Suzhou, China). Ultrapure water (18.2 MΩ) was used through this article. Rape (*Brassica campestris*) bee pollen used for the method development was collected in Hubei, China. Commercial rape bee pollen samples applied in the method application were purchased from local markets. Stock solution of standards were prepared in ACN at the concentration of 1 mg/mL. Working standard solutions were prepared by further diluted with ACN. All standard solutions were stored at 4 °C until used.

3.2. Optimal Micro Salting-Out Assisted MSPD

A bee pollen sample (0.1 g) and NaCl (0.3 g) were blended together for 30 s, and then the materials were transferred into a 10 mL tube. After the addition of 4 mL of ACN-H_2O (4:6, v/v) solution, the mixture was vortexed for 1 min. Then the mixed solution was centrifuged at 6000 rpm for 5 min to make the phase separation clear. Aliquot of the upper phase was collected and analyzed by HPLC.

3.3. Optimization of Salting-Out Parameters

A bee pollen sample (0.1 g) was spiked with 10 μL of standards solution (100 μg/mL of BPA and BPB) and 10 μL of IS solution (100 μg/mL), then stood for 30 min. The spiked sample with different masses of salts ($MgSO_4$ and NaCl with total amount from 0.3 g to 0.5 g) were blended together, then the materials were transferred into a 10 mL tube. After the addition of 4 mL of different ACN-H_2O solutions (4:6, 5:5, 6:4, and 7:3, v/v), the mixture was vortexed for 1 min. Then the mixed solution was centrifuged at 6000 rpm for 5 min, and an aliquot of the upper phase was collected and analyzed by HPLC.

3.4. Clean-Up Effect of PSA in the Micro Salting-Out Assisted MSPD

The spiked bee pollen sample with different masses of PSA (0.1, 0.2, 0.3, and 0.4 g) and 0.3 g of NaCl were blended together, then the materials were transferred into a 10 mL tube. After the addition of 4 mL of ACN-H_2O solution (4:6, v/v), the mixture was vortexed for 1 min. Then the mixed solution was centrifuged at 6000 rpm for 5 min, and an aliquot of the upper phase was collected and analyzed by HPLC.

3.5. HPLC Analysis

The HPLC system consisted of LC-20AT pump, SIL-20AC autosampler, CTO-20AC column oven, SPD-M20A DAD and RF20-AXL FLD.

For the analysis of bisphenols, an InertSustain (Shimadzu GL, Tokyo, Japan) C18 column (4.6 mm × 250 mm, 5 μm) was used for the separation. The mobile phase consisted of water (solvent A) and ACN (solvent B). Chromatographic conditions were carried out as follows: 47% solvent B at 0–14 min, 47% to 80% at 14−17 min, and maintained at 80% at 17–20 min, then post-run with 3 min for back to 47% and maintained at 47% for 10 min. The flow rate was 1 mL/min, injection volume was 10 μL, and the column temperature was 35 °C. The detection wavelength of DAD was 210 nm, and the excitation and emission wavelength of FLD were 270 nm and 305 nm, respectively.

Phenolic compounds were separated based on the previously reported method [29,38]. A WondaCract ODS-2 (Shimadzu GL, Tokyo, Japan) C18 column (4.6 mm × 150 mm, 5 μm) was applied for the separation. The mobile phase consisted of water with 0.1% (*v/v*) acetic acid (solvent A) and methanol (solvent B). Chromatographic conditions were carried as follows: 15–40% solvent B at 0–30 min, 40–55% at 30–65 min, 55–62% at 65–70 min, 62–100% at 70–80 min, then back to 15% at 80–85 min and maintained at 15% for 5 min. The flow rate was 0.8 mL/min, injection volume was 10 μL, and the column temperature was 35 °C. The detection wavelength of DAD was 280 nm.

3.6. Method Validation

Seven-level standard curves were prepared containing BPA (0.002, 0.01, 0.02, 0.05, 0.1, 0.2, and 0.4 μg/mL), BPB (0.002, 0.01, 0.02, 0.05, 0.1, 0.2, and 0.4 μg/mL), and IS (0.1 μg/mL). The ratio of peak area (analyte/IS) versus the ratio of weight (analyte/IS) was used to construct the analytical curves. The y-intercept was set to zero and a linear fit was performed. Blank bee pollen sample with the absence of analytes was used for the method validation. The limit of detection (LOD) and limit of quantification (LOQ), defined as 3×S/N (signal/noise) and 10×S/N, respectively, were investigated in spiked blank bee pollen samples, which were prepared as described in Section 3.2. Accuracy and precision studies were estimated by analyzing blank samples spiked at three levels (1×LOQ, 5×LOQ, and 10×LOQ), which were prepared as described in Section 3.2. Recovery was used to express the accuracy, and the precision was determined as relative standard deviation (RSD) to the mean recovery in repeatability (intra-day, $n = 6$) and intermediate precision (inter-day, three consecutive days, $n = 18$) analysis.

4. Conclusions

In summary, a new sample preparation method, μ-SOA-MSPD, was developed for the analysis of bisphenol contaminants in bee pollen. Salt and the ACN-H_2O mixture could be designed to achieve better recovery and signal response. In addition, the introduction of sorbent in the blending step showed excellent removal of matrix components. The proposed method was fully validated, and the results indicated that it was suitable for the reliable and sensitive detection of BPA and BPB residues in spiked bee pollen. The proposed method was simple, rapid, and provided advantages in saving time, labor, and solvents. Since the flexibility of MSPD technology in different matrices, this modified MSPD method would be valuable for the fast sample preparation of other solid and semi-solid samples.

Supplementary Materials: The following are available online, Figure S1: Effects of salts and ACN-H_2O mixture on the calculated recovery and signal response of BPB. Figure S2: Representative HPLC-FLD chromatograms of extract under different mass of PSA.

Author Contributions: Conceptualization, W.C. and X.T.; Data curation, J.Z. and F.Y.; Funding acquisition, W.C. and X.T.; Investigation, J.Z. and F.Y.; Methodology, J.Z., W.C. and X.T.; Project administration, X.T.; Resources, W.C. and X.T.; Supervision, W.C. and X.T.; Validation, Y.T. and C.D.; Writing—original draft, W.C. and X.T.; Writing—review & editing, W.Y., W.C. and X.T. All authors have read and agreed to the published version of the manuscript.

Funding: This research was funded by Natural Science Foundation of Fujian Province, grant number 2020J01535 and 2019J01409, and Natural Science Foundation of China, grant number 32072799.

Institutional Review Board Statement: Not applicable.

Informed Consent Statement: Not applicable.

Data Availability Statement: The data presented in this study are available on request from the corresponding author.

Conflicts of Interest: The authors declare no conflict of interest.

Sample Availability: Samples of the compounds are not available from the authors.

References

1. Capriotti, A.L.; Cavaliere, C.; Giansanti, P.; Gubbiotti, R.; Samperi, R.; Lagana, A. Recent developments in matrix solid-phase dispersion extraction. *J. Chromatogr. A* **2010**, *1217*, 2521–2532. [CrossRef]
2. Capriotti, A.L.; Cavaliere, C.; Foglia, P.; Samperi, R.; Stampachiacchiere, S.; Ventura, S.; Laganà, A. Recent advances and developments in matrix solid-phase dispersion. *Trends Anal. Chem.* **2015**, *71*, 186–193. [CrossRef]
3. Tu, X.; Chen, W. A Review on the Recent Progress in Matrix Solid Phase Dispersion. *Molecules* **2018**, *23*, 2767. [CrossRef]
4. Ramos, L. Use of new tailored and engineered materials for matrix solid-phase dispersion. *Trends Anal. Chem.* **2019**, *118*, 751–758. [CrossRef]
5. Barker, S.A. Matrix solid-phase dispersion. *J. Chromatogr. A* **2000**, *885*, 115–127. [CrossRef]
6. Barker, S.A. Matrix solid phase dispersion (MSPD). *J. Biochem. Biophys. Methods* **2007**, *70*, 151–162. [CrossRef]
7. Ramos, J.J.; Rial-Otero, R.; Ramos, L. Ultrasonic-assisted matrix solid-phase dispersion as an improved methodology for the determination of pesticides in fruits. *J. Chromatogr. A* **2008**, *1212*, 145–149. [CrossRef] [PubMed]
8. Aznar, R.; Albero, B.; Perez, R.A.; Sanchez-Brunete, C.; Miguel, E.; Tadeo, J.L. Analysis of emerging organic contaminants in poultry manure by gas chromatography-tandem mass spectrometry. *J. Sep. Sci.* **2018**, *41*, 940–947. [CrossRef] [PubMed]
9. Hertzog, G.I.; Soares, K.L.; Caldas, S.S.; Primel, E.G. Study of vortex-assisted MSPD and LC-MS/MS using alternative solid supports for pharmaceutical extraction from marketed fish. *Anal. Bioanal. Chem.* **2015**, *407*, 4793–4803. [CrossRef] [PubMed]
10. León-González, M.E.; Rosales-Conrado, N. Determination of ibuprofen enantiomers in breast milk using vortex-assisted matrix solid-phase dispersion and direct chiral liquid chromatography. *J. Chromatogr. A* **2017**, *1514*, 88–94. [CrossRef] [PubMed]
11. Chen, S.J.; Du, K.Z.; Li, J.; Chang, Y.X. A chitosan solution-based vortex-forced matrix solid phase dispersion method for the extraction and determination of four bioactive constituents from *Ligustri Lucidi Fructus* by high performance liquid chromatography. *J. Chromatogr. A* **2020**, *1609*, 460509. [CrossRef]
12. Chung, W.H.; Lin, J.S.; Ding, W.H. Dual-vortex-assisted matrix solid-phase dispersion coupled with isotope-dilution ultrahigh-performance liquid chromatography-high resolution mass spectrometry for the rapid determination of parabens in indoor dust samples. *J. Chromatogr. A* **2019**, *1605*, 460367. [CrossRef]
13. Gomez-Mejia, E.; Rosales-Conrado, N.; Leon-Gonzalez, M.E.; Madrid, Y. Determination of phenolic compounds in residual brewing yeast using matrix solid-phase dispersion extraction assisted by titanium dioxide nanoparticles. *J. Chromatogr. A* **2019**, *1601*, 255–265. [CrossRef]
14. Kemmerich, M.; Demarco, M.; Bernardi, G.; Prestes, O.D.; Adaime, M.B.; Zanella, R. Balls-in-tube matrix solid phase dispersion (BiT-MSPD): An innovative and simplified technique for multiresidue determination of pesticides in fruit samples. *J. Chromatogr. A* **2020**, *1612*, 460640. [CrossRef] [PubMed]
15. Chatzimitakos, T.G.; Anderson, J.L.; Stalikas, C.D. Matrix solid-phase dispersion based on magnetic ionic liquids: An alternative sample preparation approach for the extraction of pesticides from vegetables. *J. Chromatogr. A* **2018**, *1581–1582*, 168–172. [CrossRef] [PubMed]
16. Fotouhi, M.; Seidi, S.; Shanehsaz, M.; Naseri, M.T. Magnetically assisted matrix solid phase dispersion for extraction of parabens from breast milks. *J. Chromatogr. A* **2017**, *1504*, 17–26. [CrossRef]
17. Guerra, E.; Celeiro, M.; Lamas, J.P.; Llompart, M.; Garcia-Jares, C. Determination of dyes in cosmetic products by micro-matrix solid phase dispersion and liquid chromatography coupled to tandem mass spectrometry. *J. Chromatogr. A* **2015**, *1415*, 27–37. [CrossRef] [PubMed]
18. Chen, Q.; Lin, Y.; Tian, Y.; Wu, L.; Yang, L.; Hou, X.; Zheng, C. Single-Drop Solution Electrode Discharge-Induced Cold Vapor Generation Coupling to Matrix Solid-Phase Dispersion: A Robust Approach for Sensitive Quantification of Total Mercury Distribution in Fish. *Anal. Chem.* **2017**, *89*, 2093–2100. [CrossRef]
19. Deng, T.; Wu, D.; Duan, C.; Yan, X.; Du, Y.; Zou, J.; Guan, Y. Spatial Profiling of Gibberellins in a Single Leaf Based on Microscale Matrix Solid-Phase Dispersion and Precolumn Derivatization Coupled with Ultraperformance Liquid Chromatography-Tandem Mass Spectrometry. *Anal. Chem.* **2017**, *89*, 9537–9543. [CrossRef] [PubMed]
20. Xu, J.J.; Yang, R.; Ye, L.H.; Cao, J.; Cao, W.; Hu, S.S.; Peng, L.Q. Application of ionic liquids for elution of bioactive flavonoid glycosides from lime fruit by miniaturized matrix solid-phase dispersion. *Food Chem.* **2016**, *204*, 167–175. [CrossRef] [PubMed]
21. Xu, J.J.; Cao, J.; Peng, L.Q.; Cao, W.; Zhu, Q.Y.; Zhang, Q.Y. Characterization and determination of isomers in plants using trace matrix solid phase dispersion via ultrahigh performance liquid chromatography coupled with an ultraviolet detector and quadrupole time-of-flight tandem mass spectrometry. *J. Chromatogr. A* **2016**, *1436*, 64–72. [CrossRef] [PubMed]
22. Cao, J.; Peng, L.Q.; Xu, J.J.; Du, L.J.; Zhang, Q.D. Simultaneous microextraction of inorganic iodine and iodinated amino acids by miniaturized matrix solid-phase dispersion with molecular sieves and ionic liquids. *J. Chromatogr. A* **2016**, *1477*, 1–10. [CrossRef] [PubMed]

23. Guerra, E.; Llompart, M.; Garcia-Jares, C. Miniaturized matrix solid-phase dispersion followed by liquid chromatography-tandem mass spectrometry for the quantification of synthetic dyes in cosmetics and foodstuffs used or consumed by children. *J. Chromatogr. A* **2017**, *1529*, 29–38. [CrossRef]

24. Chu, C.; Jiang, L.; Mao, H.; Yan, J. A simple and environmentally-friendly method by pipette-tip matrix solid-phase dispersion microextraction coupled with high-performance liquid chromatography for the simultaneous determination of lignans and terpenes. *Sustain. Chem. Pharm.* **2021**, *20*, 100384. [CrossRef]

25. Pena-Abaurrea, M.; Ramos, L. Miniaturization of analytical methods. In *Challenges in Green Analytical Chemistry*, 1st ed.; de la Guardia, M., Garrigues, S., Eds.; Royal Society of Chemistry: Cambridge, UK, 2011; pp. 107–143.

26. Anthemidis, A.N.; Ioannou, K.-I.G. Recent developments in homogeneous and dispersive liquid-liquid extraction for inorganic elements determination. A review. *Talanta* **2009**, *80*, 413–421. [CrossRef]

27. Dmitrienko, S.G.; Apyari, V.V.; Gorbunova, M.V.; Tolmacheva, V.V.; Zolotov, Y.A. Homogeneous Liquid–Liquid Microextraction of Organic Compounds. *J. Anal. Chem.* **2020**, *75*, 1371–1383. [CrossRef]

28. Valente, I.M.; Rodrigues, J.A. Recent advances in salt-assisted LLE for analyzing biological samples. *Bioanalysis* **2015**, *7*, 2187–2193. [CrossRef] [PubMed]

29. Chen, W.; Tu, X.; Wu, D.; Gao, Z.; Wu, S.; Huang, S. Comparison of the Partition Efficiencies of Multiple Phenolic Compounds Contained in Propolis in Different Modes of Acetonitrile-Water-Based Homogenous Liquid-Liquid Extraction. *Molecules* **2019**, *24*, 442. [CrossRef]

30. Anastassiades, M.; Lehotay, S.J.; Stajnbaher, D.; Schenck, F.J. Fast and Easy Multiresidue Method Employing Acetonitrile Extraction/Partitioning and "Dispersive Solid-Phase Extraction" for the Determination of Pesticide Residues in Produce. *J. AOAC Int.* **2003**, *86*, 412–431. [CrossRef] [PubMed]

31. González-Curbelo, M.Á.; Socas-Rodríguez, B.; Herrera-Herrera, A.V.; González-Sálamo, J.; Hernández-Borges, J.; Rodríguez-Delgado, M.Á. Evolution and applications of the QuEChERS method. *Trends Anal. Chem.* **2015**, *71*, 169–185. [CrossRef]

32. Tu, X.; Wu, S.; Liu, W.; Gao, Z.; Huang, S.; Chen, W. Sugaring-Out Assisted Liquid-Liquid Extraction Combined with High-Performance Liquid Chromatography-Fluorescence Detection for the Determination of Bisphenol A and Bisphenol B in Royal Jelly. *Food Anal. Methods* **2019**, *12*, 705–711. [CrossRef]

33. Inoue, K.; Murayama, S.; Takeba, K.; Yoshimura, Y.; Nakazawa, H. Contamination of xenoestrogens bisphenol A and F in honey: Safety assessment and analytical method of these compounds in honey. *J. Food Compos. Anal.* **2003**, *16*, 497–506. [CrossRef]

34. Gałuszka, A.; Konieczka, P.; Migaszewski, Z.M.; Namiesnik, J. Analytical Eco-Scale for assessing the greenness of analytical procedures. *Trends Anal. Chem.* **2012**, *37*, 61–72. [CrossRef]

35. Valente, I.M.; Goncalves, L.M.; Rodrigues, J.A. Another glimpse over the salting-out assisted liquid-liquid extraction in acetonitrile/water mixtures. *J. Chromatogr. A* **2013**, *1308*, 58–62. [CrossRef] [PubMed]

36. Tu, X.; Sun, F.; Wu, S.; Liu, W.; Gao, Z.; Huang, S.; Chen, W. Comparison of salting-out and sugaring-out liquid-liquid extraction methods for the partition of 10-hydroxy-2-decenoic acid in royal jelly and their co-extracted protein content. *J. Chromatogr. B* **2018**, *1073*, 90–95. [CrossRef]

37. Tu, X.; Chen, W. Miniaturized Salting-Out Assisted Liquid-Liquid Extraction Combined with Disposable Pipette Extraction for Fast Sample Preparation of Neonicotinoid Pesticides in Bee Pollen. *Molecules* **2020**, *25*, 5703. [CrossRef]

38. Zhang, C.P.; Huang, S.; Wei, W.T.; Ping, S.; Shen, X.G.; Li, Y.J.; Hu, F.L. Development of high-performance liquid chromatographic for quality and authenticity control of Chinese propolis. *J. Food Sci.* **2014**, *79*, C1315–C1322.

39. Taverniers, I.; De Loose, M.; Van Bockstaele, E. Trends in quality in the analytical laboratory. II. Analytical method validation and quality assurance. *Trends Anal. Chem.* **2004**, *23*, 535–552. [CrossRef]

molecules

MDPI

Article

A Simple and Reliable Dispersive Liquid-Liquid Microextraction with Smartphone-Based Digital Images for Determination of Carbaryl Residues in *Andrographis paniculata* Herbal Medicines Using Simple Peroxidase Extract from *Senna siamea* Lam. Bark

Sam-ang Supharoek [1,2], Watsaka Siriangkhawut [3], Kate Grudpan [4] and Kraingkrai Ponhong [3,*]

1 Department of Chemistry and Center of Excellence for Innovation in Chemistry, Faculty of Science, Mahidol University, Bangkok 10400, Thailand; samang.sup@mahidol.ac.th
2 Department of Medical Science, Mahidol University, Amnatcharoen Campus, Amnat Charoen 37000, Thailand
3 Creative Chemistry and Innovation Research Unit, Department of Chemistry and Center of Excellence for Innovation in Chemistry, Faculty of Science, Mahasarakham University, Maha Sarakham 44150, Thailand; watsaka@hotmail.com
4 Department of Chemistry, Faculty of Science and Center of Excellence for Innovation in Analytical Science and Technology for Biodiversity-Based Economic and Society, Chiang Mai University, Chiang Mai 50200, Thailand; kgrudpan@gmail.com
* Correspondence: kraingkrai.p@msu.ac.th; Tel.: +66-43754-246

Citation: Supharoek, S.-a.; Siriangkhawut, W.; Grudpan, K.; Ponhong, K. A Simple and Reliable Dispersive Liquid-Liquid Microextraction with Smartphone-Based Digital Images for Determination of Carbaryl Residues in *Andrographis paniculata* Herbal Medicines Using Simple Peroxidase Extract from *Senna siamea* Lam. Bark. *Molecules* **2022**, *27*, 3261. https://doi.org/10.3390/molecules27103261

Academic Editors: Victoria Samanidou, Irene Panderi and Nuno Neng

Received: 5 April 2022
Accepted: 17 May 2022
Published: 19 May 2022

Publisher's Note: MDPI stays neutral with regard to jurisdictional claims in published maps and institutional affiliations.

Abstract: A simple and reliable dispersive liquid-liquid microextraction (DLLME) coupled with smartphone-based digital images using crude peroxidase extracts from cassia bark (*Senna siamea* Lam.) was proposed to determine carbaryl residues in *Andrographis paniculata* herbal medicines. The method was based on the reaction of 1-naphthol (hydrolysis of carbaryl) with 4-aminoantipyrine (4-AP) in the presence of hydrogen peroxide, using peroxidase enzyme simple extracts from cassia bark as biocatalysts under pH 6.0. The red product, after preconcentration by DLLME using dichloromethane as extraction solvent, was measured for blue intensity by daily life smartphone-based digital image analysis. Under optimized conditions, good linearity of the calibration graph was found at 0.10–0.50 mg·L^{-1} (r^2 = 0.9932). Limits of detection (LOD) (3SD/slope) and quantification (LOQ) (10SD/slope) were 0.03 and 0.09 mg·L^{-1}, respectively, with a precision of less than 5%. Accuracy of the proposed method as percentage recovery gave satisfactory results. The proposed method was successfully applied to analyze carbaryl in *Andrographis paniculata* herbal medicines. Results agreed well with values obtained from the HPLC-UV method at 95% confidence level. This was simple, convenient, reliable, cost-effective and traceable as an alternative method for the determination of carbaryl.

Keywords: carbaryl; cassia bark (*Senna siamea* Lam.); smartphone-based digital image analysis; 1-naphthol; peroxidase enzyme

1. Introduction

Since the outbreak of the COVID-19 virus in 2019, the application of medicinal plants to strengthen immunity and prevent viral infections has increased as another option [1–3]. *Andrographis paniculata* is an indigenous medicinal plant found in Malaysia and Thailand [1,2,4,5] that has antioxidant properties to scavenge free radicals [6,7] and stimulate the immune system [8,9] against foreign matter entering the body, thereby inhibiting the growth of cancer cells [5,10–12]. This plant is widely used for treating sore throats, flu and upper respiratory tract infections [5,13]. Therefore, the use of pesticides increases the harvest of *Andrographis paniculate* [14].

Carbaryl (1-naphthyl methylcarbamate), a synthetic insecticide in the carbamate family, causes reproductive and developmental toxicity including neurodevelopmental perturbations that impact the immune system as a possible human carcinogen. Carbaryl exhibits high toxicity to non-target organisms [15] but is only moderately toxic to aquatic organisms (LC50 values in rainbow trout and bluegill of 1.3 and 10 mg·L^{-1}, respectively) [16]. Oral LD50 of carbaryl ranges from 250 mg·kg^{-1} to 850 mg·kg^{-1} in rats, and from 100 mg·kg^{-1} to 650 mg· kg^{-1} in mice [17].

Various analytical methods have been established for the determination of carbaryl such as high-performance liquid chromatography (HPLC) [18–20], liquid chromatography-mass spectrometry (LC-MS) [21,22], liquid chromatography-tandem mass spectrometry (LC-MS/MS) [23], gas chromatography (GC) [24,25] and capillary electrophoresis [26,27]. These techniques can simultaneously determine carbaryl with other carbamate insecticides but they require highly skilled operators and are cost-consuming. Electrochemical analysis [28–31] to investigate carbaryl content has advantages in terms of specific and low detection limits. However, this method requires skilled fabrication and modification of the working electrode. Fluorescence spectroscopy has also been used to determine carbaryl [32,33], providing advantages of selectivity detection, while spectrophotometric methods [34–38] are based on carbaryl hydrolysis to 1-naphthol, which is subsequently coupled with different reagents. Spectrophotometric procedures are considered to be appropriate as they involve commonly available laboratory instruments such as a spectrometer but they still need some specific requirements [39,40].

Recently, image processing was applied for chemical analysis. Digital image acquisition devices such as a scanner, camera and smartphone built-in camera as cheap electronic components were employed to capture digital photo images [39]. Digital imaging analysis is an advanced technique to evaluate the color intensity of captured digital photo images using image processing software [41]. Pixels in red, green or blue channels with numerical values ranging from 0 to 255 can be utilized for analytical calibration [42]. Digital images provide data that can be used for fast and low-cost colorimetric detection of quantitative chemical analysis [43]. Smartphones are extensively employed in daily life as image acquisition tools because they are portable, with multiple data transmission functions and high storage capacity [44]. Digital image data acquired from image software can considerably enhance detection accuracy [45,46]. Digital image analysis has been applied in various fields including environmental monitoring, food safety and clinical analysis due to its convenience, stability, low cost and flexibility [47,48]. Digital image analysis has also been developed to determine liquid turbidity [42] and bacterial cell concentration in liquid media [49].

Enzymes are often employed as bio-accelerators because they require small amounts of substrate and utilize specific reactions that operate under mild conditions [50,51]. Peroxidase enzymatic spectrophotometry has been used to quantify carbaryl residues in vegetables by our group [52,53]. Cassia bark (*Senna simea*), an indigenous plant found in many areas including Thailand, Southern India, Sri Lanka, Myanmar (Burma), Cambodia, Malaysia and parts of Indonesia was employed as a source of peroxidase enzyme for extraction by a simple method yielding crude peroxidase enzyme. Crude peroxidase enzyme extract was used as an alternative to commercially available expensive horseradish peroxidase as a biocatalyst for the enzymatic reaction of carbaryl with 4-AP in the presence of hydrogen peroxide under optimal pH. No additional enzyme purification steps were required to avoid deterioration through enzyme degradation [54–56]. This process offered comparable analytical performance to horseradish peroxidase as a green chemical analysis.

However, a sample containing carbaryl at low levels in complicated matrices requires pretreatment before the colorimetric detection step. Sample pretreatment techniques such as dispersive liquid-liquid microextraction (DLLME) have recently attracted increased attention to concentrate the sample before analysis [52,57,58]. DLLME has advantages of simplicity of operation, rapidity, low cost, high recovery, high preconcentration factor and environmental benignity [59].

Here, we developed a simple downscaling cost effective procedure as a green chemical analysis for the colorimetric enzymatic determination of carbaryl residues in herbal medicine samples by employing crude peroxidase enzyme extracts with smartphone-based digital image analysis, using dispersive liquid-liquid microextraction for sample pretreatment.

Here, low-cost crude peroxidase enzyme extract from cassia bark was employed as biocatalysts. A down-scaled dispersive liquid-liquid microextraction technique reduced the consumption of deleterious extraction and disperser solvent to preconcentrate the analyte before daily life smartphone-based digital imaging analysis. The developed method was simple, cost-effective, reliable and tracible and produced comparable results to HPLC.

2. Results and Discussion

2.1. Activity of Peroxidase Crude-Extract

Peroxidase enzyme extracts from cassia bark were mixed with ready-to-use ABTS substrate in the presence of hydrogen peroxide. The green color of ABTS cation radical was observed, indicating that peroxidase was found in the crude extracts from cassia bark. Activity of the crude enzyme was 1.15 ± 0.05 kU 10 μL^{-1}, equivalent to 0.32 kU L^{-1} of horseradish peroxidase (HRP, Sigma-Aldrich, St. Louis, MO, USA) evaluated from HRP activity calibration in the range 0.125–2.0 kU L^{-1} [52]. Peroxide activity remained constant for 3 months at -20 °C. Activity decreased by 9% after 6 months compared with fresh extracts, while the stability of enzyme extracts declined after 8 h at room temperature, indicating that they could be used for quantification of carbaryl for only one day.

2.2. Suggested Reaction between Peroxidase Extracts and Carbaryl

The suggested reaction for the determination of carbaryl was based on the mixed solution containing 1-naphthol as the hydrolysis product of carbaryl under alkaline conditions, 4-AP and hydrogen peroxide. The mixture solution was catalyzed by crude peroxidase enzyme extracts under phosphate buffer pH 6. Maximum absorption of red product was observed at a wavelength of 500 nm (Figure 1a).

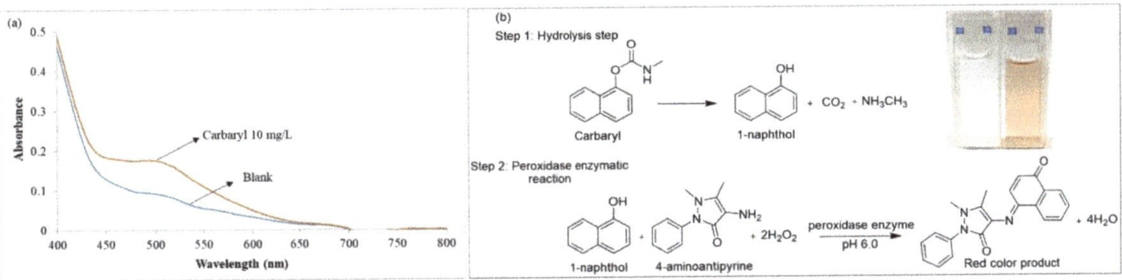

Figure 1. Determination of carbaryl based on peroxidase enzymatic reaction; (**a**) Absorption spectra 10 mg·L^{-1} carbaryl in phosphate buffer pH 6.0 and blank; (**b**) suggested mechanism for the reaction of carbaryl with 4-AP in the presence of hydrogen peroxide, exploiting crude peroxidase as a catalyst.

Figure 1b shows the reaction involving two steps as (i) hydrolysis of carbaryl under alkaline conditions yielding 1-naphthol, and (ii) enzymatic reaction of 1-naphthol with 4-AP in the presence of hydrogen peroxide using peroxidase as a catalyst. 4-AP reacted with 1-naphthol at the para-position of the aromatic ring giving a red-colored product [60]. Results indicated that carbaryl insecticide could be produced using crude peroxidase enzyme extracts from cassia bark.

2.3. Optimization of Operational Parameters for Determination of Carbaryl by Smartphone-Based Digital Images

A light-emitting diode (LED) (Yongnuo YN300 III, China) was placed in the middle of the light control box used as the illumination device. This eliminated the need for flash photography to capture images and control the uniformity of light intensity (Figure S1). The captured images of the 96-microwell plate without extraction phase and the repeatability of RGB intensities were used to assess the homogeneity of light illumination from the LEDs. Relative standard deviation (RSD) was 1.47%, indicating excellent measurement repeatability of RGB intensity under the light control box. The RSD obtained from the calibration graph in the range 0.10 to 0.50 mg·L^{-1} of standard carbaryl was less than 6% (n = 9). We concluded that light and temperature in a closed light control box were not significantly affected by the intensity of RGB color.

Image qualities obtained from the camera and smartphone were not significantly different. The RGB intensity of captured images using standard carbaryl at 0.30 mg·L^{-1} achieved using a smartphone and camera were not significantly different (t Stat = 0.123 < t Critical = 1.985 at 95% confidence level and df = 95). Moreover, the slope of the calibration graph obtained from the smartphone was not significantly different when compared with the slope of the calibration curve from the camera (t Stat = 1.14 < t Critical = 2.78 at 95% confidence level and df = 4).

After enzymatic reaction and DLLME optimization, the extraction solvent was transferred to a 96-microwell plate to capture the digital image under the light control box using a smartphone. RGB intensity of the photographed image was evaluated by ImageJ software to obtain the RGB profile as illustrated in Figure S2. For quantification purposes, the graph of RGB intensity difference (ΔI) (difference in color intensity of reagent blank zone and color intensity of standard) was plotted against carbaryl concentrations in the range 0 to 0.50 mg·L^{-1} (Figure S3). Blue (B) and green (G) intensity of the captured image related to the concentration of carbaryl but blue intensity gave the highest sensitivity. Therefore, blue intensity was utilized to determine carbaryl by the proposed method.

Sensitivity of the developed method depended on the volume of extraction phase in the microwell plate and was investigated at 100, 200 and 300 µL. The loading extraction phase at 200 µL gave good sensitivity and linearity for quantification of carbaryl. Low sensitivity was achieved at 100 µL, while 300 µL gave a narrow linear range and low sensitivity. Hence, 200 µL of the extraction phase was loaded into the 96-microwell plate. The calibration graph was constructed by plotting the change in intensity of blue color (intensity of blank − intensity of analyte) versus carbaryl concentration ranging from 0.10 to 0.50 mg·L^{-1}.

2.4. Optimized Conditions for Determination of Carbaryl Using Crude Peroxidase Enzyme

2.4.1. Effect of pH

The pH value impacts stability, conformation and activity of an enzyme. The effect of pH on the enzymatic reaction for carbaryl assay was determined using citrate-phosphate buffer (pH 3 to pH 5) and phosphate buffer (pH 6 to pH 7). Results in Figure 2a show that crude peroxidase extract provided the highest sensitivity using phosphate buffer pH 6. Therefore, the peroxidase enzyme catalytic reaction for carbaryl assay was operated at pH 6.0.

2.4.2. Effect of 4-AP Concentration

4-AP chromogenic substrate was employed for peroxidase enzymatic reaction to determine carbaryl. 4-AP acts as a hydrogen atom donor in the peroxidase catalytic reaction. Concentrations of 4-AP from 50 to 200 mg·L^{-1} were investigated. Results indicated that the analytical signal increased gradually with increasing 4-AP concentration from 50 to 150 mg·L^{-1}. The signal was not significantly different from 150 to 200 mg·L^{-1} because active sites of peroxidase were almost saturated with 4-AP (Figure 2b). Therefore, 150 mg·L^{-1} of 4-AP was selected for the next experiment.

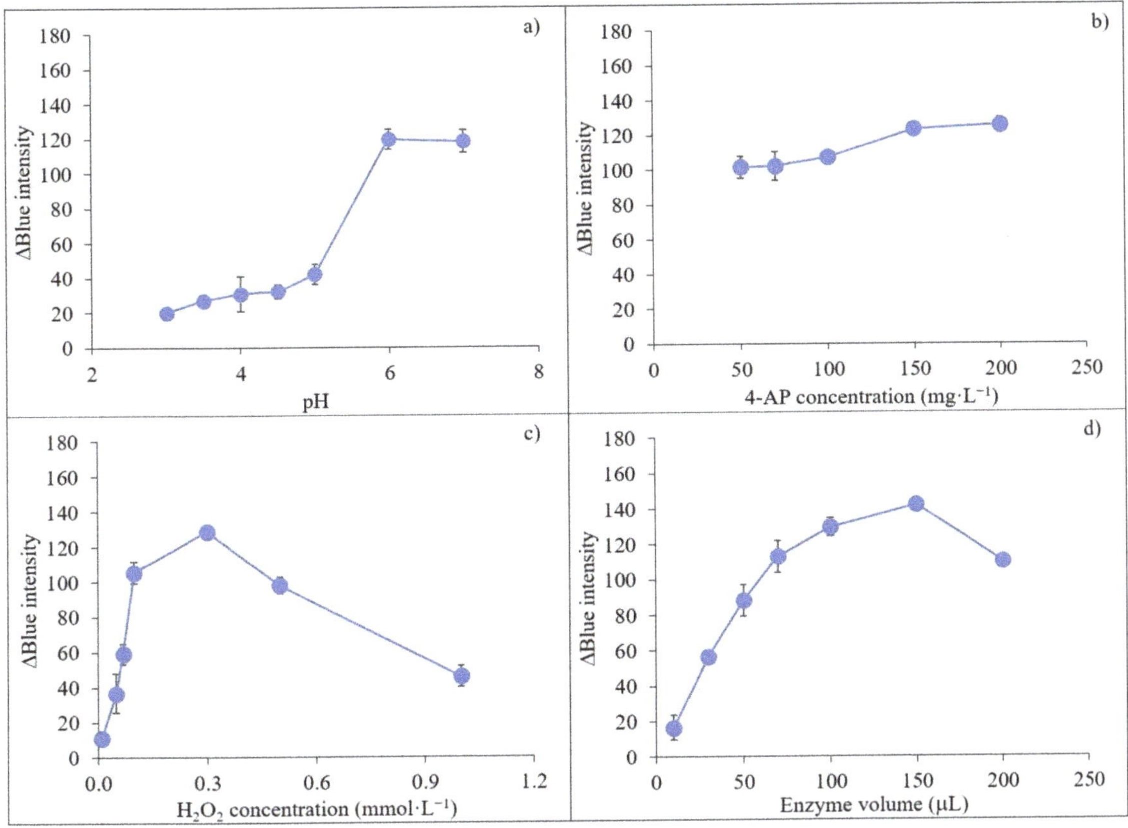

Figure 2. Investigation of various carbaryl detection parameters based on peroxidase enzymatic reaction; (**a**) effect of pH; (**b**) effect of 4-AP concentration; (**c**) effect of hydrogen peroxide concentration and (**d**) effect of enzyme volume on sensitivity of carbaryl detection by peroxidase enzymatic reaction.

2.4.3. Effect of Hydrogen Peroxide Concentration

Hydrogen peroxide acts as a hydrogen atom acceptor. Results in Figure 2c show that the signal increased sharply between 0.01 and 0.3 mmol·L^{-1} hydrogen peroxide and then dropped until 1.0 mmol·L^{-1} due to hydrogen peroxide inhibition, with 0.3 mmol·L^{-1} providing the highest value. This concentration was selected for subsequent experiments.

2.4.4. Effect of Peroxidase Enzyme Volume

The volume of peroxidase enzyme extracts impacted sensitivity by influencing enzyme activity. Here, 10–200 µL of enzyme extracts were tested. Enzyme activity increased with increasing volume of enzyme, accelerating the reaction. When increasing the volume of enzyme, the red color of the blank also increased. Results are presented in Figure 2d. The signal increased gradually from 10 µL to 150 µL and then decreased above 150 µL. Thus, 150 µL volume was selected for subsequent experiments.

2.4.5. Effect of Incubation Time

Longer incubation time increased products from the catalytic reaction. Incubation time ranging 1–20 min was studied. Sensitivity climbed continuously from 1 min to 10 min and then the signal leveled off over 10 min (data not shown). Therefore, 10 min incubation time was chosen for the next procedure giving sufficient determination sensitivity and short time of analysis.

2.5. DLLME Optimization for Carbaryl Detection

2.5.1. Effect of Types and Volume of Extraction Solvents

Types and volume of extraction solvents in DLLME impacted extraction efficiency [57,59]. Chloroform, dichloromethane, octanol and 1-dodecanol were explored, with results shown in Figure 3a. Optimal extraction efficiency for carbaryl was obtained when dichloromethane was used as the extraction solvent, and this was selected for the proposed method. Volumes of 100–700 μL extract solvent were also considered. Results showed that the signal increased from 100 μL to 500 μL and then decreased from 500 μL to 700 μL because of the dilution effect (Figure 3b). Thus, 500 μL of extraction solvent was selected for subsequent experiments.

2.5.2. Effect of Types and Volume of Dispersive Solvents

Dispersive solvents must have a good tendency between organic (extraction solvent) and aqueous phases and should be selected according to the miscibility properties of the extraction solvent and aqueous phase [61,62]. Different dispersive solvents such as acetonitrile, ethanol, methanol and acetone were investigated. Ethanol provided the highest signal compared to the other tested solvents (Figure 3c) and was selected as the dispersive solvent of DLLME. To evaluate the effect of dispersive solvent volume on extraction efficiency, a constant volume extraction solvent (dichloromethane at 500 μL) containing different volumes of ethanol from 100 μL to 700 μL was studied for the DLLME process. The signal at 300 μL ethanol showed maximum sensitivity (Figure 3d).

2.5.3. Effect of Ionic Strength

Sodium chloride was added to improve extraction efficiency through the salting-out effect by decreasing the solubility of the analyte [52,63]. Results are shown in Figure 3e. The signal sharply increased from 0.6 to 1.0% (w/v) and then reduced from 1.2 to 1.4% (w/v). Thus, 1.0% (w/v) sodium chloride was chosen for the developed method.

2.5.4. Effect of Vortex Time

Vortex mixing increases contact between the extraction solvent and the analyte to improve extraction efficiency. The analysis signal increased from 0.1 min to 1 min (data not shown), while at over 1 min the signal leveled off because equilibrium was attained. Therefore, 1 min vortex time was adopted.

2.5.5. Effect of Centrifugation Time

Centrifugation to complete phase separation between the organic and aqueous phases was examined in the range 1–10 min at 4032 g. Results showed that the signal climbed continuously from 1 to 7 min, with little increase up to 10 min (Figure 3f). Thus, centrifugation at 7 min was selected to achieve phase separation for the proposed method.

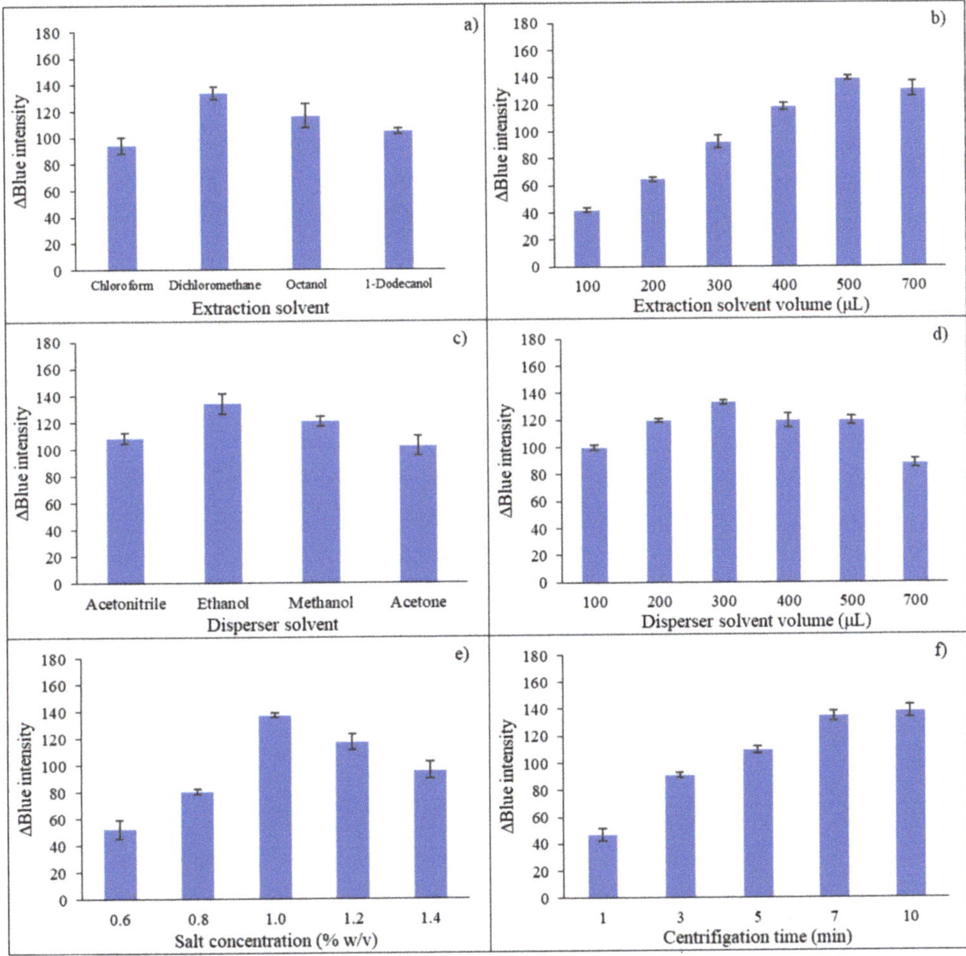

Figure 3. Investigation of DLLME parameters on carbaryl detection using smartphone-based digital image analysis; (**a**) effect of extraction solvent type; (**b**) effect of extraction solvent volume; (**c**) effect of dispersive solvent type; (**d**) effect of dispersive solvent volume; (**e**) effect of salt concentration and (**f**) effect of vortex time on the sensitivity of carbaryl determination.

2.6. Analytical Characteristics

Under optimal conditions of the proposed procedure summarized in Table S1, the linearity range of the calibration graph with DLLME for digital images based on the colorimetric method was 0.10 to 0.50 mg·L^{-1} carbaryl with good linear regression r^2 at more than 0.99 (Figure 4). LOD and LOQ were calculated by 3SD/slope and 10SD/slope, where SD is the standard deviation of the blank, at 0.03 and 0.09 mg·L^{-1}, respectively. Precision of the digital image method when analyzing carbaryl intraday at 0.30 mg·L^{-1} ($n = 7$) was 4.91%, with reproducibility 7.59% for 7 days ($n = 3 \times 7$). The calibration graph was prepared daily to minimize inherent experiment variability when using crude peroxidase enzyme extracts. A summary of the analytical characteristics of the proposed method compared to some other spectrophotometric methods/digital image colorimetry is shown in Table 1. The developed method was sensitive and simple as an alternative for the determination of carbaryl. The developed procedure was down-scaled into a microliter volume operation using 500 µL of extraction and dispersive solvent (300 µL) for the DLLME procedure compared with the previous peroxidase enzymatic spectrophotometry [52], while exploiting daily life smartphones as acquisition tools provided rapidity and traceability for carbaryl detection.

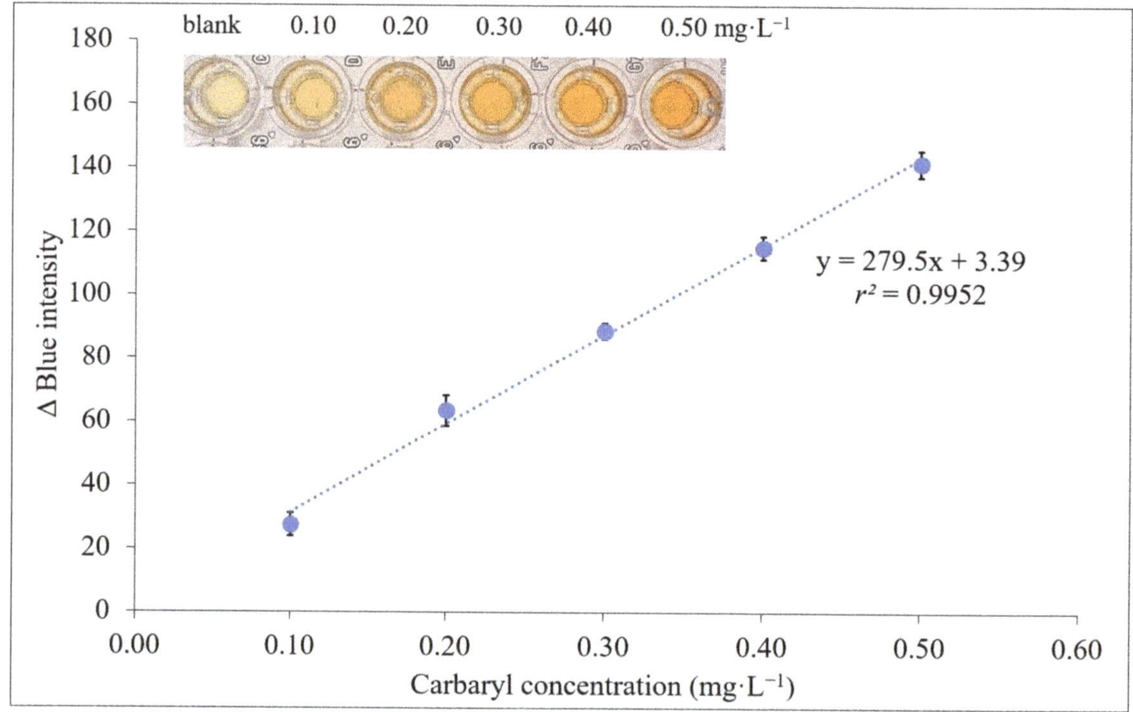

Figure 4. Calibration graph for carbaryl determination using smartphone detection plots between Δblue intensity and carbaryl concentrations in the range 0.10 to 0.50 mg·L^{-1}.

2.7. Recovery and Carbaryl Residues in Andrographis paniculata Herbal Medicines

Recovery of the developed method was studied by adding standard 0.10, 0.20 and 0.30 mg·L^{-1} carbaryl into real samples, with results presented in Table 2. Carbaryl sample recovery ranged 83–109% and 88–114% for smartphone-based digital images and HPLC-UV, respectively. RSDs of recovery were less than 10% indicating that the method showed good accuracy and good precision.

This method was also applied for carbaryl determination in 10 *Andrographis paniculate* herbal medicines. Carbaryl contents shown in Table 2 ranged 5.54 ± 0.13–16.22 ± 0.29 and 6.31 ± 0.70–15.56 ± 0.32 mg·kg^{-1} for smartphone-based digital images and HPLC, respectively. Concentrations of carbaryl residues in *Andrographis paniculate* herbal medicines obtained from both methods were not significantly different at 95% confidence level using the paired *t*-test (t Stat = 1.06, t Critical = 2.57, df = 5). High contamination of carbaryl in *Andrographis paniculate* herbal medicines was observed in sample Nos. 1, 4–7 and 10. There are no reports regarding maximum residue limit (MRL) of carbaryl in herbal medicines.

2.8. Selectivity for the Determination of Carbaryl by Peroxidase Enzymatic Reaction

Other insecticides in the carbamate family such as carbofuran, promecarb, aldicarb and isoprocarb were studied for the selectivity of the peroxidase enzyme catalytic reaction. Only carbaryl was catalyzed by peroxidase enzyme, while carbofuran, promecarb, aldicarb and isoprocarb were not observed [52,53]. Therefore, the quantification of carbaryl using peroxides enzyme extracts from cassia bark was selective since the other carbamates did not interfere with carbaryl detection.

Table 1. Comparison of peroxidase enzymatic reaction-DLLME and smartphone-based digital image method with other spectrophotometric and digital image colorimetric methods for the determination of carbaryl in various samples.

Detection Technique	Pre-concentration Method	Reagent	Linearity	[a] LOD	Recovery (%)	[b] RSD (%)	Sample	Reference
Spectrophotometry	-	Diazotized 2-aminonaphthalenesulfonic acid	$0.01–0.1$ mg·L^{-1}	-	96–98	-	Soil and insecticide	[32]
Spectrophotometry	-	p-Aminophenol, p-N,N-dimethylphenylenediamine, dihydrochloride, and 1-amino-2-naphthol-4-sulphonic acid	$0.08–1$ mg L^{-1}	0.08 mg·L^{-1}	92.0–97.5	1	Insecticide, water and grains	[35]
Spectrophotometry	-	2,6-Dibromo-4-methylaniline, 2,4,6-tribromoaniline, and 2,6-dibromo-4-nitroaniline	$0.6–10.0$ mg·L^{-1}	0.825 mg·L^{-1}	94.20–99.00	<2	Environmental Samples	[34]
Spectrophotometry	[c] CPE	Rhodamine-B	$0.04–0.4$ mg·L^{-1}	0.005 mg·L^{-1}	97.80–101.20	<2	Water and grains	[36]
Spectrophotometry	[d] DLME and [e] DMSPE	2-Naphthylamine-1-sulfonic acid	$10–100$ μg·L^{-1}	8 ng·mL^{-1}	97.3–108.1	8.5	Tap water, field water and fruit juice	[58]
Spectrophotometry	[f] SPE [g] QuEChERS and [h] DLLME	4-AP, H_2O_2 with crude rubber tree bark peroxidase extracts	$0.1–3.0$ mg L^{-1}	0.06 mg·L^{-1}	83–118	<4	Vegetable sample	[52]
Digital image colorimetry	[i] LPME	4-Methoxybenzene-diazonlum tetrafluoroborate (MBDF)	$0.03–30.0$ mg·kg^{-1}	$0.006–0.008$ mg·kg^{-1}	92.3–105.9	<5	Food sample	[41]
Smartphone-based digital image analysis	DLLME	4-AP, H_2O_2 with non-purified peroxidase extracts from Senna siamea Lam. bark	$0.10-0.50$ mg·L^{-1}	0.03 mg·L^{-1}	82.5–108.2	4.9	Pharmaceutical sample	This work

[a] LOD is limit of detection; [b] RSD is relative standard deviation; [c] CPE is could point extraction; [d] DLME is dispersive liquid microextraction; [e] DMSPE is dispersive μ-solid phase extraction, [f] SPE is solid phase extraction; [g] QuEChERS is Quick, Easy, Cheap, Effective, Rugged and Safe; [h]DLLME is dispersive liquid-liquid microextraction and [i]LPME is liquid phase microextraction.

Table 2. Mean recovery percentage of spiked standard carbaryl into real samples and concentration of carbaryl residues in *Andrographis paniculate* herbal medicines obtained by smartphone-based digital images and HPLC-UV.

Sample	Added (mg·L^{-1})	Smartphone-Based Digital Images ($n = 3$)			HPLC-UV ($n = 3$)		
		Found (mg·L^{-1} ± SD)	Mean Recovery, %(RSD)	Carbaryl Content (mg·kg^{-1} ± SD)	Found (mg·L^{-1} ± SD)	Mean Recovery, %(RSD)	Carbaryl Content (mg·kg^{-1} ± SD)
1	0.1	0.1_1 * ± 0.0_1	108(6)	9.48 ± 0.15	0.100 ± 0.005	100(7)	9.72 ± 0.30
	0.2	$0.2_0 \pm 0.0_1$	100(2)		0.177 ± 0.009	88(5)	
	0.3	$0.3_1 \pm 0.0_1$	104(9)		0.289 ± 0.014	96(5)	
2	0.1	$0.1_0 \pm 0.0_1$	101(10)	<LOD	0.100 ± 0.005	100(2)	<LOD
	0.2	$0.2_0 \pm 0.0_1$	99(3)		0.184 ± 0.009	93(5)	
	0.3	$0.2_5 \pm 0.0_1$	84(5)		0.340 ± 0.010	114(3)	
3	0.1	$0.0_9 \pm 0.0_1$	87(3)	<LOD	0.100 ± 0.003	100(1)	<LOD
	0.2	$0.1_9 \pm 0.0_0$	92(2)		0.189 ± 0.004	95(2)	
	0.3	$0.3_0 \pm 0.0_1$	101(2)		0.288 ± 0.011	96(3)	
4	0.1	$0.1_0 \pm 0.0_0$	98(5)	13.55 ± 0.34	0.089 ± 0.004	89(3)	14.83 ± 0.13
	0.2	$0.2_0 \pm 0.0_0$	98(4)		0.198 ± 0.012	99(6)	
	0.3	$0.3_2 \pm 0.0_0$	105(5)		0.285 ± 0.009	95(3)	
5	0.1	$0.0_8 \pm 0.0_0$	83(6)	6.98 ± 0.16	0.097 ± 0.007	97(5)	6.57 ± 0.11
	0.2	$0.2_0 \pm 0.0_0$	98(3)		0.184 ± 0.013	92(7)	
	0.3	$0.2_7 \pm 0.0_0$	91(4)		0.284 ± 0.009	94(3)	
6	0.1	$0.1_0 \pm 0.0_0$	103(4)	16.22 ± 0.29	0.095 ± 0.008	95(6)	15.56 ± 0.32
	0.2	$0.2_1 \pm 0.0_1$	105(2)		0.193 ± 0.005	97(3)	
	0.3	$0.3_2 \pm 0.0_0$	106(3)		0.282 ± 0.007	93(3)	
7	0.1	$0.1_0 \pm 0.0_0$	101(5)	9.42 ± 0.97	0.102 ± 0.003	102(2)	10.15 ± 0.40
	0.2	$0.2_0 \pm 0.0_0$	98(5)		0.195 ± 0.003	99(2)	
	0.3	$0.2_6 \pm 0.0_0$	87(2)		0.280 ± 0.006	93(2)	
8	0.1	$0.1_1 \pm 0.0_1$	109(5)	<LOD	0.103 ± 0.005	103(7)	<LOD
	0.2	$0.2_0 \pm 0.0_0$	99(2)		0.190 ± 0.006	95(3)	
	0.3	$0.2_9 \pm 0.0_1$	96(4)		0.307 ± 0.003	102(2)	
9	0.1	$0.0_9 \pm 0.0_1$	90(5)	<LOD	0.104 ± 0.003	103(2)	<LOD
	0.2	$0.1_9 \pm 0.0_1$	93(4)		0.212 ± 0.009	106(5)	
	0.3	$0.2_7 \pm 0.0_1$	91(3)		0.315 ± 0.006	105(2)	
10	0.1	$0.1_1 \pm 0.0_0$	108(4)	5.54 ± 0.13	0.103 ± 0.003	103(3)	6.31 ± 0.70
	0.2	$0.1_9 \pm 0.0_1$	93(5)		0.195 ± 0.002	103(3)	
	0.3	$0.2_9 \pm 0.0_1$	95(3)		0.310 ± 0.006	103(3)	

* The subscripts are the second decimal place.

3. Materials and Methods

3.1. Reagents and Chemicals

All chemicals used were analytical grade and utilized without further purification. Deionized (DI) water (Milli-Q, Millipore, Solna, Sweden) was used to prepare all the solutions, while 1000 mg·L^{-1} carbaryl stock solution (Sigma-Aldrich, Darmstadt, Germany) was prepared by weighing 0.10 g carbaryl, with volume adjusted to 100 mL with 95% ethanol (Merck, Darmstadt, Germany). Working solutions of carbaryl were freshly prepared by appropriate dilution of carbaryl stock solution by deionized water. Hydrogen peroxide 100 mmol·L^{-1} was prepared by transferring 1.02 mL 30% hydrogen peroxide solution into a 100 mL volumetric flask and adjusting the volume with deionized water. 4-AP, 1000 mg·L^{-1}, was prepared by weighing 0.10 g of 4-AP in a 100 mL volumetric flask and adjusting the volume with deionized water. Buffer solutions were prepared by mixing appropriate volumes of disodium hydrogen phosphate and citric acid, with the required pH attained by adjusting with sodium hydroxide solution.

3.2. Instruments and Apparatus

A UV-Visible spectrophotometer (UV-1800 Shimadzu, Kyoto, Japan) was utilized to evaluate enzyme activity at a wavelength of 420 nm. A Rotanta 46 R model centrifuge (Hettich Zentrifugen, Tuttlingen, Germany) was employed to achieve separation of extract solutions, yielding a clear supernatant. A pH meter (Eutech, Ayer Rajah Crescent, Singapore) was used to measure buffer pH. Reaction temperature was controlled by a water bath (Memmert, Schwabach, Germany). A vortex mixer was used to increase mass transfer of QuEChERS (Quick Easy Cheap Effective Rugged Safe) and DLLME steps. A cooking blender model EBR 2601 from Electrolux (Electrolux, Bangkok, Thailand) was utilized to homogenize the materials. iPhone model 11 Pro Max (Designed by Apple in California Assembled in China) was utilized to photograph the color products after preconcentration by DLLME under the light control box.

3.3. Light Control Box

The in-house light control box was adapted from our previous project [64] and constructed from white opaque acrylic sheet with outer dimensions 19 × 32 × 15 cm to prevent light penetration from the surroundings (Figure S1). The outer part was covered with PVC sticker sheet. A tray to place a 96-microwell plate was installed in the middle of the box. Internal illumination was provided by a LED video light with 300 high quality LED light beads of extra-large luminous chips (Yongnuo YN300 III, Shenzhen, China). An ON/OFF switch was used to control the power supply (6.5–8.5 V, 3 A). the LED light was positioned below the acrylic tray in the box used as a light diffuser. A hole 3.0 × 3.5 cm (w·l) was made in the top of the box for photography using a daily life smartphone built-in camera (iPhone 11 Pro Max, Apple, Zhengzhou, China).

3.4. Extraction of Peroxidase Enzyme from Cassia Bark

Fresh cassia bark (200 g) was collected from Amnat Charoen Province, washed with deionized water and cut into small pieces. Next, 200 g of cassia bark was weighed into a 600 mL beaker. Phosphate buffer extracted solution 100 mL, pH 6.0 was then added. The mixed contents were thoroughly blended for 5 min. The solution was then filtered using a white cloth and centrifuged at 3028× *g*, 4 °C for 30 min and 9072× *g*, 4 °C for 30 min. The supernatant was filtered using Whatman No. 1 filter paper and stored in a brown 1.5 mL microcentrifuge tube at −20 °C.

3.5. Peroxidase Enzyme Extract Activity Study

Peroxidase enzyme activity, not previously studied in cassia bark, was investigated to confirm the presence of the enzyme in the extract solution [52]. Briefly, 10 μL crude extract solution was mixed with specific peroxidase enzyme substrate ready-to-use ABTS solution (ABTS solution, Roche, Mannheim, Germany). Absorbance was immediately monitored at

420 nm for 1 min, and the initial slope of the enzymatic reaction was calculated. Enzyme activity of 1 U was defined as the amount of enzyme required to generate 0.001 absorbance of product every minute under the described conditions.

3.6. Peroxidase Enzymatic Analytical Method Synergied with DLLME for Determination of Carbaryl by Smartphone-Based Digital Image Analysis

As illustrated in Figure 5, samples (200 µL, see Section 3.10), NaOH (100 µL, 50 mmol·L^{-1}), 4-AP (1.5 mL, 1000 mg·L^{-1}), hydrogen peroxide (300 µL, 10 mmol·L^{-1}) were transferred into a 10 mL volumetric flask. Then, the enzyme extract (150 µL) was added before adjusting the volume with phosphate buffer (pH 6.0, 50 mmol·L^{-1}). The mixture was incubated at 30 °C for 15 min until a red color product was observed. The mixture was then transferred to a 15 mL centrifuge tube containing sodium chloride (0.10 g) and dichloromethane (500 µL, as extraction solvent), and rapidly injected with ethanol (300 µL, as disperser solvent) before vortexing for 1 min, followed by 7 min centrifuging (4032 g). The aqueous phase was completely withdrawn using a long needle syringe and the organic phase was diluted with 95% ethanol (500 µL). An aliquot (200 µL) of the resulting solution was transferred into a microwell plate and photographed using a smartphone in a light control box (Figure S1). Image processing was performed for RGB (red, green, blue) intensity of the captured image. The calibration graph was a plot of ΔB intensity against carbaryl concentration ranging 0.10 to 0.50 mg·L^{-1}.

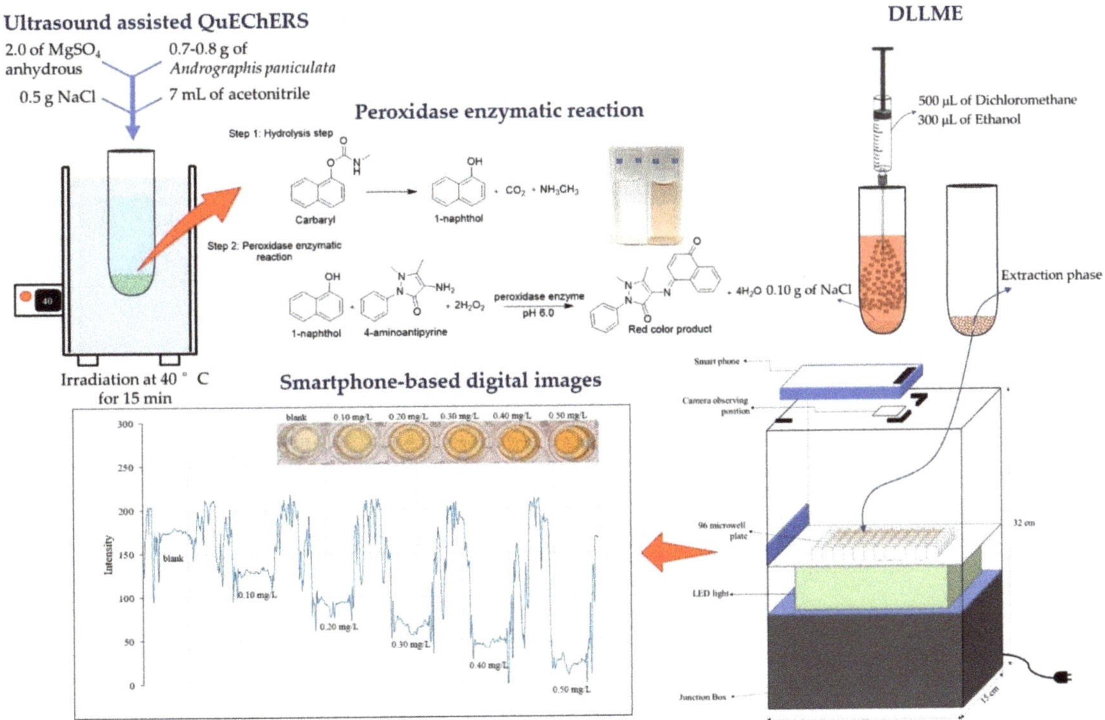

Figure 5. Illustration of DLLME synergy with smartphone-based digital images for the determination of carbaryl using enzymatic reaction of crude peroxidase enzyme extracts from cassia bark as biocatalyst.

3.7. Optimization of Carbaryl Determination Conditions Using Crude Peroxidase Enzyme

The influence of various parameters including pH, 4-AP concentration, hydrogen peroxide concentration, peroxidase crude enzyme volume and incubation time were investigated to determine the optimized conditions. The effect of pH was studied in the range 3–6, with 4-AP concentration at 50–200 mmol·L^{-1}. Hydrogen peroxide concentration was examined between 0.01 and 1 mmol·L^{-1}, while crude peroxide enzyme volumes were determined in the range 10–200 µL. Incubation time of 1–20 min was also studied.

3.8. Optimization for DLLME

Various parameters influence the DLLME procedure. Optimized conditions for DLLME were investigated for extraction solvent type, extraction solvent volume, dispersive solvent type, dispersive solvent volume, salt concentration, vortex time and centrifuge time to enhance sensitivity before the smartphone-based digital image determination step. Extraction solvent types as chloroform, dichloromethane, octanol and 1-dodecanol were employed. The effect of extraction solvent volume was explored at 100–700 µL. Dispersive solvent type was studied for acetonitrile, methanol, ethanol and acetone. Volume of dispersive solvent was explored in the range 100–700 µL. Salting out of NaCl during the DLLME process was determined between 0.6 and 1.4% (w/v) NaCl concentration. Vortex time was also tested between 0.1 and 2 min, while the influence of centrifuge time was investigated in the range 1–10 min.

3.9. Validation Methods

To evaluate method validation, linearity range, limit of detection (LOD), limit of quantification (LOQ), precision in terms of relative standard deviation (RSD) and accuracy were investigated. A linear calibration graph was studied by varying carbaryl standard in the range 0.10–0.50 mg·L^{-1} under the selected conditions. A calibration curve was constructed by plotting blue intensity difference (Y-axis) versus carbaryl concentration in mg·L^{-1} (X-axis), with the linear equation and linear regression coefficient (r^2) also evaluated. The precision of the proposed method was studied by monitoring the blue intensity of carbaryl standard at 0.10 and 0.30 mg·L^{-1} for seven replicates intraday (repeatability) and repeated for five days (reproducibility). Precision of the proposed method was reported in terms of RSD. The accuracy of this method was researched by adding various concentrations of 0.10, 0.20 and 0.30 mg·L^{-1} of carbaryl standard solution to the sample and evaluating the recovery percentage of carbaryl under optimal conditions. LOD and LOQ were calculated from 3SD/slope and 10SD/slope, respectively where SD represents the standard deviation of the blank.

3.10. Samples

Pharmaceutical preparations containing *Andrographis paniculata* were collected from different drug stores in Maha Sarakham Province, Thailand. Samples were extracted by the ultrasonication-assisted QuEChERs method [53]. Briefly, *Andrographis paniculata* powder in the capsules was weighed at 0.7–0.8 g into a 15 mL centrifuge tube. Then, 2.0 of MgSO$_4$ anhydrous and 0.5 g NaCl were added, followed by pipetting 7 mL of acetonitrile extraction solvent. The solution mixture was homogenized by a vortex mixer for 1 min and then transferred to an ultrasonic bath for irradiation at 40 °C for 15 min. The sample was then centrifuged at 4032× g for 10 min to separate the solid and organic extraction solvent. The extraction phase was filtered through Whatman filter paper No. 1 into a 10 mL volumetric flask and the volume was adjusted with acetonitrile. Next, 3 mL of the extraction phase was transferred to a 15 mL centrifuge tube containing 10 mg of activated charcoal powder for color elimination. The solution was then vortexed for 1 min and then centrifuged at 1008× g for 10 min. The organic extraction phase was filtered through a nylon filter, 0.45 µm, and kept at −4 °C until analysis via peroxidase enzymatic reaction and followed by DLLME.

3.11. Reference Method

HPLC-UV detection was proposed as the reference method to compare sample concentrations of carbaryl. Chromatographic separation was conducted using a Waters 1525 HPLC

System with a Binary Pump (Waters, Milford, MA, USA) and a LiChroCART® 150-4.6 RP-18 endcapped (4.6 × 150 mm, 5.0 μm) column (Merck, Darmstadt, Germany). An isocratic system involving 40% v/v acetonitrile in deionized water with a flow rate of 1.0 mL·min^{-1} was performed. The samples were injected manually using a Rheodyne injector with a 20 μL sample loop. Absorption of carbaryl was detected at 270 nm using a Waters 2489 UV detector (Waters, Milford, MA, USA). Breeze software version 2.0 was adopted for data acquisition and peak area integration.

4. Conclusions

A simple and reliable dispersive liquid-liquid microextraction with smartphone-based digital images for determination of carbaryl residues was developed. A simple peroxidase extract from *Senna siamea* Lam. bark served as a catalyst for the reactions at pH 6 of 4-aminoantipyrine, hydrogen peroxide and 1-naphthol, which was the hydrolysis product of carbaryl. Dispersive liquid-liquid microextraction was synergized with peroxidase enzymatic reaction to pre-concentrate the analyte. The red color product was sensed by a smartphone camera for further evaluation to quantify the carbaryl content. The developed procedure, with micro-liter volume operation, was applied for carbaryl residue assay in *Andrographis paniculata* herbal medicine. Results were not significantly different from the HPLC-UV reference method, at 95% confidence limits. The developed procedure was cost-effective, simple, reliable and down-scaled and offered traceability as an alternative for the assay of carbaryl residues in herbal medicines.

Supplementary Materials: The following supporting information can be downloaded at: https://www.mdpi.com/article/10.3390/molecules27103261/s1, Figure S1: Light control box for smartphone-based digital imaging for the determination of carbaryl, Figure S2: RGB profile plots of color intensities obtained from smartphone-based digital imaging in the enzymatic reaction and DLLME in a light control box for carbaryl in the range 0–0.50 mg·L^{-1}, Figure S3: Plots of intensity difference (delta intensity) versus carbaryl concentration: (a) delta red intensity (b) delta green intensity, and (c) delta blue intensity, (delta intensity being the intensity due to that carbaryl concentration subtracted by that of blank), Table S1: Summarized selected conditions of smartphone-based digital images with DLLME for the determination of carbaryl residues.

Author Contributions: S.-a.S.: conceptualization, investigation, methodology, validation, data curation, formal analysis, visualization, writing—original draft. W.S.: writing—review & editing. K.G.: supervision, writing—review & editing, K.P.: conceptualization, resources, investigation, methodology, validation, data curation, formal analysis, writing—original draft, writing—review & editing. All authors have read and agreed to the published version of the manuscript.

Funding: This research project was financially supported by Mahasarakham University 2021 (No. 6408038). Financial support from Mahidol University, Amnat Charoen Campus, the Center of Excellence for Innovation in Chemistry (PERCH-CIC), Ministry of Higher Education, Science, Research and Innovation is gratefully acknowledged.

Institutional Review Board Statement: Not applicable.

Informed Consent Statement: Not applicable.

Data Availability Statement: Not applicable.

Acknowledgments: We would also like to thank the Department of Chemistry, Faculty of Science, Mahasarakham University and Mahidol University, Amnat Charoen Campus for providing laboratory facilities and Chorphaka Duangta, an undergraduate student of K.P. for analytical assistance. K.G. acknowledges the Center of Excellence for Innovation in Analytical Science and Technology for Biodiversity-based Economic and Society, Chiang Mai University.

Conflicts of Interest: The authors declare no conflict of interest.

Sample Availability: Strains are available from the authors.

References

1. Babich, O.; Sukhikh, S.; Prosekov, A.; Asyakina, L.; Ivanova, S. Medicinal Plants to Strengthen Immunity during a Pandemic. *Pharmaceuticals* **2020**, *13*, 313. [CrossRef] [PubMed]
2. Khanna, K.; Kohli, S.K.; Kaur, R.; Bhardwaj, A.; Bhardwaj, V.; Ohri, P.; Sharma, A.; Ahmad, A.; Bhardwaj, R.; Ahmad, P. Herbal immune-boosters: Substantial warriors of pandemic Covid-19 battle. *Phytomedicine* **2020**, *85*, 153361. [CrossRef] [PubMed]
3. Ogunrinola, O.O.; Kanmodi, R.I.; Ogunrinola, O.A. Medicinal plants as immune booster in the palliative management of viral diseases: A perspective on coronavirus. *Food Front.* **2022**, *3*, 83–95. [CrossRef]
4. Kuchta, K.; Cameron, S.; Lee, M.; Cai, S.-Q.; Shoyama, Y. Which East Asian herbal medicines can decrease viral infections? *Phytochem. Rev.* **2021**, *21*, 219–237. [CrossRef] [PubMed]
5. Jayakumar, T.; Hsieh, C.-Y.; Lee, J.-J.; Sheu, J.-R. Experimental and Clinical Pharmacology of *Andrographis paniculata* and Its Major Bioactive Phytoconstituent Andrographolide. *Evid. Based Complementary Altern. Med.* **2013**, *2013*, 846740.
6. Adedapo, A.A.; Adeoye, B.O.; Sofidiya, M.O.; Oyagbemi, A. Antioxidant, antinociceptive and anti-inflammatory properties of the aqueous and ethanolic leaf extracts of *Andrographis paniculata* in some laboratory animals. *J. Basic Clin. Physiol. Pharmacol.* **2015**, *26*, 327–334. [CrossRef]
7. Mussard, E.; Cesaro, A.; Lespessailles, E.; Legrain, B.; Berteina-Raboin, S.; Toumi, H. Andrographolide, A Natural Antioxidant: An Update. *Antioxidants* **2019**, *8*, 571. [CrossRef]
8. Puri, A.; Saxena, R.; Saxena, K.C.; Srivastava, V.; Tandon, J.S. Immunostimulant Agents from *Andrographis paniculata*. *J. Nat. Prod.* **1993**, *56*, 995–999. [CrossRef]
9. Churiyah; Pongtuluran, O.B.; Rofaani, E.; Tarwadi. Antiviral and Immunostimulant Activities of *Andrographis paniculata*. *HAYATI J. Biosci.* **2015**, *22*, 67–72. [CrossRef]
10. Rajagopal, S.; Kumar, R.A.; Deevi, D.S.; Satyanarayana, C.; Rajagopalan, R. Andrographolide, a potential cancer therapeutic agent isolated from *Andrographis paniculata*. *J. Exp. Ther. Oncol.* **2003**, *3*, 147–158. [CrossRef]
11. Shanmugam, R.; Nagalingam, M.; Ponnanikajamideen, M.; Vanaja, M.; Chelladurai, M. Anticancer activity of *Andrographis paniculata* leaves extract against neuroblastima (IMR-32) and human colon (HT-29) cancer cell line. *World J. Pharm. Pharm. Sci.* **2015**, *4*, 1667–1675.
12. Malik, Z.; Parveen, R.; Parveen, B.; Zahiruddin, S.; Khan, M.A.; Khan, A.; Massey, S.; Ahmad, S.; Husain, S.A. Anticancer potential of andrographolide from *Andrographis paniculata* (Burm.f.) Nees and its mechanisms of action. *J. Ethnopharmacol.* **2021**, *272*, 113936. [CrossRef] [PubMed]
13. Hu, X.-Y.; Wu, R.-H.; Logue, M.; Blondel, C.; Lai, L.Y.W.; Stuart, B.; Flower, A.; Fei, Y.-T.; Moore, M.; Shepherd, J.; et al. *Andrographis paniculata* (Chuān Xīn Lián) for symptomatic relief of acute respiratory tract infections in adults and children: A systematic review and meta-analysis. *PLoS ONE* **2017**, *12*, e0181780. [CrossRef] [PubMed]
14. Zuin, V.G.; Vilegas, J.H.Y. Pesticide residues in medicinal plants and phytomedicines. *Phytother. Res.* **2000**, *14*, 73–88. [CrossRef]
15. Koshlukova, S.E.; Reed, N.R. Carbaryl. In *Encyclopedia of Toxicology*, 3rd ed.; Wexler, P., Ed.; Academic Press: Oxford, UK, 2014; pp. 668–672.
16. Harp, P.R. Carbaryl. In *Encyclopedia of Toxicology*, 2nd ed.; Wexler, P., Ed.; Elsevier: New York, NY, USA, 2005; pp. 414–416.
17. Srivastava, A.K.; Kesavachandran, C. *Health Effects of Pesticides*, 1st ed.; CRC Press: London, UK, 2019.
18. Sharma, V.; Jadhav, R.; Rao, G.; Saraf, A.; Chandra, H. High performance liquid chromatographic method for the analysis of organophosphorus and carbamate pesticides. *Forensic Sci. Int.* **1990**, *48*, 21–25. [CrossRef]
19. Fu, L.; Liu, X.; Hu, J.; Zhao, X.; Wang, H.; Wang, X. Application of dispersive liquid–liquid microextraction for the analysis of triazophos and carbaryl pesticides in water and fruit juice samples. *Anal. Chim. Acta* **2009**, *632*, 289–295. [CrossRef]
20. Pulgarín, J.A.M.; Bermejo, L.F.G.; Durán, A.C. Determination of carbamates in soils by liquid chromatography coupled with on-line postcolumn UV irradiation and chemiluminescence detection. *Arab. J. Chem.* **2018**, *13*, 2778–2784. [CrossRef]
21. Durand, G.; De Bertrand, N.; Barceló, D. Applications of thermospray liquid chromatography-mass spectrometry in photochemical studies of pesticides in water. *J. Chromatogr. A* **1991**, *554*, 233–250. [CrossRef]
22. Totti, S.; Fernández, M.; Ghini, S.; Picó, Y.; Fini, F.; Mañes, J.; Girotti, S. Application of matrix solid phase dispersion to the determination of imidacloprid, carbaryl, aldicarb, and their main metabolites in honeybees by liquid chromatography–mass spectrometry detection. *Talanta* **2006**, *69*, 724–729. [CrossRef]
23. Zhou, Y.; Guan, J.; Gao, W.; Lv, S.; Ge, M. Quantification and Confirmation of Fifteen Carbamate Pesticide Residues by Multiple Reaction Monitoring and Enhanced Product Ion Scan Modes via LC-MS/MS QTRAP System. *Molecules* **2018**, *23*, 2496. [CrossRef]
24. Nagasawa, K.; Uchiyama, H.; Ogamo, A.; Shinozuka, T. Gas chromatographic determination of microamounts of carbaryl and 1-napththol in natural water as sources of water supplies. *J. Chromatogr. A* **1977**, *144*, 77–84. [CrossRef]
25. Bagheri, H.; Creaser, C. Determination of carbaryl and 1-naphthol in english apples and strawberries by combined gas chromatography-fluorescence spectrometry. *J. Chromatogr. A* **1991**, *547*, 345–353. [CrossRef]
26. Zhang, Y.-P.; Li, X.-J.; Yuan, Z.-B. Analysis of carbaryl and other pesticides by capillary electrophoresis. *Chin. J. Chromatogr.* **2002**, *20*, 341–344.
27. Zhang, C.; Ma, G.; Fang, G.; Zhang, Y.; Wang, S. Development of a Capillary Electrophoresis-Based Immunoassay with Laser-Induced Fluorescence for the Detection of Carbaryl in Rice Samples. *J. Agric. Food Chem.* **2008**, *56*, 8832–8837. [CrossRef]

28. Wang, M.; Huang, J.; Wang, M.; Zhang, D.; Chen, J. Electrochemical nonenzymatic sensor based on CoO decorated reduced graphene oxide for the simultaneous determination of carbofuran and carbaryl in fruits and vegetables. *Food Chem.* **2014**, *151*, 191–197. [CrossRef]

29. Liu, B.; Xiao, B.; Cui, L. Electrochemical analysis of carbaryl in fruit samples on graphene oxide-ionic liquid composite modified electrode. *J. Food Compos. Anal.* **2015**, *40*, 14–18. [CrossRef]

30. Salih, F.E.; Achiou, B.; Ouammou, M.; Bennazha, J.; Ouarzane, A.; Younssi, S.A.; El Rhazi, M. Electrochemical sensor based on low silica X zeolite modified carbon paste for carbaryl determination. *J. Adv. Res.* **2017**, *8*, 669–676. [CrossRef]

31. Rahmani, T.; Bagheri, H.; Behbahani, M.; Hajian, A.; Afkhami, A. Modified 3D Graphene-Au as a Novel Sensing Layer for Direct and Sensitive Electrochemical Determination of Carbaryl Pesticide in Fruit, Vegetable, and Water Samples. *Food Anal. Methods* **2018**, *11*, 3005–3014. [CrossRef]

32. Jia, G.; Li, L.; Qiu, J.; Wang, X.; Zhu, W.; Sun, Y.; Zhou, Z. Determination of carbaryl and its metabolite 1-naphthol in water samples by fluorescence spectrophotometer after anionic surfactant micelle-mediated extraction with sodium dodecylsulfate. *Spectrochim. Acta Part A Mol. Biomol. Spectrosc.* **2007**, *67*, 460–464. [CrossRef]

33. Zhang, C.; Cui, H.; Cai, J.; Duan, Y.; Liu, Y. Development of Fluorescence Sensing Material Based on CdSe/ZnS Quantum Dots and Molecularly Imprinted Polymer for the Detection of Carbaryl in Rice and Chinese Cabbage. *J. Agric. Food Chem.* **2015**, *63*, 4966–4972. [CrossRef]

34. Sastry, C.S.P.; Vijaya, D.; Mangala, D.S. Spectrophotometric determination of carbaryl and propoxur using aminophenols and phenylenediamine. *Analyst* **1987**, *112*, 75–78. [CrossRef]

35. Mathew, L.; Reddy, M.L.P.; Rao, T.P.; Iyer, C.S.P.; Damodaran, A.D. Simple spectrophotometric method for the determination of carbaryl in soil and insecticide formulations. *Analyst* **1995**, *120*, 1799–1801. [CrossRef]

36. Kumar, K.S.; Suvardhan, K.; Chiranjeevi, P. Preparation of Reagents for the Sensitive Spectrophotometric Determination of Carbaryl in Environmental Samples. *Anal. Lett.* **2005**, *38*, 697–709. [CrossRef]

37. Gupta, N.; Pillai, A.K.; Parmar, P. Spectrophotometric determination of trace carbaryl in water and grain samples by inhibition of the rhodamine-B oxidation. *Spectrochim. Acta Part A Mol. Biomol. Spectrosc.* **2015**, *139*, 471–476. [CrossRef]

38. Handa, S.K.; Dikshit, A.K. Spectrophotometric method for the determination of residues of carbaryl in water. *Analyst* **1979**, *104*, 1185–1188. [CrossRef]

39. Haifa, B.A.; Bacârea, V.C.; Iacob, O.-A.; Călinici, T.D.; Schiopu, A. Comparison between Digital Image Processing and Spectrophotometric Measurements Methods. Application in Electrophoresis Interpretation. *Appl. Med. Inf.* **2011**, *28*, 29–36.

40. Cal, E.; Güneri, P.; Köse, T. Comparison of digital and spectrophotometric measurements of colour shade guides. *J. Oral Rehabil.* **2006**, *33*, 221–228. [CrossRef]

41. Jing, X.; Wang, H.; Huang, X.; Chen, Z.; Zhu, J.; Wang, X. Digital image colorimetry detection of carbaryl in food samples based on liquid phase microextraction coupled with a microfluidic thread-based analytical device. *Food Chem.* **2020**, *337*, 127971. [CrossRef]

42. Golovanov, V.I.; Golovanov, S.V.; Varganov, M.S. Use of Contrast of Digital Photo Images for the Determination of the Turbidity of Liquids. *J. Anal. Chem.* **2018**, *73*, 667–673. [CrossRef]

43. Firdaus, M.L.; Alwi, W.; Trinoveldi, F.; Rahayu, I.; Rahmidar, L.; Warsito, K. Determination of Chromium and Iron Using Digital Image-based Colorimetry. *Procedia Environ. Sci.* **2014**, *20*, 298–304. [CrossRef]

44. Yang, R.; Cheng, W.; Chen, X.; Qian, Q.; Zhang, Q.; Pan, Y.; Duan, P.; Miao, P. Color Space Transformation-Based Smartphone Algorithm for Colorimetric Urinalysis. *ACS Omega* **2018**, *3*, 12141–12146. [CrossRef] [PubMed]

45. Lin, B.; Yu, Y.; Cao, Y.; Guo, M.; Zhu, D.; Dai, J.; Zheng, M. Point-of-care testing for streptomycin based on aptamer recognizing and digital image colorimetry by smartphone. *Biosens. Bioelectron.* **2018**, *100*, 482–489. [CrossRef] [PubMed]

46. Granica, M.; Tymecki, Ł. Analytical aspects of smart (phone) fluorometric measurements. *Talanta* **2019**, *197*, 319–325. [CrossRef] [PubMed]

47. Wongniramaikul, W.; Limsakul, W.; Choodum, A. A biodegradable colorimetric film for rapid low-cost field determination of formaldehyde contamination by digital image colorimetry. *Food Chem.* **2018**, *249*, 154–161. [CrossRef]

48. Zhao, W.; Tian, S.; Huang, L.; Liu, K.; Dong, L.; Guo, J. A smartphone-based biomedical sensory system. *Analyst* **2020**, *145*, 2873–2891. [CrossRef]

49. Lahuerta Zamora, L.; Pérez-Gracia, M.T. Using digital photography to implement the McFarland method. *J. R. Soc. Interface* **2012**, *9*, 1892–1897. [CrossRef]

50. Wong, C.-H.; Shen, G.-J.; Pederson, R.; Wang, Y.-F.; Hennen, W. Enzymatic catalysis in organic synthesis. In *Methods in Enzymology*; Academic Press: Cambridge, MA, USA, 1991; Volume 202, pp. 591–620.

51. Bisswanger, H. Enzyme assays. *Perspect. Sci.* **2014**, *1*, 41–55. [CrossRef]

52. Supharoek, S.-A.; Ponhong, K.; Siriangkhawut, W.; Grudpan, K. A new method for spectrophotometric determination of carbaryl based on rubber tree bark peroxidase enzymatic reaction. *Microchem. J.* **2018**, *144*, 56–63. [CrossRef]

53. Didpinrum, P.; Ponhong, K.; Siriangkhawut, W.; Supharoek, S.-A.; Grudpan, K. A Cost-Effective Spectrophotometric Method Based on Enzymatic Analysis of Jackfruit Latex Peroxidase for the Determination of Carbaryl and Its Metabolite 1-Napthol Residues in Organic and Chemical-Free Vegetables. *Food Anal. Methods* **2019**, *13*, 433–444. [CrossRef]

54. Ponhong, K.; Supharoek, S.-A.; Siriangkhawut, W.; Grudpan, K. Employing peroxidase from Thai indigenous plants for the application of hydrogen peroxide assay. *J. Iran. Chem. Soc.* **2016**, *13*, 1307–1313. [CrossRef]

55. Vieira, I.; Fatibello-Filho, O.; Angnes, L. Zucchini crude extract-palladium-modified carbon paste electrode for the determination of hydroquinone in photographic developers. *Anal. Chim. Acta* **1999**, *398*, 145–151. [CrossRef]

56. de Carvalho, C.C.C.R. Enzymatic and whole cell catalysis: Finding new strategies for old processes. *Biotechnol. Adv.* **2011**, *29*, 75–83. [CrossRef] [PubMed]

57. Rezaee, M.; Yamini, Y.; Faraji, M. Evolution of dispersive liquid–liquid microextraction method. *J. Chromatogr. A* **2010**, *1217*, 2342–2357. [CrossRef] [PubMed]

58. Mukdasai, S.; Thomas, C.; Srijaranai, S. Enhancement of sensitivity for the spectrophotometric determination of carbaryl using dispersive liquid microextraction combined with dispersive μ-solid phase extraction. *Anal. Methods* **2012**, *5*, 789–796. [CrossRef]

59. Al-Saidi, H.; Emara, A.A. The recent developments in dispersive liquid–liquid microextraction for preconcentration and determination of inorganic analytes. *J. Saudi Chem. Soc.* **2014**, *18*, 745–761. [CrossRef]

60. Aly, O.M. Spectrophotometric Determination of Sevin in Natural Waters. *J. Am. Water Work. Assoc.* **1967**, *59*, 906–912. [CrossRef]

61. Rezaee, M.; Assadi, Y.; Hosseini, M.R.M.; Aghaee, E.; Ahmadi, F.; Berijani, S. Determination of organic compounds in water using dispersive liquid–liquid microextraction. *J. Chromatogr. A* **2006**, *1116*, 1–9. [CrossRef]

62. Quigley, A.; Cummins, W.; Connolly, D. Dispersive Liquid-Liquid Microextraction in the Analysis of Milk and Dairy Products: A Review. *J. Chem.* **2016**, *2016*, 4040165. [CrossRef]

63. Psillakis, E.; Kalogerakis, N. Developments in liquid-phase microextraction. *TrAC Trends Anal. Chem.* **2003**, *22*, 565–574. [CrossRef]

64. Didpinrum, P.; Siriangkhawut, W.; Ponhong, K.; Chantiratikul, P.; Grudpan, K. A newly designed sticker-plastic sheet platform and smartphone-based digital imaging for protein assay in food samples with downscaling Kjeldahl digestion. *RSC Adv.* **2021**, *11*, 36494–36501. [CrossRef]

molecules

Article

Fluorimetric Analysis of Five Amino Acids in Chocolate: Development and Validation

Maria S. Synaridou [1], Vasilis Tsamis [1], Georgia Sidiropoulou [1], Constantinos K. Zacharis [1], Irene Panderi [2] and Catherine K. Markopoulou [1,*]

1   Laboratory of Pharmaceutical Analysis, Department of Pharmaceutical Technology, School of Pharmacy, Aristotle University of Thessaloniki, 54124 Thessaloniki, Greece; msynarid@pharm.auth.gr (M.S.S.); vasilistsm@gmail.com (V.T.); sgeorgiae@gmail.com (G.S.); czacharis@pharm.auth.gr (C.K.Z.)
2   Laboratory of Pharmaceutical Analysis, Department of Pharmaceutical Analysis, Faculty of Pharmacy, National and Kapodistrian Universityof Athens, Panepistimiopolis, 15771 Athens, Greece; ipanderi@pharm.uoa.gr
*   Correspondence: amarkopo@pharm.auth.gr; Tel.: +30-2310-997665

Abstract: Amino acids present ergogenic action, helping to increase, protect, and restore the muscular system of young athletes. Moreover, the encapsulation of five relevant amino acids in chocolate pellet form will appeal to them, facilitating their daily consumption. A reliable HPLC fluorimetric method was developed to detect and quantitatively determine L-Leucine, L-Isoleucine, L-Histidine, L-Valine, and β-Alanine in chocolate using aniline as an internal standard. Experimental design methodology was used to investigate and optimize the clean-up procedure of the samples. Therefore, three extraction techniques (solid-phase extraction (by two different SPE cartridges) and liquid–solid extraction (LSE)) were compared and evaluated. The LOQ values in chocolate varied from 24 to 118 ng/g (recovery 89.7–95.6%, %RSD < 2.5). Amino acids were pre-column derivatized with o-phthalaldehyde (OPA), while derivatization parameters were thoroughly investigated by experimental design methodology. The analysis was performed by HPLC-fluorescence (emission: $\lambda = 455$ nm, excitation: $\lambda = 340$ nm) method using a C_{18} column and a mixture of phosphate buffer (pH = 2.8; 20 mM)-methanol as a mobile phase in gradient elution. The method was validated ($r^2 > 0.999$, %RSD < 2, LOD: 10 ng mL^{-1} for histidine and leucine, 2 ng mL^{-1} for alanine and valine, and 4 ng mL^{-1} for Isoleucine) according to the International Conference on Harmonization guidelines.

Keywords: amino acids; chocolate; derivatization; HPLC; fluorescence

Citation: Synaridou, M.S.; Tsamis, V.; Sidiropoulou, G.; Zacharis, C.K.; Panderi, I.; Markopoulou, C.K. Fluorimetric Analysis of Five Amino Acids in Chocolate: Development and Validation. *Molecules* 2021, 26, 4325. https://doi.org/10.3390/molecules 26144325

Academic Editor: Marcello Locatelli

Received: 28 June 2021
Accepted: 15 July 2021
Published: 16 July 2021

Publisher's Note: MDPI stays neutral with regard to jurisdictional claims in published maps and institutional affiliations.

1. Introduction

Dietary supplements are constantly gaining supporters, particularly in the field of sports. These supplements are concentrated sources of nutrients with great nutritional value; they are intended to supplement normal diets, according to each person's needs. There are many types of food supplements on the market, such as tablets, pastilles, ampoules containing liquids, and powders, designed to be taken in measured, small-unit quantities. Alternatively, they are usually consumed with nutritious food, such as cereal bars or milk drinks [1]. Chocolate is a food product consumed worldwide, by different populations, due to its desirable sensory characteristics [2]. Dark chocolate, in particular, with its high cacao content, is one of the most promising functional foods, owing to its high levels of bioactive compounds, including flavonoids and phenolic acids [3].

Amino acids (AAs) are basic dietary supplements for those who exercise. They are also referred to as the building blocks of life due to their important roles in ribosomal protein synthesis. Protein AAs can be divided into two categories: those that cannot be synthesized by the organism (essential), and must be supplied from external sources, and non–essential amino acids that are synthesized by the organism [4].

The use of dietary supplements is determined by the amino acid content. For example, formulations containing branched chain AAs (consisting of leucine, isoleucine, and valine 2: 1: 1) are among the most widely used by athletes as they contribute to muscle recovery after training.

The scientific community has conducted a lot of research in this field, analyzing amino acids in various matrices, such as environmental samples [5], food [6,7], plants [8], or herbal raw material [9] and biological fluids [10,11]. Particular emphasis has been placed on the determination of amino acids in nutritional substrates and the purity of the sample [12].

The extraction of protein-bound amino acids usually precedes the hydrolysis stage, which is a complex process [13,14] followed by a clean-up step. For most sample matrices, solid-phase extraction (SPE) is optional, though it can greatly improve the quality of a sample by cleaning and concentrating the overall analytical outcomes [15]. In AA analysis, ion-exchange SPE is typically employed. Being zwitterionic, both cation and anion exchange SPE can be used to selectively retain amino acids. In contrast, cartridges with C_{18} are not suggested since it was reported that they had poor retention of the underivatized polar amino acids and low selectivity [16–18].

A variety of analytical techniques have been reported in literature regarding the determination of AAs, including capillary electrophoresis (CE) [19], thin layer chromatography (TLC) [20], liquid chromatography (HPLC) [21], or gas chromatography (GC). In most cases, HPLC is coupled with a fluorimetric (FLD) [22] or a mass spectrometric (MS) detector [13,23,24].

In general, the analysis of low molecular-weight amino acids is considered to be a challenging analytical task due to their polar nature, small size, and their relative low ability to absorb ultraviolet radiation (lack of chromophore groups). The derivatization of such molecules with appropriate reagents often helps to effectively overcome these obstacles. O-phthalaldehyde (OPA) is one of the most widely used reagents for the derivatization of compounds with a primary amino group [25].

The present study initially aimed to incorporate five amino acids (L-Leucine, L-Isoleucine, L-Histidine, L-Valine, and β-Alanine) in dark chocolate (pellet form) in order to prepare a complete dietary supplement. The selection of amino acids and their composition for young athletes was based on their properties to reduce fatigue (before training), and to rebuild their muscles. Thus, since the daily intake of AAs is considered essential, their formulation into dark chocolate pellets facilitates their oral intake.

According to the declaration of the European Parliament Directive (2002/46 / EC) on food supplements (Article 8–9), the values of nutrients or substances with nutritional or physiological effects contained in the product must be indicated on the label. In addition, their average values based on the product analysis, by the manufacturer, must be reported. Thus, following the process, a reliable analytical method was developed for the quantitative determination of the AAs in dark chocolate to test the suitability of the proposed formulation. Due to the complexity of the substrate, two different extraction techniques, a solid-phase extraction (SPE) using two different cartridges (NH_2, Diol) and a liquid–solid extraction (LSE) were studied, compared, and evaluated. In order to increase the sensitivity of the method, amino acids were derivatized (pre-column) with OPA, separated by HPLC and determined by fluorescence.

The proposed method, due to its efficiency, reliability and ease of use, has a dual field of application in both routing and trace analysis.

2. Results

2.1. Optimization of the HPLC Conditions

Focusing on finding and applying a reliable chromatographic method for the determination of the amino acids, the optimization stage is considered essential. The retention time and characteristics of their chromatographic peaks (e.g., resolution and tailing factor) are directly influenced by parameters, such as the kind of the solvent or the concentration of the buffers, the flow rate, column type, and the temperature.

The biggest problem for the selection of the optimum stationary phase was the low separation of alanine and valine and the coelution of leucine and isoleucine (isomers). Since the problem was not solved using any conventional RP column, the short C_{18} (15 cm × 4.6 mm, 5.0 μm, Supelco, Bellefonte, United States) analytical column was replaced by a C_{18} DB (25 cm × 4.6 mm, 5.0 μm, Supelco, Bellefonte, PA, USA) analytical column providing better resolutions. As far as the column temperature is concerned, the increase in temperature implies a reduction in elution time, and as a result, the incomplete separation of leucine and isoleucine. Thus, among the three values studied (25, 30, and 40 °C), 25 °C was chosen as the ideal.

By using the DB analytical column, the isocratic elution did not improve the separation of the analytes; thus, a binary gradient elution was selected. Various mixtures of water with or without buffers (mobile phase A) and organic modifiers, including acetonitrile (ACN) and/or methanol (MeOH) (mobile phase B), were tested. Initially, in order to find out the most suitable mobile phase composition, a wide gradient elution system was employed (mobile phase B: 20 to 80%). The procedure was repeated in triplicate using methanol 100%, acetonitrile 100%, or methanol/acetonitrile mixture at a ratio 1:1 v/v as phase B. Under these conditions, it was observed that MeOH (as mobile phase B) and H_2O (as mobile phase A) gave sharper peaks and, hence, were selected. Thereafter, for the final configuration of the gradient elution program, emphasis was placed on the separation of the two isomers considering their behavior data from respective isocratic mobile phases.

For further improvement of the peaks' widths and resolution, two additional factors were examined: (a) pH adjustment using an aqueous buffer and (b) the buffer concentration. Knowing that the isoelectric point of the amino acids is about 6.0 and the pKa of their acidic group 2.3, the water was replaced with a sodium dihydrogen phosphate buffer, and studied at 2.8, 5.0, 6.0, and 7.0 pH values. The value of 2.8 was chosen as the most appropriate since the separation of the analytes improved with the reduction of the pH value. Taking into account that the high concentration of phosphate buffer, during gradient elution, should cause precipitation of salt and high back pressure in the system [22,26], three relatively low levels of buffer concentrations (10, 20, and 30 mM) were also studied. Since there were no significant improvements in the chromatographic behavior of the analytes, 20 mM was considered the ideal value. The flow rate was chosen to be low enough (0.6 and 0.7 mL min^{-1}) to achieve better peak resolutions and low back pressures. Results of the system suitability parameters are depicted in Table 1.

Table 1. System suitability.

Parameters	Retention Time (min)	k′	Resolution	Tailing Factor
Histidine	6.45	-	-	1.217
Alanine	16.78	1.599	31.518	1.209
Valine	23.75	2.683	20.922	1.203
Isoleucine	27.60	3.280	9.486	1.193
Leucine	29.07	3.505	2.831	1.207
IS	30.65	3.749	2.695	1.198

2.2. Optimization of the OPA-Amino Acids Derivatization Conditions

Critical parameters affecting the amino acid derivatization, such as the time, OPA concentration, and pH of the borate buffer, were studied and optimized. Thus, with the aid of the Box-Behnken experimental design, 18 experimental combinations were performed and their results were evaluated.

The derivatization reaction of organic compounds, containing a primary amino group in their molecule, with OPA, is rapid [2,27], and leads to the formation of a fluorescence derivative due to the presence of chromophores and the rigidity of the molecule (Figure 1).

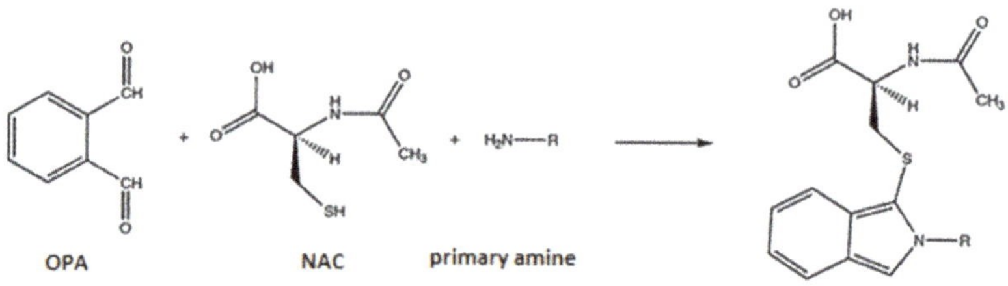

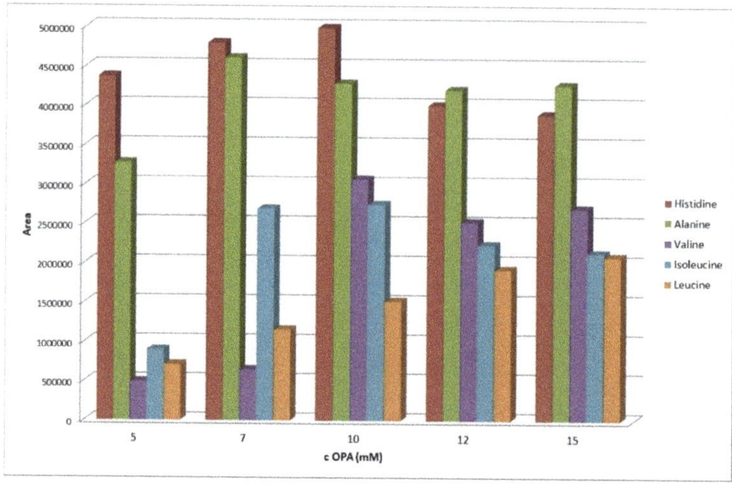

Figure 1. Derivatization reaction of primary amines with OPA.

In the present experimental investigation, in order to determine the required reaction time (at ambient temperature), values ranging from 1 to 10 min were studied. As can be seen from the results of the experimental design (Supplementary Material of Figure S1), the ideal time, set as the reaction time, was 8 min.

As expected, the concentration of the OPA is also one of the most critical factors that determine the effectiveness of the reaction, and is directly related to its stoichiometry. Preliminary experiments were performed using a standard solution of AAs (histidine 5 µg mL^{-1}, alanine 2 µg mL^{-1}, valine 10 µg mL^{-1}, isoleucine 10 µg mL^{-1}, leucine 20 µg mL^{-1}, and aniline 10 µg mL^{-1} as IS), while the OPA concentration varied from 1 mmol L^{-1} to 20 mmol L^{-1}. As shown in Figure 2, there is an almost linear increase in the signal from 5 to 10 mmol L^{-1} and a negligible decrease from that point on.

Figure 2. Effect of OPA concentration on the effectiveness of the reaction.

Hence, the ideal OPA concentration, for the given concentration of analytes in the sample, is 10 mmol L^{-1}, which was selected for future experiments. The pKa values of the amino groups of amino acids used in the chocolate preparation range from 9 to 9.7 (Table S1—Supplementary Materials). The OPA derivatization reaction takes place under alkaline conditions to prevent ionization of the secondary amino groups. Therefore, the pH of the borate buffer was studied in the range 8.5–11.0. The ideal value indicated by the software was pH 10.5. Figure S1 (Supplementary Material) depicts the total value limits studied and the ideal conditions of the selected parameters.

2.3. Stability of Amino Acids-OPA Derivatives

According to the literature, a major disadvantage of OPA amines-derivatives is their short shelf life [2]. For this reason, their stability was tested at 0, 12, and 24 h. More specifically, three different concentrations of derivatized mixtures of amino acids were prepared and measured (three replications at 25 °C) by HPLC. The within days %RSD of AAs derivatives was <2.0%, which indicates their stability up to 24 h.

2.4. Extraction Recovery–Preliminary Experiments

Chocolate is a complex matrix and, moreover, API's determination is difficult, especially when amino acids behave in the same way as the substrate. Thus, a purification step, which would aim at the precipitation of proteins/saturated fats, and improve the efficiency of the method, is crucial. Therefore, the appropriate experimental conditions that will promote the best recoveries should be selected after investigation.

Two extraction techniques, LSE and SPE, were used in the present study. In the case of liquid/solid extraction, acetonitrile was chosen as the protein precipitant because it offers good recovery and removes most of the impurities. However, in order to achieve the complete release of the analytes, as well as for their dissolution, the presence of water in the diluent is considered necessary. Hence, after a series of experiments, it was found that the ratio of extraction solvents with the best recovery results was H_2O-ACN 30:70% v/v. It must be noticed that the freezing of the sample for at least 45 min after its centrifugation is one more determinant parameter for the precipitation of impurities.

LSE is carried out in a single step avoiding significant losses of the analytes (Table 2) [28]. The method displays high recoveries with good reproducibility, while its relative low sensitivity in trace analysis could be considered a drawback.

Table 2. Accuracy—LSE.

Standard	C (Added) (μg mL^{-1})	Observed (ng mL^{-1})	Recovery (%)
		Mean $\pm$ SD	Mean $\pm$ SD
Histidine	0.09	0.09 $\pm$ 0.00	97.8 $\pm$ 0.2
	3.00	2.97 $\pm$ 0.01	99.0 $\pm$ 0.4
	5.00	4.92 $\pm$ 0.06	98.3 $\pm$ 1.1
	Average (n = 9)		98.4 $\pm$ 0.9
	Confidence interval (95%)		97.5–99.3
Alanine	0.02	0.02 $\pm$ 0.00	98.5 $\pm$ 0.5
	1.00	1.01 $\pm$ 0.01	100.6 $\pm$ 0.7
	2.00	2.00 $\pm$ 0.04	99.9 $\pm$ 1.2
	Average (n = 9)		99.6 $\pm$ 1.7
	Confidence interval (95%)		97.9–101.4
Valine	0.03	0.03 $\pm$ 0.00	99.1 $\pm$ 0.5
	5.00	4.92 $\pm$ 0.02	98.4 $\pm$ 0.4
	8.00	7.86 $\pm$ 0.01	98.2 $\pm$ 0.1
	Average (n = 9)		98.6 $\pm$ 0.6
	Confidence interval (95%)		98.0–99.2
Isoleucine	0.03	0.03 $\pm$ 0.00	98.3 $\pm$ 0.4
	5.00	4.94 $\pm$ 0.01	98.8 $\pm$ 0.1
	8.00	7.90 $\pm$ 0.04	98.8 $\pm$ 0.5
	Average (n = 9)		98.6 $\pm$ 0.5
	Confidence interval (95%)		98.2–99.1
Leucine	0.06	0.06 $\pm$ 0.00	99.1 $\pm$ 0.2
	10.00	10.05 $\pm$ 0.09	100.5 $\pm$ 0.9
	15.00	15.17 $\pm$ 0.11	101.1 $\pm$ 0.7
	Average (n = 9)		100.2 $\pm$ 1.1
	Confidence interval (95%)		99.1–101.4

For the SPE procedure, two types of cartridges, Diol and NH_2, were tested. In this case too, special attention must be paid to the stage of protein/impurities precipitation.

Thus, the amino acid extraction from the chocolate matrix with the appropriate mixture of solvents (15 mL water-35 mL ACN) should be carried out as a preliminary step. Then, after appropriate dilutions, the sample loaded into the cartridge should contain the minimum amount of water (H_2O-ACN, 5:95% v/v). The recovery results of the analytes, at three concentration levels/three replicates, are presented in Table 3. From a chromatographic point of view, the samples obtained from the Diol cartridge were better, as there were no additional peaks in the chromatogram (see Section 2.6.1).

Table 3. Accuracy—PE (Diol and NH_2).

Standard	C (Added) (μg mL^{-1})	SPE/Diol			SPE/NH$_2$	
		Observed (ng mL^{-1})	Recovery (%)		Observed (ng mL^{-1})	Recovery (%)
		Mean ± SD	Mean ± SD		Mean ± SD	Mean ± SD
Histidine	0.09	0.09 ± 0.00	98.5 ± 0.7		0.09 ± 0.00	96.1 ± 0.2
	3.00	2.93 ± 0.02	97.7 ± 0.5		2.92 ± 0.01	97.2 ± 0.2
	5.00	4.87 ± 0.01	97.4 ± 0.2		4.88 ± 0.01	97.5 ± 0.2
Average (*n* = 9) Confidence interval (95%)			97.8 ± 0.7 97.1–98.6		Average (*n* = 9) Confidence interval (95%)	96.9 ± 0.7 96.2–97.6
Alanine	0.02	0.02 ± 0.00	98.9 ± 0.1		0.02 ± 0.00	96.7 ± 0.4
	1.00	1.01 ± 0.01	100.9 ± 0.8		1.01 ± 0.01	101.3 ± 1.3
	2.00	2.03 ± 0.00	101.7 ± 0.2		2.06 ± 0.03	102.9 ± 1.3
Average (*n* = 9) Confidence interval (95%)			100.5 ± 1.3 99.2–101.8		Average (*n* = 9) Confidence interval (95%)	100.3 ± 3.0 97.3–103.3
Valine	0.03	0.03 ± 0.00	99.1 ± 0.3		0.03 ± 0.00	97.1 ± 0.2
	5.00	4.96 ± 0.03	99.2 ± 0.5		4.95 ± 0.03	99.1 ± 0.5
	8.00	7.99 ± 0.02	99.9 ± 0.2		8.03 ± 0.08	100.4 ± 0.9
Average (*n* = 9) Confidence interval (95%)			99.4 ± 0.5 98.9–100.0		Average (*n* = 9) Confidence interval (95%)	98.9 ± 1.6 97.3–100.4
Isoleucine	0.03	0.03 ± 0.00	98.0 ± 0.1		0.03 ± 0.00	97.0 ± 0.6
	5.00	4.90 ± 0.02	98.0 ± 0.5		4.85 ± 0.02	96.9 ± 0.3
	8.00	7.83 ± 0.05	97.9 ± 0.6		7.73 ± 0.02	96.7 ± 0.2
Average (*n* = 9) Confidence interval (95%)			98.0 ± 0.5 97.5–98.4		Average (*n* = 9) Confidence interval (95%)	96.9 ± 0.5 96.4–97.3
Leucine	0.06	0.06 ± 0.00	100.4 ± 1.3		0.06 ± 0.00	96.5 ± 0.4
	10.00	10.07 ± 0.08	100.7 ± 0.8		9.82 ± 0.04	98.2 ± 0.4
	15.00	15.10 ± 0.23	100.6 ± 1.5		14.85 ± 0.06	99.0 ± 0.4
Average (*n* = 9) Confidence interval (95%)			100.6 ± 1.3 99.2–101.9		Average (*n* = 9) Confidence interval (95%)	97.9 ± 1.2 96.7–99.1

2.5. Improvement of Trace Analysis

For the trace analysis of AAs in the chocolate substrate, the combination of the two most effective purification procedures (LSE and SPE with Diol cartridge) were used. Due to the large number of parameters that probably affect the recovery of the analytes and, consequently, the large number of experiments needed, the study was divided into two steps; one for the LSE and the other for the Diol-SPE. In both cases, the efficiency of the extraction procedure was investigated, taking into account the influence of four parameters. For the LSE technique, there were the chocolate weight, the centrifugation, freezing, and ultrasonication time (Table S2A,B–Supplementary Material), whereas for the Diol cartridge there were the diluent's initial volume, the pH of the initial diluent, the load volume of the sample, and the elution volume (Table S3A,B–Supplementary Material). All parameters were evaluated using experimental design methodology (Design Expert 11 software) and by applying the two-level factorial mode. In each case, (two models) 16 experiments were performed, and their results were evaluated in terms of %Recovery values for each amino acid (responses). The proposed solutions for the two models were combined and determined the final conditions of the overall experiment (Table 4), which were tested at three concentration levels for each amino acid (three replications). The %Recoveries and %RSD values of the experiments revealed that the proposed method is robust and can be used for the quantitation of the amino acids in chocolate at very low concentrations (Table 5). The %Prediction Error was studied in order to evaluate the predictive capability of the experimental design models (Table 6) and was found to be < 10.7. The LOQ values in chocolate were found to be 24 ng/g for histidine and alanine, 35 ng/g for valine, 118 ng/g for isoleucine, and 59 ng/g for leucine.

Table 4. Final experimental conditions.

Parameter	Optimal Value	Selected Value
Chocolate weight (mg)	848.2	850.0
Centrifugation time (min)	18.3	20.0
Freezing time (min)	43.3	45.0
Ultrasonic time (min)	20.8	20.0
Initial volume of diluent (mL)	6.5	5.0
pH (initial) diluent	5.0	5.0
Volume of sample on SPE (mL)	2.2	2.0
Elution volume on SPE (µL)	300.0	300.0

Table 5. Accuracy—LSE/ SPE after experimental design.

Standard	C (Added) (ng mL^{-1})	Observed (ng mL^{-1}) Mean $\pm$ SD	Recovery (%) Mean $\pm$ SD
Histidine	20.0	18.7 $\pm$ 0.1	93.7 $\pm$ 0.6
	100.0	93.3 $\pm$ 0.4	93.3 $\pm$ 0.4
	500.0	460.2 $\pm$ 2.5	92.0 $\pm$ 0.5
Average ($n = 9$)			93.0 $\pm$ 0.9
Confidence interval (95%)			92.1–93.9
Alanine	20.0	18.2 $\pm$ 0.1	91.2 $\pm$ 0.4
	100.0	90.3 $\pm$ 0.6	90.3 $\pm$ 0.6
	500.0	446.2 $\pm$ 0.9	89.2 $\pm$ 0.2
Average ($n = 9$)			90.2 $\pm$ 1.0
Confidence interval (95%)			89.3–91.2
Valine	30.0	27.2 $\pm$ 0.3	90.8 $\pm$ 0.8
	200.0	179.9 $\pm$ 1.2	90.0 $\pm$ 0.6
	1000.0	885.0 $\pm$ 7.5	88.5 $\pm$ 0.7
Average ($n = 9$)			89.7 $\pm$ 1.3
Confidence interval (95%)			88.5–91.0
Isoleucine	100.0	95.1 $\pm$ 0.3	95.1 $\pm$ 0.3
	200.0	187.5 $\pm$ 0.4	93.7 $\pm$ 0.2
	1000.0	938.3 $\pm$ 0.9	93.8 $\pm$ 0.1
Average ($n = 9$)			94.2 $\pm$ 0.7
Confidence interval (95%)			93.5–94.9
Leucine	50.0	48.0 $\pm$ 0.1	96.0 $\pm$ 0.2
	200.0	191.5 $\pm$ 1.4	95.7 $\pm$ 0.7
	1000.0	949.7 $\pm$ 1.7	95.0 $\pm$ 0.2
Average ($n = 9$)			95.6 $\pm$ 0.7
Confidence interval (95%)			94.9–96.2

Table 6. Prediction error.

	%Recovery				
	Histidine	**Alanine**	**Valine**	**Isoleucine**	**Leucine**
Predicted Value–LSE	90.6	89.5	86.4	87.0	89.4
Predicted Value–SPE	87.1	84.0	84.8	81.2	86.6
Predicted value (average)	88.9	86.8	85.6	84.1	88.0
Experimental value (average)	93.0	90.2	89.7	94.2	95.6
%Prediction Error	4.4	3.8	4.6	10.7	7.9

2.6. Method Validation

2.6.1. Selectivity of the Developed Chromatographic Method

Chromatograms of drug-free chocolate samples (blank) were compared with three corresponding samples spiked with histidine at 0.18 µg mL^{-1}, alanine at 0.3 µg mL^{-1}, valine and isoleucine both at 0.6 µg mL^{-1}, leucine at 1.2 µg mL^{-1}, and aniline (IS) at 3 µg mL^{-1}. Both series of samples were processed using three different clean-up methods and derivatized with OPA. As can be observed in the chromatograms shown in Figure 3B–D, there are no co-eluted matrix interferences at the retention times of the analytes. Of course, in the case of the SPE with the -NH$_2$ cartridge, there is an additional peak near alanine, which essentially makes its quantification difficult.

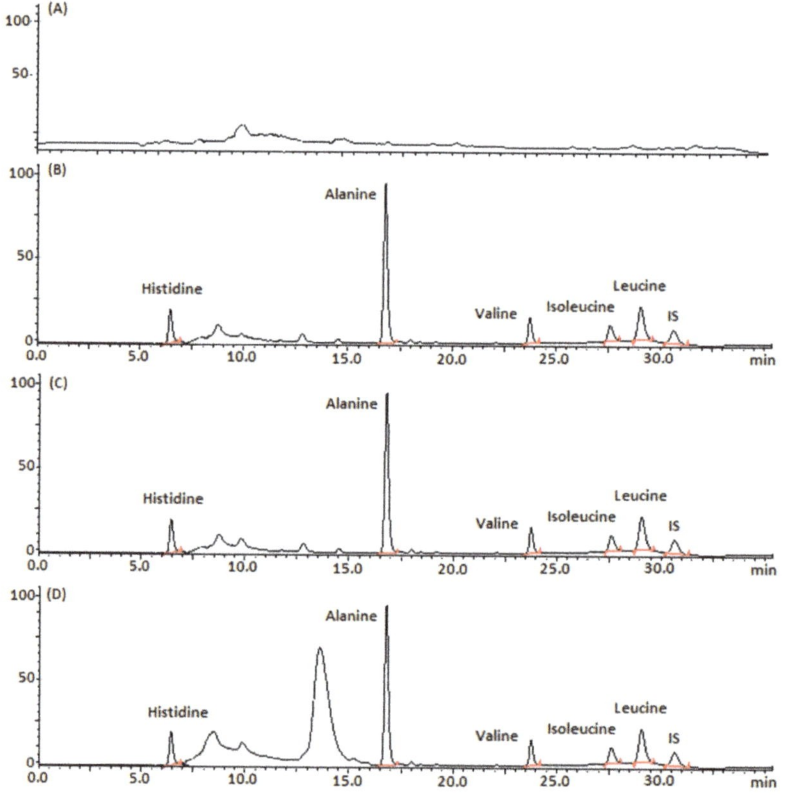

Figure 3. Chromatograms of the analysis of (**A**) blank chocolate sample; (**B**) spiked sample after LSE; (**C**) spiked sample after SPE-Diol; and (**D**) spiked sample after SPE-NH$_2$.

2.6.2. Linearity, LODs, and LOQs

The linearity of the method was studied using five standard solutions with different concentration ranges for each amino acid (three replicates). Results of linearity regression, correlation coefficient, LOD, and LOQ for the amino acids are presented in Table 7.

Table 7. Results of linearity regression, correlation coefficient, LOD, and LOQ for amino acids.

Standard	Calibration Range (μg mL^{-1})	Regression Equation	r^2	LOD (μg mL^{-1})	LOQ (μg mL^{-1})
Histidine	0.02–5.0	27.9x − 0.23	0.9992	0.01	0.02
Alanine	0.01–2.0	95.96x − 1.89	0.9991	0.002	0.01
Valine	0.01–8.0	18.67x − 0.06	0.9996	0.002	0.01
Isoleucine	0.01–8.0	18.96x − 0.13	0.9997	0.004	0.01
Leucine	0.03–15.0	10.15x − 0.04	0.9991	0.01	0.03

2.6.3. Accuracy and Precision

To evaluate accuracy, spiked placebo samples were analyzed in triplicate at three different concentration levels (0.09–5.0 μg mL^{-1} for histidine, 0.2–2.0 μg mL^{-1} for alanine, 0.3–8.0 μg mL^{-1} for valine and isoleucine, and 0.6–15.0 μg mL^{-1} for leucine. The %Recovery and %RSD values were used as evaluation criteria for each purification procedure. As indicated in Table 2; Table 3, recovery values were within the acceptable limits (98.4–100.2 for LSE, 97.8–100.6 for SPE-Diol, and 96.9–100.3 for SPE-NH$_2$), whereas the %RSD values ranged from 0.01 to 0.02 for LSE, 0.46 to 1.32 for SPE-Diol, and 0.47 to 3.01 for SPE-NH$_2$, respectively. However, considering that during the extraction of the AAs with the SPE-NH$_2$ cartridge, an additional peak appears in the chromatogram (close to alanine), the method was rejected, while both LLE and SPE-Diol were further examined with experimental design, and their combination was used for the trace analysis of amino acids in chocolate.

To evaluate the precision of the method, repeatability and intermediate precision were examined. Therefore, a standard solution of amino acids (3.0 μg mL^{-1} for histidine, 1.0 μg mL^{-1} for alanine, 5.0 μg mL^{-1} for valine and isoleucine, and 10.0 μg mL^{-1} for leucine) was analyzed (six replicates) on the same day. The %RSD values were found to be <2%. Regarding the intermediate precision, samples in three different concentrations (0.02–5.0 μg mL^{-1} for histidine, 0.01–2.0 μg mL^{-1} for alanine, 0.01–8.0 μg mL^{-1} for valine and isoleucine, and 0.03–15.0 μg mL^{-1} for leucine) were prepared and analyzed on three consecutive days (three replicates). In all cases, %RSD values were found to be < 2% (Table S4—Supplementary).

2.6.4. Robustness

The Plackett–Burman experimental design was used to evaluate method robustness [29] and to limit the number of tests needed (11 tests) (Table S5—Supplementary Table). The parameters studied were: the concentration of the buffer (sodium dihydrogen phosphate, mobile phase A) at a range of 18–20 mM, the pH values of the solution (2.7 to 2.9), the temperature of the column (from 24 to 26 °C), the initial methanol content (mobile phase B: 39 to 40% v/v), as well as the injection volume (from 19 to 21 μL). The evaluation of the experimental results was carried out based on the intensity ratios (amino acid area/internal standard), as well as on the resolution of the chromatographic peaks.

As we can see from the Pareto diagrams (Figure S2 in the Supplements), none of the parameters studied at the specific value range significantly affect ($p < 0.05$) the ratio of peak areas and the resolution. Therefore, the proposed chromatographic method is robust.

3. Materials and Methods

3.1. Chemicals, Materials and Reagents

All amino acids (L-Leucine, L-Isoleucine, L-Histidine, L-Valine, and β-Alanine) were Pharma grade (purity 98.5–101.5%) and were purchased from Merck (Darmstadt, Germany).

Aniline ($\geq$99.5%), hydrochloric acid (37% w/v), N-acetyl-cysteine (NAC) ($\geq$99.0%), o-phthalaldehyde (OPA) ($\geq$99.0%), sodium dihydrogen phosphate, boric acid, and sodium hydroxide pellets were provided by Sigma-Aldrich (Steinheim, Germany). Acetonitrile (ACN), methanol (MeOH) and water of HPLC grade were supplied by VWR chemicals (Vienna, Austria), whereas phosphoric acid, which met USP specifications, was purchased from Merck (Darmstadt, Germany). Dark chocolate (70% cacao) was obtained from local markets (Greece). SPE tubes (Diol and NH_2) were purchased from Supelco (Sigma Aldrich, Steinheim, Germany).

Standard stock solutions of amino acids were prepared on a daily basis by dissolving the corresponding amount of each amino acid to HCl 0.1 N to obtain a 1000 μg mL^{-1} solution. Working standards were made by appropriate dilutions of the stock solutions in d. water. The OPA derivatization reagent (10 mmol L^{-1}) was prepared by dissolving 5 mg of o-phthalaldehyde in 0.5 mL MeOH followed by dilution with 9.5 mL water and the NAC solution (10 mmol L^{-1}) by dissolving 10 mg of the reagent in 0.5 mL water. Both solutions were stable at 4 °C for three working days since they were protected from light. For the preparation of the borate buffer (10 mmol L^{-1}, pH 10.5), 7 mg of boric acid was dissolved in 2 mL water, and the pH of the solution was adjusted using NaOH (in pellets). Finally, the pH of the phosphate buffer (20 mmol L^{-1}) was adjusted to the desired value of 2.8 by adding drops of concentrated H_3PO_4.

3.2. Instrumentation and Chromatographic Conditions

Chromatographic separations of amino acids were performed on a Shimadzu HPLC system consisting of two LC-20AD isocratic pumps, a DGU-14A degasser, an SIL-10AD autosampler and a fluorescence detector (RF-535) (Shimadzu, Tokyo, Japan). The analytical column was a reversed phase LC-C_{18} DB (250 $\times$ 4.6 mm, 5.0 μm, Supelco (Bellefonte, United States). LC-solution® software version 1.25 SP4 was utilized for hardware control and data manipulation. All separations were accomplished using a binary gradient elution program. The mobile phases A and B were 20 mM phosphate buffer (pH 2.8) and methanol, respectively. Their initial ratio was 40% v/v of B and the flow rate was set at 0.7 mL min^{-1}. The ratio of the mobile phase B was linearly increased to 65% in 15 min and kept constant for 9 min. Then, it was further increased to 80% in 1 min, while the flow was decreased at 0.6 mL min^{-1} and stayed constant for up to 30 min. Then, it reverted to its initial conditions (40% B and flow 0.7 mL min^{-1}) in 5 min and kept constant for up to 40 min to obtain reproducible separations. The injection volume was set at 20 μL (for the analysis of the product) and 100 μL (to determine the LOD value on the chocolate substrate). The column was thermostated at 25 °C. Amino acid-OPA derivatives were detected spectrofluorimetrically at λex/λem = 340/455 nm.

3.3. Derivatization Procedure

A 100 μL aliquot of standard with amino acids or sample solution was mixed with 100 μL OPA (10 mmol L^{-1}), 100 μL NAC solution, and 700 μL borate buffer (10 mmol L^{-1}, pH 10.5) in a 1.5 mL autosampler glass vial. After vortex mixing for 10 s, the derivatization mixture was allowed to react for 8 min at ambient temperature prior to injection into the HPLC system. Blank samples were prepared without amino acids.

3.4. Preparation of the Chocolate Formulation

According to international guidelines [30,31], the required daily amounts of amino acids to be consumed by a child 10–12 years old are: 42–48 mg/kg for leucine, 18–22 mg/kg for isoleucine, 23–28 mg/kg for valine, 13–17 mg/kg for histidine, and 750–1800 mg for alanine.

Taking the above into consideration, a 12-year-old child weighing an average of 40 kg should take 4.6 gr of chocolate pellets per day (divided into 2–3 doses before and after training) containing the amounts of amino acids listed in Table 8.

Table 8. Recommended daily dosage and composition of the formulation.

Amino Acids	Recommended Dosage (mg)/Day	Quantity (mg)/ Formulation
Leucine	1700	540
Isoleucine	850	270
Valine	850	270
Histidine	520	175
Alanine	750	250
Milk chocolate (excipient.)	-	4.6 g

For the preparation of the pellets, precisely weighed quantities of amino acids corresponding to 30 formulations were transferred to a stainless-steel bowl and blended for 10 min using a Turbula T2F shaker (WAB, Muttenz, Switzerland). Then, the required amount of chocolate (slightly melted, 30–40 °C) was added and mixing was continued for another 5 min. The mixture was placed in a silicone mold and allowed to cool in a refrigerator for 1 h.

3.5. Sample Purification Procedures

Four different procedures for the purification of chocolate samples and the quantification of amino acids on such a substrate are described below. The latter focuses on trace analysis.

3.5.1. Liquid–Solid Extraction (LSE)

A total of 500 mg of the formulation (chocolate pellets enriched with amino acids), which was previously dissolved in 10 mL H_2O, was placed in a water bath set at 30 °C. The sample was sonicated for 10 min and stirred for another 10 min. A volume of 2 mL of the supernatant was transferred to a centrifuge tube to which 3 mL of ACN and 5 mL of water was added. The sample was centrifuged for 15 min at 4000 rpm and then was left in the freezer for 60 min. An aliquot of 1 mL of the supernatant was diluted with water to 100 mL. Derivatization (according to the procedure) and analysis in HPLC followed.

3.5.2. Solid-Phase Extraction (SPE)

A solid-phase extraction method using two types of cartridges (Diol and NH_2) was developed to clean-up the sample and extract the amino acids. More specifically, 500 mg chocolate pellets were dissolved in a mixture of 15 mL water–35 mL can, followed by sonication for 10 min (30 °C), and stirring for another 10 min. The sample was centrifuged for 15 min at 4000 rpm; 1 mL of supernatant was diluted up to 100 mL with H_2O-ACN (5–95% *v/v*) (Sample 1); 1 mL of sample passed through the column (Diol and NH_2, respectively) after first being conditioned with 1 mL H_2O-ACN (5–95% *v/v*) and 1 mL ACN. The column was then rinsed with 1 mL of ACN and the amino acids were eluted with 1 mL of water. Derivatization was performed (as mentioned) and the samples were analyzed in HPLC.

3.5.3. Trace Analysis of AAs in Chocolate

A total of 850 mg of chocolate were dissolved in 5 mL of ACN-H_2O mixture 95–5% *v/v*. The sample was sonicated for 20 min (30 °C) and stirred for 10 min. The sample was centrifuged for 15 min at 4000 rpm and then put into a freezer for another 45 min; 2 mL of supernatant passed through the Diol column after firstly being conditioned with 1 mL H_2O-ACN (5–95% *v/v*) and 1 mL ACN. The column was then rinsed with 1 mL ACN and the amino acids were eluted with 0.3 mL water. Derivatization was performed (as mentioned) and the samples were analyzed in HPLC.

3.6. Validation of the Proposed Method

The proposed method was validated according to ICH and U.S. Food Drug guidelines [29–33]. The parameters assessed were specificity, linearity, precision, accuracy, LOD/LOQ, and robustness.

3.6.1. Specificity

The retention times of the amino acids, as well as the corresponding peaks of blank and spiked placebos samples, were compared in order to investigate the ability of the proposed method to unequivocally assess the analytes.

3.6.2. Linearity

For the assessment of linearity, standard solutions were prepared by appropriate dilution of a stock standard at the ranges of 0.02–5.0 µg mL^{-1} for histidine, 0.01–2.0 µg mL^{-1} for alanine, 0.01–8.0 µg mL^{-1} for valine and isoleucine, and 0.03–15.0 µg mL^{-1} for leucine (five concentrations for each analyte). Each standard was analyzed in triplicate. A linear regression line was calculated for each amino acid using the concentrations versus peak areas ratio (AAs/internal standard). The correlation coefficient (r^2) of the regression equation was calculated to validate the linearity parameter.

3.6.3. Limit of Detection (LOD) and Limit of Quantification (LOQ)

The LOD, LOQ limits [31] of the analytes were calculated based on the residual standard deviation (σ) and the slope (S) of the regression equation according to the following.

$$\text{LOD} = 3.3\sigma/S \tag{1}$$

$$\text{LOQ} = 10\sigma/S \tag{2}$$

3.6.4. Precision and Accuracy

For the evaluation of the precision of the method, repeatability (intra-day) and intermediate precision (inter-day) were assessed. A standard mix solution of an intermediate concentration (0.02–5.0 µg mL^{-1} for histidine, 0.01–2.0 µg mL^{-1} for alanine, 0.01–8.0 µg mL^{-1} for valine and isoleucine and 0.03–15.0 µg mL^{-1} for leucine), as well as a series of 3 standard solutions at three different concentrations, were analyzed on the same day and on three different days, respectively. The %Relative standard deviation (%RSD) was used to assess the precision.

To assess accuracy, chocolate samples spiked with three different concentrations (0.09–5.0 µg mL^{-1} for histidine, 0.2–2.0 µg mL^{-1} for alanine, 0.3–8.0 µg mL^{-1} for valine and isoleucine, and 0.6–15.0 µg mL^{-1} for leucine) were prepared and pretreated using the three proposed techniques (LSE and SPE). Each sample was analyzed in triplicate. The %Recovery (Equation (3)) of the analyte, the %RSD and the confidence interval (95%) were used to evaluate the methods.

$$\text{Recovery, \%} = \text{ x 100} \tag{3}$$

3.6.5. Robustness

Small, deliberate changes to selected crucial parameters of the proposed method were conducted in order to assess the robustness of the HPLC method. Specifically, the concentration (18–20 mM) and the pH (2.7–2.9) of the phosphate buffer (mobile phase A), the column temperature (24–26 °C), the %Methanol at the beginning of the gradient elution (mobile phase B) (39.0 to 40% *v*/*v*) and the injection volume (19 to 21 µL) were investigated using the Plackett–Burman experimental design. The results were an evaluation based on peak response (area/IS area) and the resolution [34].

4. Conclusions

In the present study, a reliable HPLC fluorometric method was developed for the simultaneous quantitative and qualitative determination of five amino acids (L-leucine, L-isoleucine, L-histidine, L-valine, and β-alanine) encapsulated in chocolate pellets used as a dietary supplement. Pre-column derivatization with an OPA reagent was applied before the analysis for the fluorimetric detection of the analytes, whereas their chromatographic separation was performed using a C_{18} column and a mixture of the phosphate buffer (pH = 2.8; 20 mM)-methanol as mobile phase in gradient elution. For the purification of the samples and the quantitative recovery of amino acids from the substrate, three extraction techniques were developed and compared: solid-phase extraction, using two different SPE cartridges (NH_2, Diol), and liquid–solid extraction (LSE). The optimization of both the derivatization parameters and the extraction method were evaluated using experimental design. The proposed method was validated according to ICH guidelines and demonstrated high precision, accuracy, and sensitivity. This procedure can be successfully applied for the determination of the five amino acids in chocolate pellet formulations (routine analysis) or to trace them on a chocolate substrate (trace analysis).

Supplementary Materials: The following are available online: Figure S1: Ideal values of the parameters studied during the optimization of the OPA-amino acids' derivatization conditions. Figure S2: Pareto charts for the robustness test of HPLC instrumental parameters (ratio & resolution) for A) Histidine, B) Alanine, C) Valine, D) Isoleucine, E) Leucine. Table S1: Molecular structure, molecular weight, pKa and isoelectric point of amino acids, Table S2A: Design experiments and requirements used in the experimental design of LSE method, Table S2B: Conducted experiments for the optimization of LSE method, Table S3A: Design experiments and requirements used in the experimental design of SPE method, Table S3B: Conducted experiments for the optimization of SPE method, Table S4: Intermediate precision, Table S5: Conducted experiments with Box-Behnken design for the evaluation of the robustness.

Author Contributions: Conceptualization, C.K.M. and M.S.S.; methodology, C.K.M., C.K.Z. and I.P; software, C.K.M. and C.K.Z.; validation, M.S.S., V.T. and G.S.; formal analysis, M.S.S., V.T. and G.S.; investigation, C.K.M. and M.S.S.; data curation, M.S.S., C.K.M. and C.K.Z.; writing—original draft preparation, M.S.S.; writing—review and editing, C.K.M.; C.K.Z. and I.P.; visualization, C.K.M. and M.S.S.; supervision, C.K.M. All authors have read and agreed to the published version of the manuscript.

Funding: This research received no external funding.

Institutional Review Board Statement: Not applicable.

Informed Consent Statement: Not applicable.

Conflicts of Interest: The authors declare no conflict of interest.

Sample Availability: Samples of the compounds are not available from the authors.

References

1. The European Parliament and the Council of the European Union. Regulation (EU) No 609/2013 of the European Parliament and of the Council of 12 June 2013 on food intended for infants and young children, food for special medical purposes, and total diet replacement for weight control and repealing Council Directive 92/52/EEC, Commission Directives 96/8/EC, 1999/21/EC, 2006/125/EC and 2006/141/EC, Directive 2009/39/EC of the European Parliament and of the Council and Commission Regulations (EC) No 41/2009 and (EC) No 953/2009. *Off. J. Eur. Union* **2013**, *181*, 35–56.
2. Oracz, J.; Nebesny, E.; Zyzelewicz, D.; Budryn, G.; Luzak, B. Bioavailability and metabolism of selected cocoa bioactive compounds: A comprehensive review. *Crit. Rev. Food Sci. Nutr.* **2020**, *60*, 1947–1985. [CrossRef]
3. Oracz, J.; Nebesny, E.; Żyżelewicz, D. Identification and quantification of free and bound phenolic compounds contained in the high-molecular weight melanoidin fractions derived from two different types of cocoa beans by UHPLC-DAD-ESI-HR-MSn. *Food Res. Int.* **2019**, *115*, 135–149. [CrossRef] [PubMed]
4. Ferré, S.; González-Ruiz, V.; Guillarme, D.; Rudaz, S. Analytical strategies for the determination of amino acids: Past, present and future trends. *J. Chromatogr. B Anal. Technol. Biomed. Life Sci.* **2019**, *1132*, 121819. [CrossRef] [PubMed]
5. Gao, S.; Xu, B.; Zheng, X.; Wan, X.; Zhang, X.; Wu, G.; Cong, Z. Developing an analytical method for free amino acids in atmospheric precipitation using gas chromatography coupled with mass spectrometry. *Atmos. Res.* **2021**, *256*, 105579. [CrossRef]

6. Peace, R.W.; Gilani, G.S. Chromatographic determination of amino acids in foods. *J. AOAC Int.* **2005**, *88*, 877–887. [CrossRef]

7. Dala-Paula, B.M.; Deus, V.L.; Tavano, O.L.; Gloria, M.B.A. In vitro bioaccessibility of amino acids and bioactive amines in 70% cocoa dark chocolate: What you eat and what you get. *Food Chem.* **2021**, *343*, 128397. [CrossRef]

8. Dziagwa-Becker, M.M.; Marin Ramos, J.M.; Topolski, J.K.; Oleszek, W.A. Determination of free amino acids in plants by liquid chromatography coupled to tandem mass spectrometry (LC-MS/MS). *Anal. Methods* **2015**, *7*, 7574–7581. [CrossRef]

9. Przybylska, A.; Gackowski, M.; Koba, M. Application of capillary electrophoresis to the analysis of bioactive compounds in herbal raw materials. *Molecules* **2021**, *26*, 2135. [CrossRef]

10. Fekkes, D.; Voskuilen-Kooyman, A.; Jankie, R.; Huijmans, J. Precise analysis of primary amino acids in urine by an automated high-performance liquid chromatography method: Comparison with ion-exchange chromatography. *J. Chromatogr. B Biomed. Sci. Appl.* **2000**, *744*, 183–188. [CrossRef]

11. Teerlink, T.; Van Leeuwen, P.A.M.; Houdijk, A. Plasma amino acids determined by liquid chromatography within 17 min. *Clin. Chem.* **1994**, *40*, 245–249. [CrossRef]

12. Samardzic, K.; Rodgers, K.J. Cytotoxicity and mitochondrial dysfunction caused by the dietary supplement l-norvaline. *Toxicol. Vitr.* **2019**, *56*, 163–171. [CrossRef] [PubMed]

13. Violi, J.P.; Bishop, D.P.; Padula, M.P.; Steele, J.R.; Rodgers, K.J. Considerations for amino acid analysis by liquid chromatography-tandem mass spectrometry: A tutorial review. *TrAC-Trends Anal. Chem.* **2020**, *131*, 116018. [CrossRef]

14. Moore, S.; Stein, W.H. Photometric ninhydrin method for use in the chromatography of amino acids. *J. Biol. Chem.* **1948**, *176*, 367–388. [CrossRef]

15. How, Z.T.; Busetti, F.; Linge, K.L.; Kristiana, I.; Joll, C.A.; Charrois, J.W.A. Analysis of free amino acids in natural waters by liquid chromatography-tandem mass spectrometry. *J. Chromatogr. A* **2014**, *1370*, 135–146. [CrossRef]

16. Wang, L.; Xu, R.; Hu, B.; Li, W.; Sun, Y.; Tu, Y.; Zeng, X. Analysis of free amino acids in Chinese teas and flower of tea plant by high performance liquid chromatography combined with solid-phase extraction. *Food Chem.* **2010**, *123*, 1259–1266. [CrossRef]

17. Tang, X.; Gu, Y.; Nie, J.; Fan, S.; Wang, C. Quantification of amino acids in rat urine by solid-phase extraction and liquid chromatography/electrospray tandem mass spectrometry: Application to radiation injury rat model. *J. Liq. Chromatogr. Relat. Technol.* **2014**, *37*, 951–973. [CrossRef]

18. Armstrong, M.; Jonscher, K.; Reisdorph, N.A. Analysis of 25 underivatized amino acids in human plasma using ion-pairing reversed-phase liquid chromatography/time-of-flight mass spectrometry. *Rapid Commun. Mass Spectrom.* **2007**, *21*, 2717–2726. [CrossRef]

19. Bonvin, G.; Schappler, J.; Rudaz, S. Capillary electrophoresis-electrospray ionization-mass spectrometry interfaces: Fundamental concepts and technical developments. *J. Chromatogr. A* **2012**, *1267*, 17–31. [CrossRef]

20. Bhawani, S.A.; Mohamad Ibrahim, M.N.; Sulaiman, O.; Hashim, R.; Mohammad, A.; Hena, S. Thin-layer chromatography of amino acids: A review. *J. Liq. Chromatogr. Relat. Technol.* **2012**, *35*, 1497. [CrossRef]

21. D'Atri, V.; Fekete, S.; Clarke, A.; Veuthey, J.L.; Guillarme, D. Recent Advances in Chromatography for Pharmaceutical Analysis. *Anal. Chem.* **2019**, *91*, 210–239. [CrossRef]

22. Corleto, K.A.; Singh, J.; Jayaprakasha, G.K.; Patil, B.S. A sensitive HPLC-FLD method combined with multivariate analysis for the determination of amino acids in L-citrulline rich vegetables. *J. Food Drug Anal.* **2019**, *27*, 717–728. [CrossRef]

23. Danielsen, M.; Nebel, C.; Dalsgaard, T.K. Simultaneous determination of L- And D-amino acids in proteins: A sensitive method using hydrolysis in deuterated acid and liquid chromatography–tandem mass spectrometry analysis. *Foods* **2020**, *9*, 309. [CrossRef]

24. Zhou, W.; Wang, Y.; Yang, F.; Dong, Q.; Wang, H.; Hu, N. Rapid determination of amino acids of nitraria tangutorum bobr. From the qinghai-tibet plateau using HPLC-FLD-MS/MS and a highly selective and sensitive pre-column derivatization method. *Molecules* **2019**, *24*, 1665. [CrossRef] [PubMed]

25. Rutherfurd, S.M.; Gilani, G.S. Amino Acid Analysis. *Curr. Protoc. Protein Sci.* **2009**. [CrossRef] [PubMed]

26. Dong, M.W. *Modern HPLC for Practicing Scientists*; John Wiley & Sons, Inc: Hoboken, NJ, USA, 2006; pp. 1–286.

27. Christofi, M.; Markopoulou, C.K.; Tzanavaras, P.D.; Zacharis, C.K. UHPLC-fluorescence method for the determination of trace levels of hydrazine in allopurinol and its formulations: Validation using total-error concept. *J. Pharm. Biomed. Anal.* **2020**, *187*, 113354. [CrossRef] [PubMed]

28. AlRabiah, H.; Attia, S.M.; Al-Shakliah, N.S.; Mostafa, G.A.E. Development and validation of an HPLC-UV detection Assay for the determination of clonidine in mouse plasma and its application to a pharmacokinetic study. *Molecules* **2020**, *25*, 4109. [CrossRef]

29. ICH. *Validation of Analytical Procedures: Text and Methodology (Q2 R1)*; European Medicines Agency: Amsterdam, The Netherlands, 1995; CPMP/ICH/381/95.

30. National Research Council. Amino Acids. In *Recommended Dietary Allowances*, 10th ed.; The National Academies Press: Washington, DC, USA, 1989; Chapter 6. [CrossRef]

31. Reeds, P.J.; Garlick, P.J. Protein and amino acid requirements and the composition of complementary foods. *J. Nutr.* **2003**, *133*, 2953–2961. [CrossRef]

32. U.S. Department of Health and Human Services Food and Drug Administration. *Analytical Procedures and Methods Validation for Drugs and Biologics Guidance for Industry Analytical Procedures and Methods Validation for Drugs and Biologics Guidance for Industry*; FDA: Silver Spring, MD, USA, 2015.

33. Shabir, G.A. Validation of high-performance liquid chromatography methods for pharmaceutical analysis. *J. Chromatogr. A* **2003**, *987*, 57–66. [CrossRef]
34. Miller, J.N.; Miller, J.C. *Statistics and Chemometrics for Analytical Chemistry*, 6th ed.; Pearson Education: London, UK, 2005.

 *molecules*

Article

Measuring Bismuth Oxide Particle Size and Morphology in Film-Coated Tablets

Stefani Fertaki [1,2], Georgios Bagourakis [3], Malvina Orkoula [1] and Christos Kontoyannis [1,2,*]

[1] Department of Pharmacy, University of Patras, GR-26504 Rio Achaias, Greece; sfertaki@iceht.forth.gr (S.F.); malbie@upatras.gr (M.O.)
[2] Institute of Chemical Engineering Sciences, Foundation of Research and Technology-Hellas (ICE-HT/FORTH), GR-26504 Platani Achaias, Greece
[3] ELPEN S.A. Pharmaceutical Industry, 5 Marathonos Ave., GR-19009 Pikermi Athens, Greece; gbagourakis@elpen.gr
* Correspondence: kontoyan@upatras.gr; Tel.: +30-2610-962328

Abstract: The assessment of active pharmaceutical ingredient (API) particle size and morphology is of great importance for the pharmaceutical industry since it is expected to significantly affect physicochemical properties. However, very few methods are published for the determination of API morphology and particle size of film-coated (FC) tablets. In the current study we provide a methodology for the measurement of API particle size and morphology which could be applied in several final products. Bismuth Oxide 120 mg FC Tabs were used for our method development, which contain bismuth oxide (as tripotassium dicitratobismuthate (bismuth subcitrate)) as the active substance. The sample preparation consists of partial excipient dissolution in different solvents. Following this procedure, the API particles were successfully extracted from the granules. Particle size and morphology identification in Bismuth Oxide 120 mg FC Tabs was conducted using micro-Raman mapping spectroscopy and ImageJ software. The proposed methodology was repeated for the raw API material and against a reference listed drug (RLD) for comparative purposes. The API particle size was found to have decreased compared to the raw API, while the API morphology was also affected from the formulation manufacturing process. Comparison with the RLD product also revealed differences, mainly in the API particle size and secondarily in the crystal morphology.

Keywords: bismuth oxide; API particle size; API morphology; film-coated tablets; Raman spectroscopy; ImageJ; tablet disintegration

Citation: Fertaki, S.; Bagourakis, G.; Orkoula, M.; Kontoyannis, C. Measuring Bismuth Oxide Particle Size and Morphology in Film-Coated Tablets. *Molecules* **2022**, *27*, 2602. https://doi.org/10.3390/molecules27082602

Academic Editors: Victoria Samanidou and Irene Panderi

Received: 14 March 2022
Accepted: 17 April 2022
Published: 18 April 2022

Publisher's Note: MDPI stays neutral with regard to jurisdictional claims in published maps and institutional affiliations.

1. Introduction

Particle size of pharmaceutical powders can have a significant effect on the safety and efficacy of the drug product. The particle size distribution (PSD) and morphology of APIs have a profound impact on the drug product performance including dissolution, bioavailability, and formulation stability [1–4]. Additionally, the PSD of the API powders may also influence the drug flowability, blend uniformity, compactibility, or other characteristics at almost every step of the manufacturing process, including mixing, granulation, and compression [4–6]. Therefore, the particle sizes and morphology of pharmaceutical powders should be evaluated at different pharmaceutical development phases and primarily in the final product.

A number of methods are available for determining particle size of which the most common include sieve analysis, laser diffraction, dynamic light scattering, and direct imaging techniques [7]. The most common method for particle size distribution is laser diffraction measurement [8]. However, such techniques are used for APIs' PSD prior to any manufacturing processes and tableting. Very few studies have been published regarding the measurement of API particle size and morphology in pharmaceutical final products. Michal Šimek et al. have used hot-stage microscopy in order to melt the excipients present in tablets containing tadalafil or meloxicam API. After successful disintegration of the tablets using 0.5 mL of water, the remaining powder was placed under the microscope and onto the hot-stage where it was heated until complete melt of the excipients. The API particles were then observed and measured using image analysis [9]. This procedure exhibits some limitations since the API melting point should be higher than the rest of the excipients. Usually though, the temperature of API melt is not very high (~200 °C) in contrast to the excipients. Atsushi Kuriyama and Yukihiro Ozaki managed to identify the PSD of Ebastine API in Ebastel tablets using micro-Raman chemical imaging and polystyrene microsphere size standards as reference. They have applied a specific binarization threshold value compared to the reference tablets containing microspheres in order to calculate the PSD of the API particles. However, this approach can work only if API particles are well dispersed into the formulations and the histogram indicates no API-specific distribution, and all pixels are a mixture of components with the API maximum intensity prior to formulation not being retained. Additionally, the morphology of the API crystals cannot be clearly observed [10].

A few additional publications used Raman imaging for the determination of API particle size distribution in pharmaceutical formulations [11–13]. However, this technique by itself could be more useful for the identification of the 'domain' particle size since the discrimination of frequently observed agglomerated particles cannot be easily performed.

In the present work, the bismuth oxide API particle size and morphology in Bismuth Oxide 120 mg FC Tabs were studied. This medicine is used for the treatment of peptic ulcer and gastroesophageal reflux disease and also shows antibacterial activity against Helicobacter pylori [14]. Several solvents were used to dissolve the main excipients and release the API from the granules. The sample was spread in a highly reflective sample carrier at 40 °C in order to dry while micro-Raman mapping was performed to identify all the remaining particles. The identified API particles were then measured with ImageJ software. The particles' morphology was also studied. Comparison with an RLD product was also performed. The effect of the manufacturing process and sample preparation in the API particle size and morphology was analyzed as reference.

2. Results

2.1. Sample Preparation for Bismuth Oxide 120 mg Film-Coated (FC) Tabs Disintegration and API De-Aggregation

The tablet core of Bismuth Oxide 120 mg FC Tabs consists of API, maize starch, povidone K29/32, polacrilin potassium, macrogol 6000, and magnesium stearate (Table 1). The API content is approximately 30% *w*/*w*.

Table 1. Ingredients and content of excipients in placebo used for Bismuth Oxide 120 mg FC Tabs as received from ELPEN SA.

Sample Information	Content (%)	Function	PSD (D[4,3] μm)
Maize starch	58.00	Diluent/disintegrant	-
Macrogol 6000	5.25	plasticizer/lubricant	-
Povidone K-29/39	14.00	Binder	-
Polacrilin potassium	19.25	disintegrant	66.6
Magnesium stearate	3.50	lubricant	10.1

The tablet preparation was performed using dry granulation. The API along with maize starch, macrogol 6000, and PVP were premixed and granulated using a roller compactor. The produced compacts were sized and mixed with extragranular PEG 6000, polacrilin potassium, and magnesium stearate in order to prepare the lubricated blend. Finally, the blend was compressed to tablets before the coating step.

In order to eliminate the excipients' presence, the film coating was initially removed with a scalpel. The main difficulty of the current task was the successful API extraction from the granules without damaging the API crystals. Literature revealed several solvents that could potentially dissolve some of the excipients, whereas the API remained insoluble. Using both literature and a trial-and-error process, it was observed that acetonitrile managed to dissolve povidone K30 and macrogol 6000. The images obtained from the suspended Bismuth Oxide 120 mg FC Tabs in acetonitrile revealed some remaining agglomerates and possible granule presence though (Figure 1B). Additionally, the dispersion was dominated with maize starch presence (spherical particles noted with arrow in Figure 1B).

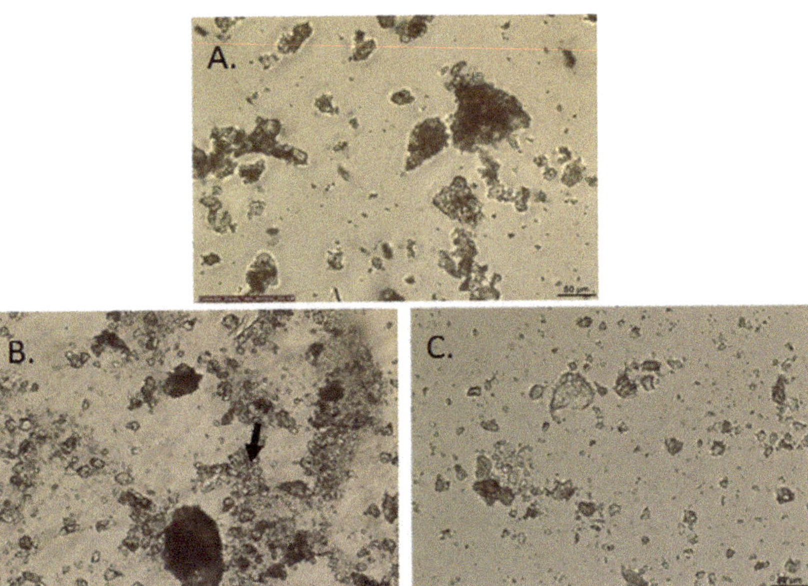

Figure 1. Optic microphotographs of Bismuth Oxide 120 mg FC Tabs as received after film removal (**A**), after treatment with acetonitrile (**B**), and after treatment with acetonitrile and subsequent treatment with heated DMSO (**C**) using 20× magnification objective lens.

Maize starch was therefore considered as the next excipient target since it is a granulation binder. Heated dimethylsulfoxide (DMSO) can dissolve maize starch and form a viscous colloidal solution. The latter was verified when pure maize starch was diluted in DMSO at 80 °C while the API was once again unaffected. As seen in Figure 1C, the agglomerates were almost absent. Polacrilin potassium and Mg stearate could not be dissolved with any of the known solvents without affecting the API.

Considering the previous information, the following procedure was performed as the most adequate in terms of sample preparation for tablet disintegration and consequent API extraction from the granules (also presented in Scheme 1 as flow chart).

After film removal, the tablet was cut into eight smaller pieces using a scalpel in order to avoid direct crushing and consequently damaging the API particles. Then, approximately 100 mg of the sample (four pieces) was dispersed in 6 mL acetonitrile in falcon tubes. The dispersion was vortexed for 5 min at 2500 rpm until complete powdering of the tablet pieces. The sample dispersion was then filtered and left to dry at 25 °C for 1 h. The remaining powder was then dispersed in heated DMSO (80 °C) for 2 h in a closed vial. The dispersion was then centrifuged in order to remove the DSMO solution. The remaining powder was re-dispersed in fresh DMSO in order to obtain a homogeneously dispersed mixture of the sample and to assist in obtaining clearer images. A few drops of the dispersion were added on a highly reflective gold glass slide which was placed on a hot-stage at 60 °C in order to dry the particles for the following Raman study, since the presence of liquid can affect the Raman signal and result in unsuccessful discrimination of the particles.

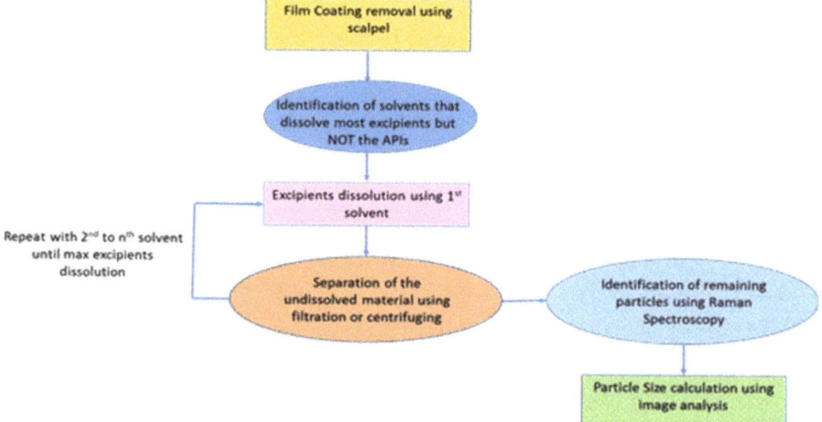

Scheme 1. Flow chart describing the methodology used for API particle size and morphology identification.

2.2. Particle Identification Using Micro-Raman Mapping Spectroscopy

The obtained particles from the previously stated procedure were further investigated using micro-Raman spectroscopy in order to be identified.

Raman spectra of pure samples of bismuth oxide API and the excipients are shown in Figure 2. Their spectral differences are distinct, therefore, Raman spectra can be used for discrimination among the remaining particles.

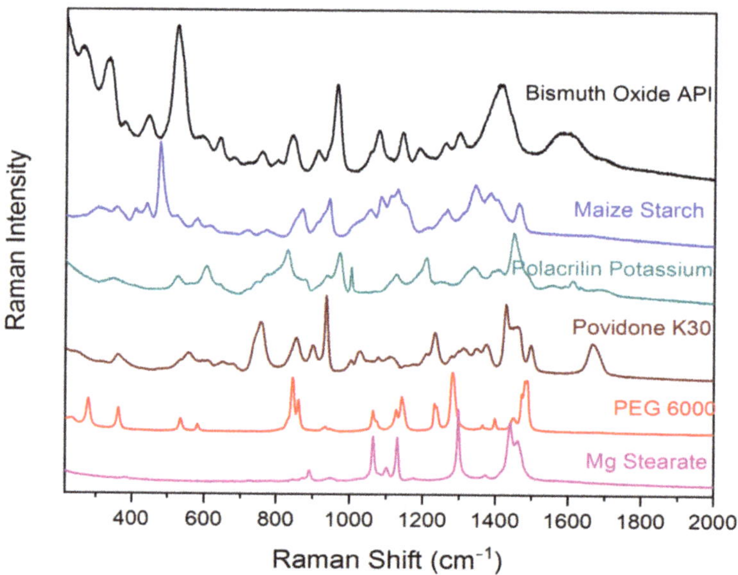

Figure 2. Raman spectra of bismuth oxide API, maize starch, polacrilin potassium, povidone K30, PEG 6000, and magnesium stearate.

In Figure 3, the spectra of the initial tablet before any treatment were compared with the respective spectra obtained from the tablets dispersed in acetonitrile (ACN) and after DMSO dispersion in order to verify the extinction of the dissolved excipients at each step. As expected, the Raman spectra of the initial tablets were combined by bismuth oxide API, maize starch (blue lines in Figure 3), polacrilin (cyan dotted lines in Figure 3), and PEG 6000. After dispersion in acetonitrile, the resulting Raman spectra were dominated by API, some maize starch, and some polacrilin. Finally, after DMSO dispersion, the Raman spectra of the tablets were dominated by API. In some parts, some polacrilin and some remaining maize starch peaks were noticed. It was thus assumed that, indeed, the respective sample preparation could result in partial excipient dissolution and API extraction from the granules.

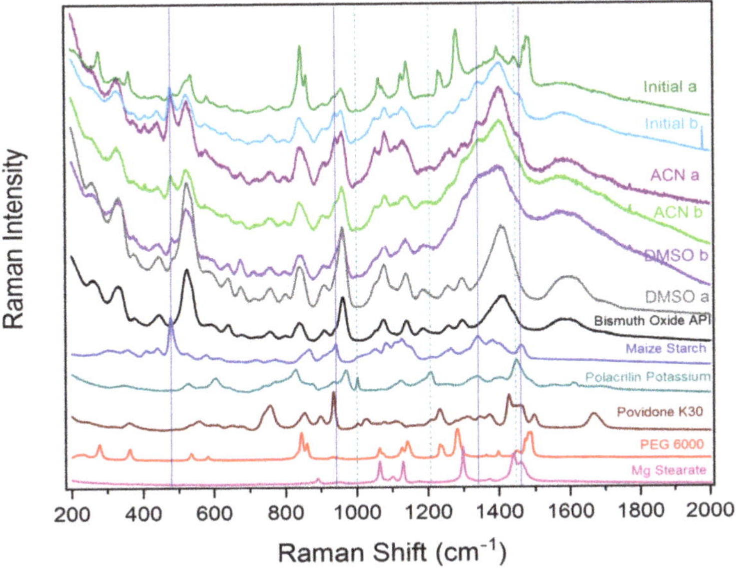

Figure 3. Raman spectra of bismuth oxide API, excipients, initial tablet, tablet dispersed in acetonitrile, and tablet further dispersed and centrifuged in heated DMSO.

A point-to-point mapping procedure was performed in order to scan a total number of at least 400 particles. The data from each pixel were analyzed using direct classical least squares (DCLS) method using API as the reference spectrum. Visualized score maps were thus generated with white and light blue pixels indicating high threshold values (higher than 80% score) corresponding to the API, and the blue pixels (lower than 80% score) indicating low values corresponding to other components (Figure 4). The Raman spectra from all pixels were compared against the API and the excipients (Figure 4). The analysis was performed in duplicate for statistical purposes.

From Figure 5, it is apparent that all white and light blue spots correspond to API particles whereas most of the other spots (Figure 4) are attributed to polacrilin potassium particles. Magnesium stearate was not observed, probably because it exhibits a small particle size while at the same time it is present at a very low content in the final products (Table 1). The API particles were predominant in all scanned areas. Particles of irregular shape were observed exhibiting a different morphology from the API particles initially added in the formulation (see Section 2.4).

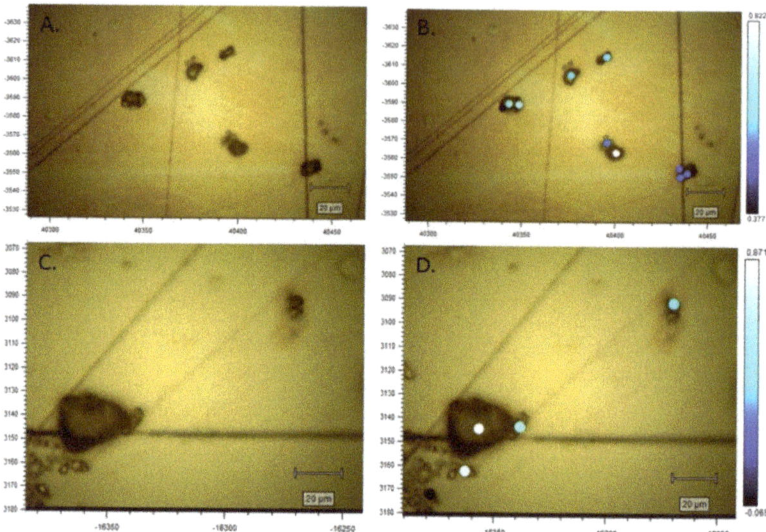

Figure 4. Images of the first and second random mapped area of Bismuth Oxide 120 mg FC Tabs after dispersion in original mode (**A,C**) and after processing with direct classical least squares (DCLS) method (**B,D**) using 50× magnification objective lens.

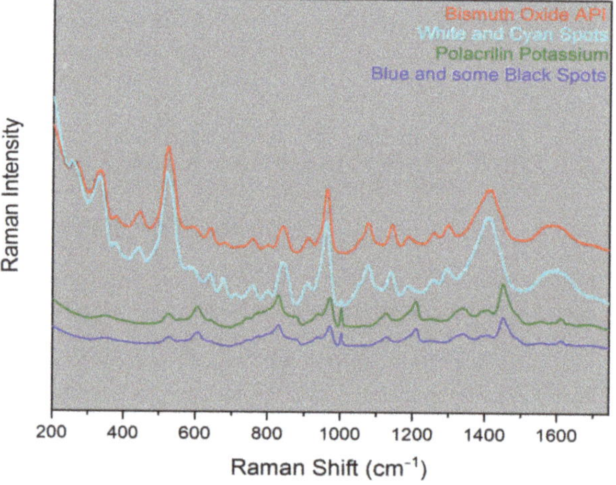

Figure 5. Typical Raman spectra of the particles marked with white and cyan spots (cyan line), blue spots (blue line), and black spots of Figure 4 (black line), colloidal bismuth tripotassium dicitrate (red line), polacrilin potassium (green line), and DMSO (grey line).

2.3. API Particle Size Determination Using ImageJ

The particle size of the already identified API particles with Raman spectroscopy was measured using ImageJ software. The area of each individual API particle was calculated manually using free-hand selection and drawing the particle's perimeter (Figure 6). The software automatically calculated the area of each particle and particle's diameter. The procedure was repeated using all areas previously scanned with Raman spectroscopy until at least 400 API particles were measured in duplicate. The D[0.1], D[0.5], and D[0.9] values of the sample were finally determined using the appropriate mathematic equations (Table 2).

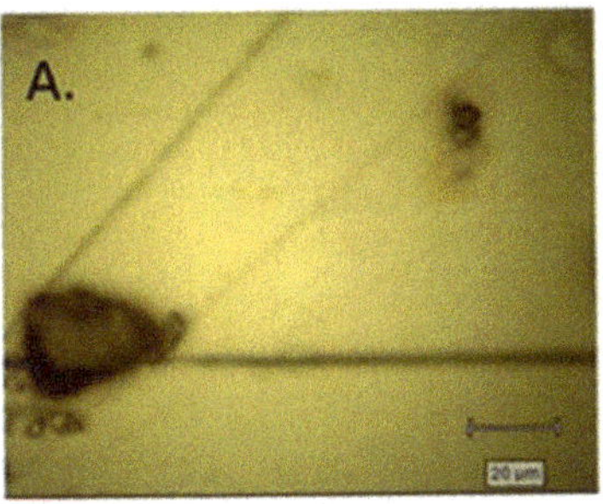
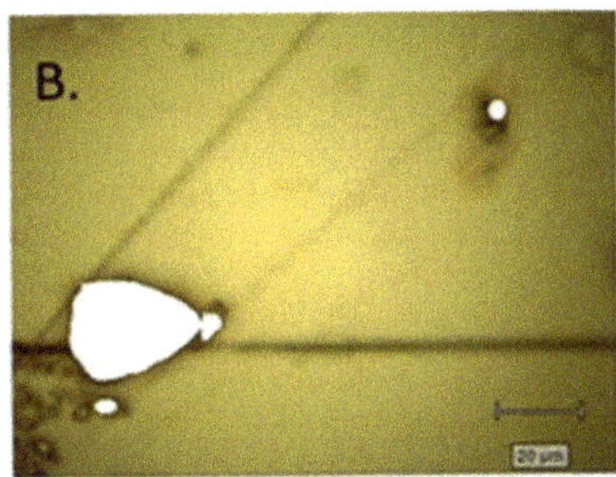

Figure 6. Images of one random mapped area of Bismuth Oxide 120 mg FC Tabs after dispersion in original mode (**A**) and after image analysis method (**B**) using 50× magnification objective lens.

Table 2. Average cumulative PSD values of bismuth oxide API particles in 120 mg FC Tabs after disintegration of the tablet.

D[0.1]	D[0.5]	D[0.9]
14.35 ± 0.07 μm	27.00 ± 1.41 μm	43.25 ± 0.35 μm

2.4. Effect of Sample Preparation in Bismuth Oxide API

The effect of the current manufacturing process in the morphology and particle size of the pure API was also studied. Initially, the morphology of the API was studied from the original powder and after the application of the sample preparation method described in Section 2.1 using optical microscopy and LAS image analysis (Figure 7) on the pure API particles. The particles appeared almost identical and mostly spherical in both cases indicating that the suggested sample preparation scheme does not affect the API's crystal morphology.

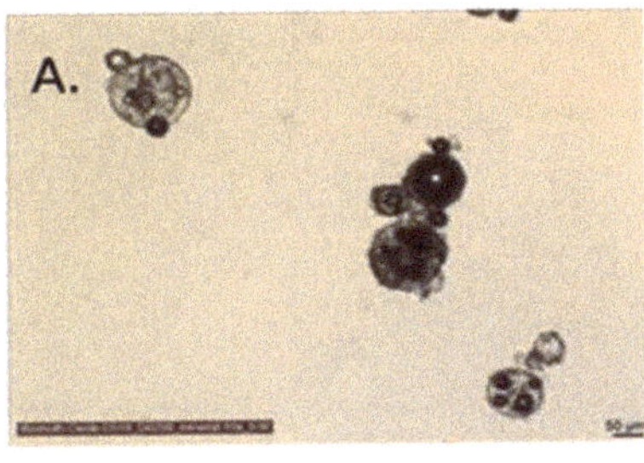
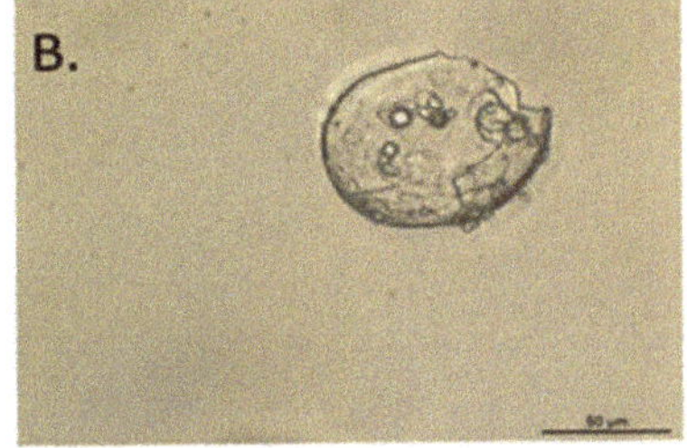
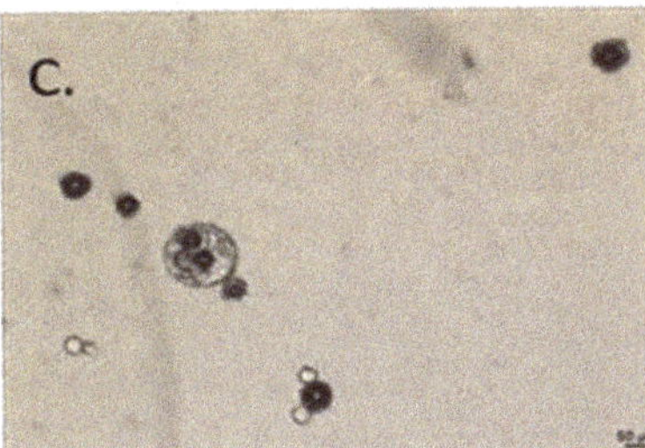
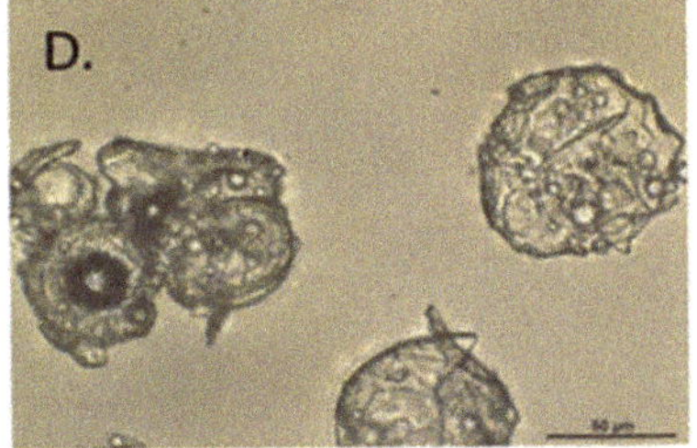

Figure 7. Optic microphotographs of original bismuth oxide API using 10× magnification objective lens (**A**) and 40× magnification objective lens (**B**), and after treatment with acetonitrile and subsequently with DMSO particles using 10× magnification objective lens (**C**) and 40× magnification objective lens (**D**).

In order to verify that the particle size of the API was not affected from the sample preparation process, the PSD of the API was measured before and after the application of the proposed method using laser diffraction technique. From Table 3 and according to ICH guidelines [15] it was concluded that the PSD of the API was not significantly affected by the respective sample preparation method.

Table 3. PSD values of bismuth oxide API before and after being processed according to sample treatment.

Sample Information	D[0.1]	D[0.5]	D[0.9]
Original bismuth oxide API	32.8 ± 5.1 μm	70.3 ± 7.5 μm	123.0 ± 13.5 μm
API particles treated with acetonitrile and subsequently with DMSO	30.6 ± 3.7 μm	66.3 ± 6.2 μm	121.0 ± 14.1 μm
RSD%	6.71	5.69	1.62

2.5. Effect of Final Formulation Manufacturing Process on Bismuth Oxide API Particle Size

A significant difference was noticed in the API particle size calculated in the final formulation compared to the initial as-received API. According to the study performed in Section 2.4, the proposed sample treatment did not affect either the morphology or the particle size of the API. It was therefore suspected that the compression applied during the manufacturing process of Bismuth Oxide 120 mg FC Tabs could damage the API particles and consequently yield decreased particle size values. For this purpose, the morphology and particle size of the API were analyzed before and after pressure application at 1 ton and 10 tons. From Figure 8 and Table 4 it was verified that bismuth oxide API is significantly sensitive under any amount pressure, justifying the decreased API particle size observed in the final products.

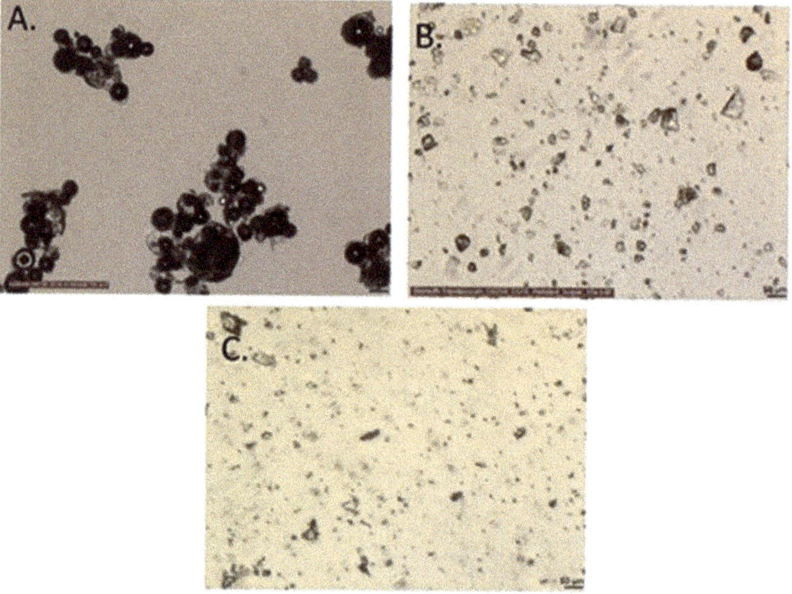

Figure 8. Optic microphotographs of original bismuth oxide API (**A**), pressed at 1 ton API (**B**) and pressed at 10 tons API (**C**) using 10× magnification objectives.

Table 4. PSD values (with laser diffraction) of bismuth oxide API before and after pressure at 10 tons.

Sample Information	D[0.1]	D[0.5]	D[0.9]
Original bismuth oxide API	35.8 ± 5.1 µm	71.3 ± 7.5 µm	125.0 ± 13.5 µm
Pressed bismuth oxide API	4.66 ± 1.8 µm	27.7 ± 2.5 µm	54.3 ± 4.8 µm

2.6. Comparison against an RLD Product

The proposed sample preparation method was repeated for Ulcamed ® 120 mg Fc Tabs (reference listed drug product) in order to identify any possible differences that could result in potential failure of the in vitro equivalence study. The API particle size of the RLD product was found increased in regards to D[0.1] and D[0.1] values according to ICH guidelines Q2R1 compared to the respective Bismuth Oxide 120 mg FC Tabs value (Table 5). Additionally, Ulcamed ® 120 mg tablet disintegration was performed faster than the respective test product. Both these findings indicate possible lower pressure application in the RLD product.

Table 5. Average PSD values of bismuth oxide API using micro-Raman spectroscopy in Ulcamed ® 120 mg FC Tabs against tested Bismuth Oxide 120 mg FC Tabs.

Sample Information	D[0.1]	D[0.5]	D[0.9]
Ulcamed ® 120 mg FC Tabs	14.30 ± 1.41 µm	33.15 ± 0.21 µm	54.85 ± 2.33 µm
Bismuth Oxide 120 mg FC Tabs	14.35 ± 0.07 µm	27.00 ± 1.41 µm	43.25 ± 0.35 µm
RSD%	0.25	14.46	16.72

3. Discussion

Identification of particle size and morphology of the components present in pharmaceutical formulations is of crucial importance since it is expected to significantly affect the API dissolution rate, bioequivalence, etc. The methodology and sample preparation developed in this study have managed to successfully calculate bismuth oxide API particle size while simultaneous identification of the crystal morphology of the API in prepared Bismuth Oxide 120 mg FC Tabs was also performed.

The experimental procedure was based on the increased API content by dissolving most of the formulation excipients. Attention should be paid to solvent selection because the API must be insoluble in the chosen solvent. Careful transformation from a tablet to powder was also performed, along with proper dispersion of the remaining component particles. The use of a highly reflective sample carrier (gold-plated) favored the API particles identification of the dried suspension using point-to-point mapping Raman spectroscopy. Particle size calculation was finally achieved using image analysis software. Performance of the same procedure on pure API particles revealed that the analysis is harmless for the API particles and therefore appropriate to be used for bismuth oxide pharmaceutical formulations.

Significant changes were noticed in the API particles of the final product compared with the original API particles in terms of size and morphology. Tablet compression seemed responsible for the decreased API particle size and the different crystal shape.

Difference in the API particle size of an RLD product was also confirmed.

The respective method development and sample preparation could be used in the pharmaceutical industry as a routine method for API particle size and morphology determination in film-coated tablets and for other formulations.

4. Materials and Methods

4.1. Reagents

The commercial film-coated tablets (Ulcamed® 120 mg Filmtabletten), bismuth tripotassium dicitrate, the corresponding placebo, and all the excipients were kindly provided by

ELPEN SA (Pikermi, Attica, Greece). The excipients of Ulcamed® 120 mg were similar to our prepared tablet [14].

4.2. Preparation of Bismuth Oxide 120 mg Film-Coated Tablets

The process used for the production of Bismuth Oxide 120 mg FC Tabs by ELPEN SA was conventional and adopted standard manufacturing techniques and equipment. Bin mixer (150 L) and roller compactor (Alexanderwerk) were used for the granulation step. Granules compressed using a KILIAN E150 rotary tablet press machine with 10 mm concave punches resulted in tablets with resistance to crushing of approximately 120 N (testing equipment: hardness tester ERWEKA TBH220TD), which were finally coated using Chamunda Pealcoat Ci 900 coater system with a perforated pan.

4.3. Dispersions

For the disintegration of a tablet with 120 mg API strength, a dispersion in acetonitrile (HPLC grade) was prepared by dispersing 100 mg of the tablet in 6 mL of acetonitrile. The sample was then vortexed at 2500 rpm using an IKA MS2 Minishaker. Filtration through 0.22 μm GSWP nitrocellulose membrane filters (Merck Millipore Ltd., Cork, Ireland) and a vacuum pump (KNF Neuberger Inc. Laboport, Trenton, NJ, USA) followed. A second dispersion of the remaining undissolved solid in DMSO (80 °C) was prepared by dispersing the undissolved solid in 5 mL DMSO followed by vigorous swirling. The sample was placed in an ultrasonic cleaner (Elmasonic P, P30H, Singen (Hohentwiel), Germany) heated at 80 °C for 2 h.

4.4. Raman Spectroscopy

A Raman spectrometer (InVia Reflex Raman spectrometers, Renishaw, UK) equipped with an optical microscope (Research Grade, Leica DMLM microscope) and a laser with a 785 nm excitation line was used. The laser line was focused through a $50\times$ objective lens on the sample. The system was equipped with a CCD detector (Peltier cooled, near-infrared enhanced). The power of the incident laser was 250 mW. The typical spectral resolution was $2\,\text{cm}^{-1}$. Windows-based software was used (WiRE© 4.2) to obtain the spectra. Instrument response (laser power and the wavenumber) was checked by recording the spectrum of Si.

For Raman spectra collection, the samples were placed on a highly reflective sample carrier (EMF Corporation, Angola, IN, USA). Quantities of a few milligrams for the solids were used. The recorded spectra were the sum of 3 scans of 10 s exposure time and the acquired region was 250–$2000\ \text{cm}^{-1}$.

For Raman spectra collection of the solutions, 3 drops of the sample were placed on a highly reflective sample carrier (EMF Corporation, Angola, IN, USA) and dried for 3 h in a Leica TPX-Type D hot-stage. A $350 \times 500\,\mu\text{m}$ area was captured each time and each particles' spectrum was acquired using a point-to-point mapping until at least 600 particles were measured. The recorded spectra were the sum of 2 scans of 5 s exposure time and the acquired region was 1240–$1700\ \text{cm}^{-1}$.

4.5. Particle Size Distribution (PSD)

Laser analyzer Mastersizer 3000 (Malvern Instruments, Malvern, UK) with Hydro SV adapter was used to determine the PSD of the samples. The measurement range of the apparatus is 0.02–200 μm. Red light of 633 nm wavelength was the source. The stirrer speed was set at 1800 rpm, obscuration range was set at 10–30%, and the particle size was calculated on a volume basis using the Mie theory. Isopar G was used as the dispersant.

4.6. Optical Microscopy

An optical microscope (Leica DM 2500M, Leica Microsystems LtD., Heerbrugg, Switzerland) equipped with a video camera (Leica DFC420 C, Leica Microsystems LtD., Heerbrugg, Switzerland) was used to obtain pictures of the API crystals in transmittance mode and brightfield illumination, using the $10\times$ Leica objective lenses; for solutions, the $20\times$ Leica objective lenses were used.

Approximately 3 mg of the powder samples was dispersed in 3 mL of mineral oil (ACRŌS ORGANICS, Geel, Belgium) and the dispersions were analyzed as prepared. One or two drops of the aforementioned samples were then placed on a 76 mm $\times$ 26 mm $\times$ 1 mm microscope slide (Paul Marienfeld GmbH and Co. KG, Lauda-Königshofen, Germany), covered with a 22 mm $\times$ 22 mm cover slip (Paul Marienfeld GmbH and Co. KG, Lauda-Königshofen, Germany). Windows-based software (LAS© V4.11) was used for image acquisition and analysis.

4.7. ImageJ

ImageJ software was used for the calculation of API particle size. The Raman obtained images were filtered using enhance local contrast. The area of each particle was automatically calculated after manually drawing each particle perimeter using free-hand selection. The diameter of each particle was acquired from the each measured particle area.

4.8. Centrifuge

The samples were placed in a 10 mL falcon tube and were centrifuged at 8000 rpm and 25 °C for 23 min using a refrigerated centrifuge (Heraeus Biofuge Stratos, Kendro, Osterode, Germany). The supernatant, which was approximately 4 mL, was removed and the precipitate was collected.

Author Contributions: Conceptualization, S.F., M.O. and C.K.; methodology, S.F., G.B. and C.K.; validation, S.F.; investigation, S.F. and C.K.; data curation, S.F. and G.B.; writing—original draft preparation, S.F.; writing—review and editing, M.O. and C.K.; supervision, C.K.; project administration, C.K. All authors have read and agreed to the published version of the manuscript.

Funding: No external funding was received.

Institutional Review Board Statement: Not applicable.

Informed Consent Statement: Not applicable.

Data Availability Statement: Not applicable.

Conflicts of Interest: The authors declare no conflict of interest.

References

1. Iacocca, R.G.; Burcham, C.L.; Hilden, L.R. Particle engineering: A strategy for establishing drug substance physical property specifications during small molecule development. *J. Pharm. Sci.* **2010**, *99*, 51–75. [CrossRef] [PubMed]
2. Mosharraf, M.; Nyström, C. The effect of particle size and shape on the surface specific dissolution rate of micronized practically insoluble drugs. *Int. J. Pharm.* **1995**, *122*, 35–47. [CrossRef]
3. Always, B.; Sangchantra, R.; Stewart, P.J. Modelling the dissolution of diazepam in lactose interactive mixtures. *Int. J. Pharm.* **1996**, *130*, 213–224. [CrossRef]
4. Shekunov, B.Y.; Chattopadhyay, P.; Tong, H.H.Y.; Chow, A.H. Particle size analysis in pharmaceutics: Principles, methods and applications. *Pharm. Res.* **2007**, *24*, 203–227. [CrossRef] [PubMed]
5. Fayed, M.E.; Otten, L. *Handbook of Powder Science & Technology*, 2nd ed.; Chapman & Hall: London, UK, 1997.
6. Hlinak, A.J.; Kuriyan, K.; Morris, K.R.; Reklaitis, G.W.; Basu, P.K. Understanding critical material properties for solid dosage form design. *J. Pharm. Innov.* **2006**, *1*, 12–17. [CrossRef]
7. Eitel, K.; Bryant, G.; Schöpe, H.J. A Hitchhiker's Guide to Particle Sizing Techniques. *Langmuir* **2020**, *36*, 10307–10320. [CrossRef] [PubMed]
8. Hildegard, B. Particle Characterisation in Excipients, Drug Products and Drug Substances. *Life Sci. Tech. Bull.* **2008**, *1*, 1–6.
9. Simek, M.; Grünwaldová, V.; Kratochvíl, B. Hot-Stage Microscopy for Determination of API Particles in a Formulated Tablet. *BioMed Res. Int.* **2014**, *2014*, 832452. [CrossRef] [PubMed]
10. Kuriyama, A.; Ozaki, Y. Assessment of Active Pharmaceutical Ingredient Particle Size in Tablets by Raman Chemical Imaging Validated using Polystyrene Microsphere Size Standards. *AAPS PharmSciTech* **2014**, *15*, 375–387. [CrossRef] [PubMed]
11. Henson, M.J.; Zhang, L. Drug Characterization in Low Dosage Pharmaceutical Tablets Using Raman Microscopic Mapping. *Appl. Spectrosc.* **2006**, *60*, 1247–1255. [CrossRef] [PubMed]

12. Frosch, T.; Wyrwich, E.; Yan, D.; Popp, J.; Frosch, T. Fiber-Array-Based Raman Hyperspectral Imaging for Simultaneous, Chemically-Selective Monitoring of Particle Size and Shape of Active Ingredients in Analgesic Tablets. *Molecules* **2019**, *24*, 4381. [CrossRef] [PubMed]
13. Farkas, A.; Nagy, B.; Marosi, G. Quantitative Evaluation of Drug Distribution in Tablets of Various Structures via Raman Mapping. *Period. Polytech. Chem. Eng.* **2018**, *62*, 1–7. [CrossRef]
14. European Medicines Agency. *Ulcamed 120 mg Filmtabletten Bismuth Subcitrate*; Public Assessment Report, AT/H/0598/001/DC, CMDh/223/2005, February 2014; European Medicines Agency: Amsterdam, The Netherlands, 2014.
15. European Medicines Agency. *ICH Topic Q 2 (R1) Validation of Analytical Procedures: Text and Methodology*; June 1995 CPMP/ICH/381/95; European Medicines Agency: Amsterdam, The Netherlands, 1995.

 *molecules* 

Article

Comparative Study of Sample Carriers for the Identification of Volatile Compounds in Biological Fluids Using Raman Spectroscopy

Panagiota Papaspyridakou [1], Michail Lykouras [1], Christos Kontoyannis [1,2] and Malvina Orkoula [1,*]

1 Department of Pharmacy, University of Patras, GR-26504 Rio, Achaias, Greece; gioulipap@upatras.gr (P.P.); lykourasm@upnet.gr (M.L.); kontoyan@upatras.gr (C.K.)
2 Institute of Chemical Engineering Sciences, Foundation of Research and Technology-Hellas (ICE-HT/FORTH), GR-26504 Platani, Achaias, Greece
* Correspondence: malbie@upatras.gr; Tel.: +30-2610-962342

Abstract: Vibrational spectroscopic techniques and especially Raman spectroscopy are gaining ground in substituting the officially established chromatographic methods in the identification of ethanol and other volatile substances in body fluids, such as blood, urine, saliva, semen, and vaginal fluids. Although a couple of different carriers and substrates have been employed for the biochemical analysis of these samples, most of them are suffering from important weaknesses as far as the analysis of volatile compounds is concerned. For this reason, in this study three carriers are proposed, and the respective sample preparation methods are described for the determination of ethanol in human urine samples. More specifically, a droplet of the sample on a highly reflective carrier of gold layer, a commercially available cuvette with a mirror to enhance backscattered radiation sealed with a lid, and a home designed microscope slide with a cavity coated with gold layer and covered with transparent cling film have been evaluated. Among the three proposed carriers, the last one achieved a quick, simple, and inexpensive identification of ethanol, which was used as a case study for the volatile compound, in the biological samples. The limit of detection (LoD) was found to be 1.00 μL/mL, while at the same time evaporation of ethanol was prevented.

Keywords: raman spectroscopy; carriers; sample holders; gold layer; cuvette; ethanol; urine; volatile compounds; biological fluids

Citation: Papaspyridakou, P.; Lykouras, M.; Kontoyannis, C.; Orkoula, M. Comparative Study of Sample Carriers for the Identification of Volatile Compounds in Biological Fluids Using Raman Spectroscopy. *Molecules* **2022**, *27*, 3279. https://doi.org/10.3390/molecules27103279

Academic Editors: Victoria Samanidou and Irene Panderi

Received: 17 March 2022
Accepted: 18 May 2022
Published: 20 May 2022

Publisher's Note: MDPI stays neutral with regard to jurisdictional claims in published maps and institutional affiliations.

1. Introduction

It is commonly accepted that consumption of alcoholic beverages and drinks (wine, beer, etc.) helps human interaction, socialization, and euphoria. However, excessive use may have undesired or even dangerous consequences. Alcohol's effects on humans vary depending on the levels in the body. When the alcohol concentration in blood is up to 1.27 μL/mL, it simply helps people to relax, but, when it exceeds this level, symptoms like car driving impairment appear. Intoxication is observed at levels above 2.53 μL/mL, while alcohol greater than 5.07 μL/mL in blood may lead to death [1]. There are multiple studies for the identification of alcohol also in other body fluids, such as saliva and urine. However, the intoxication levels are different. The ratio of alcohol in urine to blood was found approximately equal to 1.3, but it may be also enhanced according to the time of alcohol consumption and the amount of urine in the bladder [2–5].

The identification of volatile compounds in human body fluids could be very challenging. Concerning alcohol, the employed method is important to be suitable for analyte detection at least at the highest acceptable concentration for driving (1.27 μL/mL in blood and 1.65 μL/mL in urine). Among the established methods for determining the concentration of ethanol in a body is the breath alcohol test, which measures the concentration of ethanol in one's breath using the breathalyzers [6]. The accuracy and precision of the breathalyzers, however, is controversial, as there are studies indicating uncertainty in the estimation of breath alcohol concentrations due to lack of sensitivity and specificity of the method [7,8]. For higher accuracy and precision, ethanol is determined in human blood through various types of gas chromatography (GC), including direct injection GC and headspace gas chromatography (HS-GC) [9]. The GC methods, also, can achieve very low limits of detection (LoDs) of alcohol in blood reaching as low as 10 μg/mL or 12.5 nL/mL [10] while using rather low quantity (100–500 μL) of the biological sample [10–12]. Although the GC methods are the most reliable in measuring the alcohol levels in human body fluids, they often suffer from sample preparation issues [13]. Moreover, GC is a destructive technique for the sample, while experienced personnel are required for the preparation of the samples and their analysis.

In order to overcome these issues, vibrational spectroscopies could be employed for determining alcohol levels in individuals' bodies. Raman spectroscopy and infrared (IR) spectroscopy offer a rapid, easy in use, environmentally friendly, non-destructive, and non-invasive solution in biochemical analysis. The basic principle of Raman spectroscopy is based on the inelastic scattering of radiation by the examined sample [14]. The sample is irradiated by a monochromatic light source. Most of the photons are scattered by the sample at the same wavelength as the incident photons, a process known as Rayleigh scattering. Nevertheless, there are a few photons which are scattered at a shifted wavelength from that of the incoming radiation, which is referred to as inelastic scattering [15]. Inelastic scattering or Raman scattering, which is named after C.V. Raman, who discovered this phenomenon in 1928 [16], can be further divided into Stokes and anti-Stokes scattering [15], depending on the initial and the final vibrational state of the molecules.

Recently, Raman spectroscopy and other vibrational spectroscopic techniques have been employed for the analysis of body fluids, such as blood, serum, plasma, urine, saliva, semen, or vaginal fluids [17]. A variety of applications of Raman spectroscopy are described in literature. Raman spectroscopy has gained ground in the characterization of bone malfunctions, different types of cancer and other diseases, counterfeit drugs in bottles, as well as in the quantification of pharmaceuticals and their polymorphs in solid [18,19] and liquid formulations [20]. Among the applications of Raman spectroscopy, the determination of volatile essential and vegetable oils is included [21], while the identification of ethanol and methanol in alcoholic beverages [22] is also noteworthy. Lately, Raman spectroscopy was used for the determination of alcohol levels in the blood of individuals who consumed high quantities of ethanolic drinks [23]. However, there are still obstacles in the qualitative and quantitative analysis of such volatile compounds in biological fluids through Raman spectroscopy that should be surmounted.

Although the required sample preparation for Raman spectroscopy is quick and easy and the technique is sensitive for many biochemical applications, Raman scattering can prove inefficient because of the used substrate or sample holder [24]. The application of the wrong substrate can result in lower intensities of the studied sample or in fluorescent effects. A variety of different substrates and holders, as well as different sample preparation methods, have been used for studying biological fluids and volatile compounds via Raman spectroscopy. The most popular approach in Raman analysis of body fluids is the placement of a drop on a glass microscope slide which has been covered with aluminum foil in order to reduce the fluorescence from glass [25–32]. Another substrate that has been commonly used is a glass circular slide, which is designed appropriately so that it could be used with the mapping stage at the same time [33,34]. CaF$_2$ has been also employed as substrate for the Raman analysis of variable biological fluids, such as serum and blood [35,36], while the application of silicon wafer surfaces [31] and silicon nanocrystal substrates has also been tested [37,38]. For forensic purposes, Raman spectroscopy for the analysis of bloodstains or other biological fluids has been applied on more sophisticated substrates, such as fabric clothes, denim from blue jeans, un-dyed cotton swatches, glass, metallic surfaces, walls, and a pale-yellow bathroom tile [39,40]. In most of the cases, the body fluids' drops had been allowed to dry before their Raman spectra were acquired [25–30,33,34,40]. Concerning the Raman spectroscopic analysis of liquids and other fluid compounds, glass or quartz cuvettes [36,41–43] and well-sealed glass vessels [44] have been mentioned as sample holders in a couple of studies. The usage of such substrates could be demonstrated to be beneficial for the effective analysis of volatile compounds, such as ethanol, methanol, acetone, or hydrogen peroxide [42]. In addition, Raman spectra of liquid drug samples have also been acquired directly from their vials, ampoules, or bottles using a fiber optic probe [43,45].

Despite the benefits of the numerous different substrates for Raman spectroscopy that have been tested for biological and chemical applications, they are characterized by some weaknesses when volatile compounds in biological fluids are analyzed. The risks of volatility of these compounds, as well as the background noise due to the substrate, should be taken into consideration when the most appropriate substrate or sample holder carrier for Raman analysis of volatile compounds is chosen. Moreover, the required volume of the sample should be considered in the selection of the sample preparation method so that it will be at least comparable to the volume required in GC methods. At the same time, the developed method should certify a LoD below the highest acceptable levels by national laws.

In the present study, different substrates and sample holders have been tested, and the respective sample preparation methods have been developed for the identification of volatile compounds in body fluids. More specifically, the effectiveness of a glass microscope slide coated with gold layer creating a highly reflective carrier, a commercially available quartz cell for Raman analysis with a mirror on the back side, and a home-designed glass microscope slide with a cavity coated with gold layer, in the determination of ethanol in human urine, have been examined. Furthermore, the advantages and disadvantages of each carrier and sample holder have been studied.

2. Results

2.1. Identification of the Characteristic Peaks of Ethanol and Urine

The Raman spectra of ethanol, urine, and a sample of 5 μL/mL ethanol in human urine were recorded. The spectra of the biological fluids were acquired from a droplet of each sample which was placed on a highly reflective slide (glass microscope slide coated with gold substrate). However, the Raman spectrum of the volatile alcohol could not be recorded from a droplet, as it evaporated before the spectrum acquisition was completed. For this reason, the spectrum of ethanol was obtained using a commercially available cuvette for Raman spectroscopy.

The most prominent peak of urine was observed at 1003 cm^{-1} (Figure 1a), which is attributed to the symmetric C–N stretching of urea [46]. Although for solid urea the symmetric C–N stretching vibration is detected at 1010 cm^{-1}, in solution it is shifted to lower wavenumbers [47]. Concerning ethanol, the most intense peak at 883 cm^{-1} is due to the symmetric stretching of the C–C bond. The Raman spectrum of ethanol is also characterized by less intense peaks at 434 cm^{-1} (bending vibration of C–C–O), 1053 cm^{-1} (asymmetric stretching of the C–O bond), 1096 cm^{-1} (rock vibrations of CH$_3$), 1277 cm^{-1} (torsion and rotational vibrations of CH$_2$), 1455 cm^{-1} (bending vibrations of CH$_3$ and CH$_2$), and 1482 cm^{-1} (bending vibrations of CH$_3$) (Figure 1a) [48].

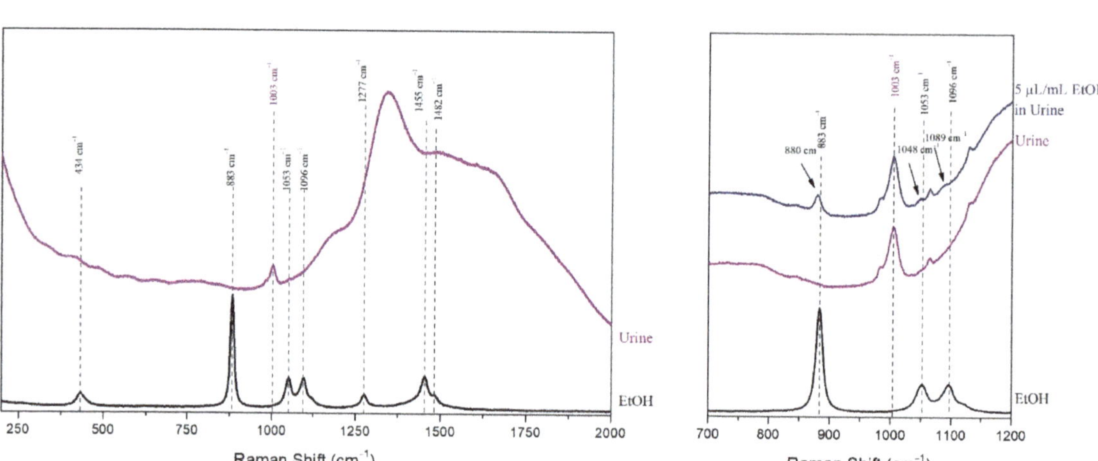

Figure 1. The Raman spectra of ethanol, human urine, and the 5 μL/mL ethanolic sample in human urine were acquired. (**a**) The characteristic peaks of ethanol and human urine are noted with dashed lines; (**b**) the characteristic peaks of ethanol were shifted in the sample of 5 μL/mL ethanol in human urine, and the shifted peaks are noted with arrows.

In the Raman spectrum of the 5 μL/mL ethanolic sample in human urine, the peak of urea at 1003 cm^{-1} was clearly detected, as well as the most prominent peak of ethanol. However, the peak of ethanol, which is attributed to the symmetric stretching of the C–C bond, was shifted to 880 cm^{-1} from the initial 883 cm^{-1}. Similarly, the peaks of ethanol at 1053 cm^{-1} and 1096 cm^{-1} were shifted to 1048 cm^{-1} and 1089 cm^{-1}, respectively (Figure 1b). These slight shifts to lower wavenumbers can be explained by the intermolecular interactions of ethanol and water. The molecules of water form a three-dimensional grid of hydrogen bonds among the hydroxylic groups. However, the addition of ethanol in concentrations up to 15–20% w/w leads to a structural rearrangement of the hydrogen bond network, as one molecule of ethanol is inserted for every five molecules of water. In this case, the hydrogen bonds are stronger than in water, resulting in weakening of the neighboring C–C and C–H bonds [49]. This is due to charge transfer leading to significant changes in the polarizability of the bonds. This change in polarizability is depicted in the Raman spectrum as a red shift, i.e., the peaks corresponding to the vibrations of the specific bonds are shifted to lower wavenumbers [48,49].

2.2. Method of Droplet on a Gold-Coated Glass Microscope Slide

The aim of this study was to develop a simple method requiring a minimum amount of sample (a few μL) and minimum sample preparation and offering low LoD of the analyte in the biological samples, while evaporation of the volatile compounds would be avoided. For this purpose, the determination of 100 μL/mL ethanol in human urine was used. The first method considered was that of placing a sample droplet on a highly reflective carrier (droplet method). The volume required was approximately 15 μL, and a gold-coated glass microscope slide was selected as a highly reflective sample carrier.

2.2.1. Focus Optimization

The Raman spectra acquired from the droplet showed a dependency on the spot of the droplet on which the Raman laser beam was focused. Thus, three different spots of the droplet were investigated; the first (A) was on the side; the second (B) was in the middle and the third one (C) on the top of the droplet (Figure 2a). The Raman spectra acquired from the three different positions exhibited a deviation of the relative intensities of the two analytes (Figure 2b). This can be attributed to the absence of uniformity concerning the concentrations of the analytes in the droplet.

a.

b.

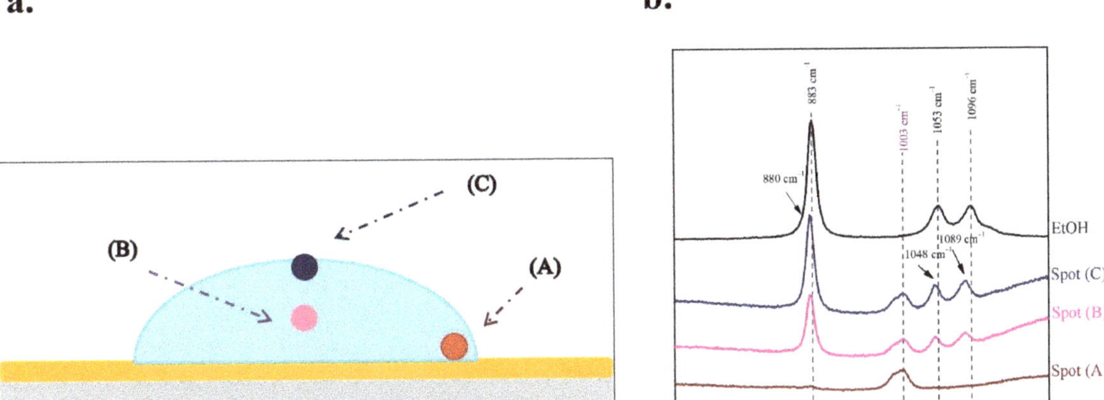

Figure 2. The method of sample droplet on a glass microscope slide coated with a gold highly reflective substrate was applied for the determination of 100 µL/mL ethanol in human urine. (**a**) Schematic illustration of the focus on three different spots of the sample droplet; (**b**) the respective Raman spectra from the three different spots of the droplet were recorded.

The characteristic peak of urine at 1003 cm^{-1} was easily detectable in all three Raman spectra. However, significant variations of the intensity of the most prominent peak of ethanol at 880 cm^{-1} were observed depending on the spot of the droplet on which the Raman laser beam was focused. More specifically, the peak of ethanol was barely distinguished from noise level in the Raman spectrum of the droplet's spot (A), while the intensity of the ethanol peak is the maximum possible in the Raman spectrum of the spot (C) (Figure 2b). Thus, the spot (C) of the droplet is the optimum position to focus when the Raman spectrum is acquired, as the ratio of the peak intensities (ethanol to urine) is maximum. This is the outcome of the non-uniformity of ethanol concentration inside the droplet and can be attributed to mass and thermal convection phenomena [50]. Higher evaporation flux near the contact line leads to minimal ethanol concentration at the edge of the droplet (spot A). On the contrary, many ethanol molecules are trapped in the center near the top of the droplet where concentration is maximized [50].

2.2.2. LoD Determination

For the LoD determination of the volatile compound in the droplet method, samples of human urine spiked with 0.25 µL/mL, 0.50 µL/mL, 2.00 µL/mL, 3.50 µL/mL, and 5.00 µL/mL ethanol were prepared. The Raman spectra of these samples were recorded after placing a droplet of each sample on a highly reflective gold-coated glass slide. The LoD of ethanol in human urine was determined, subsequently, by visual evaluation and signal-to-noise ratio as well [51]. The ethanolic peak at 880 cm^{-1} could not be detected in the 0.25 µL/mL sample Raman spectrum, but it was barely detected in the spectrum of the 0.50 µL/mL sample and easily detected in the spectrum of the 2.00 µL/mL sample (Figure 3). The signal-to-noise ratio was indeed higher than 3:1 [51] in the 0.50 µL/mL urine sample spiked with ethanol, while its value was below the minimum acceptable value (3:1) for the determination of the LoD (Table 1). Therefore, the LoD of ethanol in human urine when using the droplet method was found to be approximately 0.50 μL/mL.

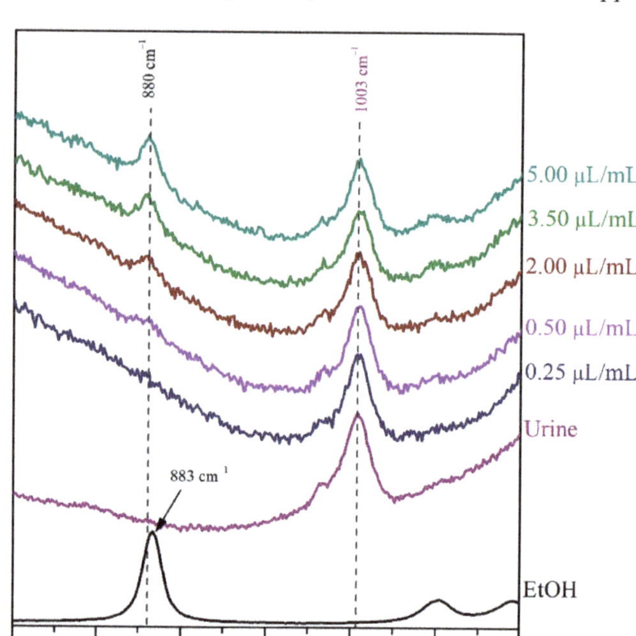

Figure 3. The LoD of ethanol in human urine using the droplet method was determined by visual evaluation.

Table 1. The LoD of ethanol in human urine for the droplet method was determined based on the signal-to-noise ratio.

Sample	Signal-to-Noise Ratio
Urine	1.00
0.25 μL/mL	2.40
0.50 μL/mL	3.62
2.00 μL/mL	4.99
3.50 μL/mL	8.49
5.00 μL/mL	11.52

2.2.3. Kinetic Study of Ethanol Evaporation

The kinetics of ethanol evaporation from the droplet was investigated in a sample of 5 μL/mL ethanol in human urine. The Raman laser was focused on the top of the droplet (spot (C)), and sequential Raman spectra of the droplet were recorded. The time of each measurement was 57 s. The intensity of the ethanol peak was reduced with time. In the last Raman spectrum recorded after 10 min and 49 s since the placement of the droplet on the carrier, the intensity of the ethanolic peak was half of the initial one (Supplementary Materials, Figure S1). The reduction rate of the ethanolic peak at 880 cm⁻¹ with respect to the peak of urine at 1003 cm⁻¹ was calculated for each Raman spectrum (0 min and 0 s, 0 min and 57 s, 1 min and 50 s, and 10 min and 49 s) (Supplementary Materials, Table S1). These results suggest that the evaporation of ethanol in the method of the droplet cannot be ignored despite the reduction in the scan time for the acquisition of the Raman spectra. Therefore, the rapid evaporation of the volatile compound appears to be the major drawback of the method, which outweighs the advantage of the small sample volume required and will result in great quantitative errors.

2.3. Method of Cuvette

As ethanol volatility proved to be a major obstacle in analysis in the previous section, another method which would prevent evaporation was sought. For this purpose, a commercially available cuvette for Raman spectra acquisition was employed. This cuvette was quite similar to the UV-Vis cells. It was a 5 mm thick synthetic quartz cuvette with a mirror on its back side so that the scattered radiation would be reflected for signal enhancement. The cuvette was fully filled with the appropriate volume of the sample (1.75 mL), plugged with a lid, and shielded with Parafilm M® so as to exclude the air above the sample and inhibit evaporation.

2.3.1. Focus Optimization

The optimal focus of the Raman laser beam on the cuvette was investigated. Five different positions were tested: the front side of the cuvette (A), various positions in the interior of the cuvette towards the back side (B)–(D), and the back side of the cuvette on the mirror (E) (Figure 4a). Due to the fact that the cuvette was not easily cleaned of the biological fluid, Xerostom® was used instead for the study. Thus, the Raman spectra of Xerostom® from the five different spots of the cuvette as well as its Raman spectrum, using the method of droplet, were recorded (Figure 4b). In the Raman spectra acquired from the front (A) and back (E) side of the cuvette, the most prominent peaks of the contents of the sample were absent. Although the peaks of Xerostom® were detected in the Raman spectrum from the spot (B), their intensities were rather low. The Raman spectra from the positions (C) and (D) resulted in higher intensities of the characteristic peaks of the sample due to the longer distance of the radiation in the sample. The intensity ratio of the peaks in the Raman spectra of spots (B), (C), and (D) are practically identical to each other (Figure 4). However, the spots with the highest peak intensities are preferred, as lower limits of detection (LoDs) are expected in these cases. Therefore, focusing on the interior of the commercially available cuvette near its back side (spot (D)) is suggested for the Raman spectra acquisition when the method of cuvette is employed.

2.3.2. LoD Determination

The ethanolic spiked samples of human urine (0.25 μL/mL, 0.50 μL/mL, 2.00 μL/mL, 3.50 μL/mL, and 5.00 μL/mL) were also analyzed using the method of cuvette so that the LoD of this method would be determined. After visual evaluation [40] of the respective Raman spectra, the LoD of ethanol in urine was determined to be equal to 2.00 μL/mL, as ethanol could not be detected in lower concentrations (Figure 5). Except for the visual evaluation, the LoD was also determined based on the signal-to-noise ratio method [51]. This method resulted in the same LoD as the one determined by the visual evaluation (Table 2). Thus, the method of cuvette results in a rather high LoD of ethanol in the specific body fluid.

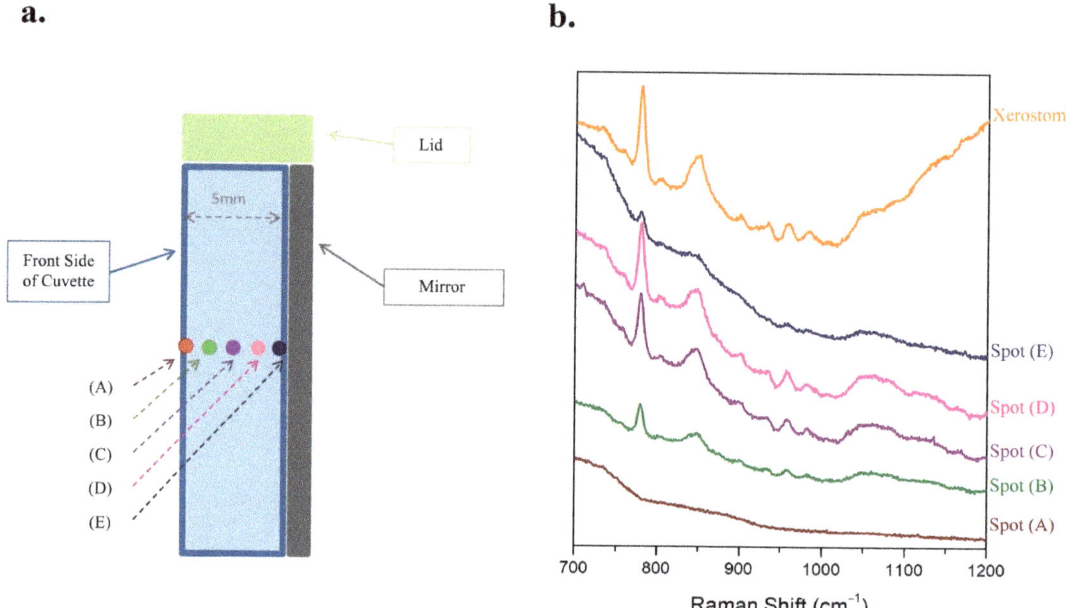

Figure 4. The method of the commercially available cuvette was applied on a liquid sample of Xerostom®. (**a**) Schematic illustration of the focus on five different spots of the cuvette; (**b**) the respective Raman spectra of Xerostom® from the five different focuses on the cuvette were recorded.

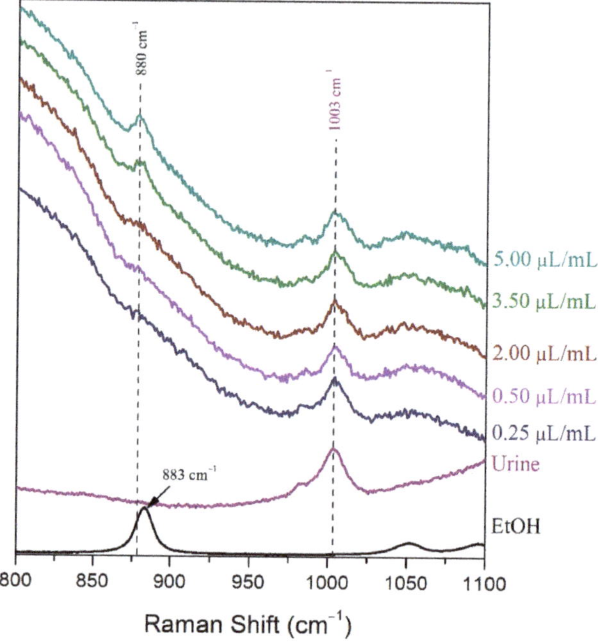

Figure 5. The LoD of ethanol in human urine using the method of cuvette was determined by visual evaluation.

Table 2. The LoD of ethanol in human urine for the method of cuvette was determined based on the signal-to-noise ratio.

Sample	Signal-to-Noise Ratio
Urine	1.00
0.25 μL/mL	2.70
0.50 μL/mL	2.90
2.00 μL/mL	6.72
3.50 μL/mL	10.95
5.00 μL/mL	14.20

2.3.3. Kinetic Study of Ethanol Evaporation

The rate of ethanol evaporation from the sample in the cuvette was studied. A sample of 5 μL/mL ethanol in human urine was used for this study as in the method of the droplet. The Raman laser was focused on the interior of the cuvette close to the mirror on the back side (spot (D)), and sequential Raman spectra of the sample were acquired from the moment the sample was inserted and shielded in the cuvette. Each scan's duration was adjusted to 5 min and 17 s. Only a slight decrease in the intensity of the peak of ethanol with time was observed (Supplementary Materials, Figure S2). The rate of change of the height of the peak of ethanol at 880 cm^{-1} with respect to the height of the peak of urine at 1003 cm^{-1} was determined for four different time intervals (0 min and 0 s, 5 min and 17 s, 15 min and 52 s, and 31 min and 42 s). After approximately half an hour, the reduction of ethanol's concentration was slightly higher than 12% (Supplementary Materials, Table S2). On the contrary, when the method of the droplet was applied, the concentration of ethanol was reduced to half after 10 min (Supplementary Materials, Table S1). Consequently, for the method of cuvette, the evaporation of the volatile constituent is obstructed to a large extent, but the significant quantity of the sample needed (1.75 mL) accompanied by the rather high LoD and the difficult cleaning of the biological fluid remain the most serious handicaps.

2.4. Method of a Gold-Coated Glass Slide with Cavity

2.4.1. Gold-Coated Glass Slide with Cavity

In order to overcome the issues of large amount of sample of the cuvette method and the volatility of ethanol in the droplet method, a novel carrier was designed and constructed. More specifically, a glass microscopy slide with a cavity in the center was adopted. The slide, as well as the cavity, was coated with a gold substrate (EMF Corporation, Ithaca, NY, USA) offering high reflectivity to the carrier (Figure 6a). The volatile sample was deposited in the cavity. A quantity of 150 μL was required for complete filling of the cavity. An appropriate cover was also necessary to prevent evaporation of the volatile constituent which would allow the laser radiation to pass through at the same time (Figure 6b). Various transparent media have been tested and quoted below.

a.

b.

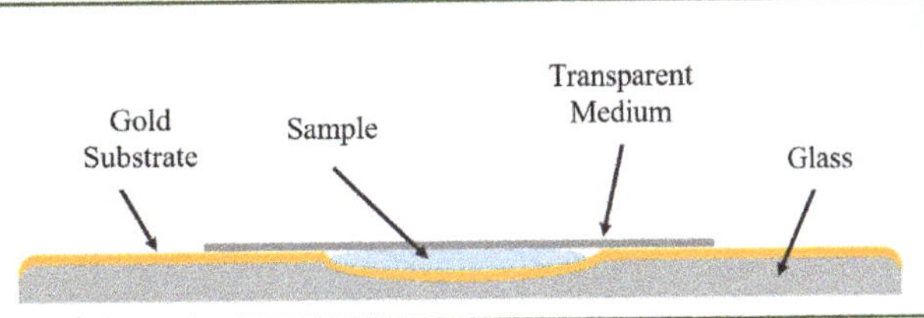

Figure 6. (**a**) The microscope slide with the cavity in the center of the slide coated with gold was used for recording the Raman spectra of volatile compounds in biological fluids; (**b**) schematic illustration of the glass microscope slide with the cavity coated with gold and covering the sample in the cavity with a transparent medium.

2.4.2. Cavity Covering: Microscope Cover Slip

The first transparent medium tested for covering the sample was the microscope cover slip. A sample of 5 µL/mL ethanol in human urine was deposited in the cavity of the slide and covered with the glass cover slip. The laser was focused on the cover slip, and a Raman spectrum was collected which appeared identical to the one acquired when a sample's droplet was caged between the gold coated glass slide and the cover slip. The peak of urine at 1003 cm^{-1} was barely detected, whereas no characteristic peak of ethanol could be observed (Figure 7). When the laser was focused in the main volume of the sample (under the cover slip), the peak of urea at 1003 cm^{-1} was clearly detectable; however, the most prominent peak of alcohol at 880 cm^{-1} could not be observed (Figure 7). Last, the cover was removed, and the spectrum of the sample was recorded. Both peaks of the urine at 1003 cm^{-1} and ethanol at 880 cm^{-1} were clearly detected (Figure 7). It can be, thus, concluded that the glass microscope cover slip may protect the sample from evaporation, but no clear signal from the sample is collected. The minor constituent is not detected.

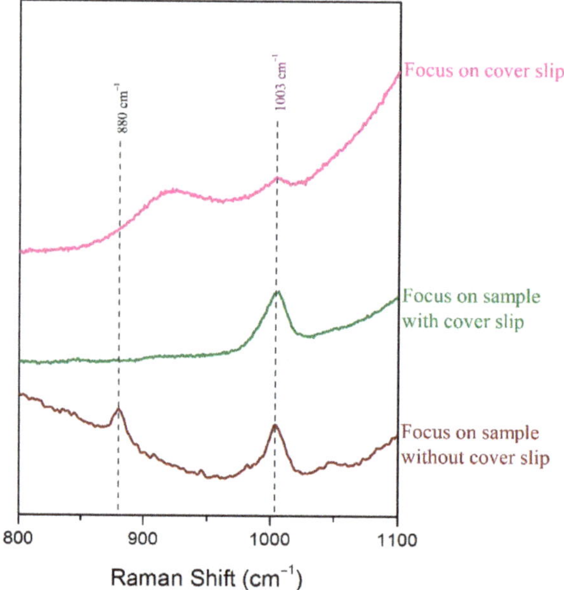

Figure 7. The Raman spectrum of 5 µL/mL ethanol in human urine after placing the sample in the cavity of the gold-coated glass slide was recorded without and with being covered with a microscope cover slip. Two different focus levels were tested when the cover slip was applied.

2.4.3. Cavity Covering: Transparent Membrane

Since the glass of the microscope cover slip had a significant effect on the acquired Raman spectrum, a transparent membrane (food cling film) was tested as an alternative medium for covering the cavity.

The material of cling films is crucial. It varies among the different suppliers as well. Two different cling films were tested; the first one was a 0.006 mm thick poly(vinyl chloride-vinyl acetate-vinyl alcohol) transparent membrane (Sanitas®), while the second one was a 0.006 mm thick polyethylene low-density transparent cling film (Vileda Freshmate®). The Raman spectra of the two membranes were recorded and found to differ significantly. In the spectrum of the Sanitas® transparent membrane only a few peaks were observed, none of them in the area between 880 cm^{-1} and 1003 cm^{-1}. However, multiple humps were observed; a broad significant hump was indeed observed at 883 cm^{-1}, which interferes with the peak of ethanol (Figure 8a). On the contrary, in the Vileda Freshmate® film's

Raman spectrum a few intense peaks appeared in the spectral area 1060–1500 cm^{-1}. None of them would interfere with peaks of ethanol and urine at 883 cm^{-1} and 1003 cm^{-1}, while only a broad peak of very low intensity was observed at 883 cm^{-1} (Figure 8a).

a.　　　　　　　　　　　　　　　　　　　　　　**b.**

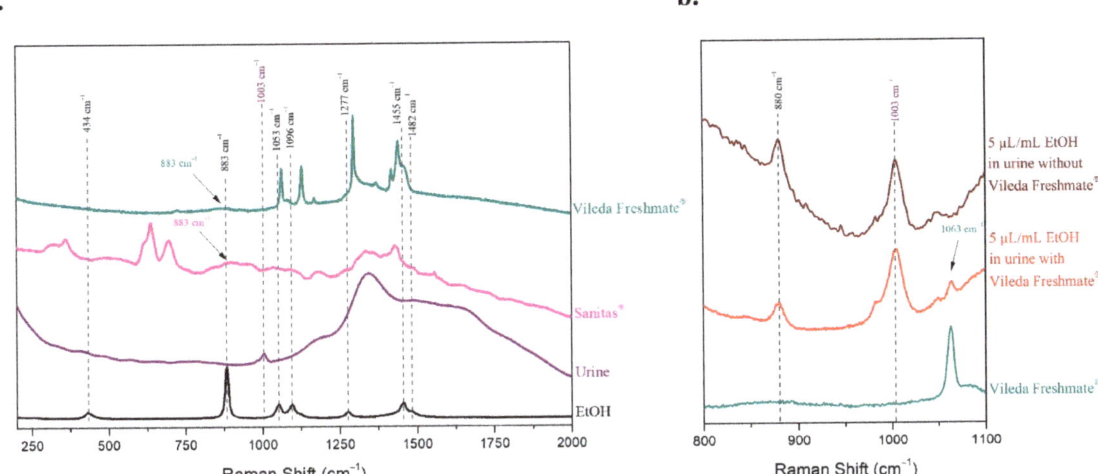

Figure 8. (**a**) The Raman spectra of ethanol, human urine, and the two transparent cling films (Sanitas® and Vileda Freshmate®) were recorded, and the characteristic peaks of ethanol and urine are noted with dashed lines; (**b**) the Raman spectrum of the 5 µL/mL ethanol in human urine sample was recorded without and with being covered with the Vileda Freshmate® membrane.

The effect of Vileda Freshmate® cling film's small broad peak at 883 cm^{-1} on the detection of the ethanol peak at the same Raman shift was further investigated. A sample of 5 µL/mL ethanol in human urine filled the cavity, and a piece of the membrane covered it, carefully taking care to be completely in touch with the sample so that no air is trapped in between. Except for the cavity, the membrane covered a surface of the gold-coated glass slide around the cavity in order to prevent leaking of the liquid sample out of the well. The Raman spectrum of the sample was recorded before and after covering with the transparent membrane. Both peaks of urine at 1003 cm^{-1} and ethanol at 880 cm^{-1} were clearly detectable, while the extra peaks of the membrane did not affect the determination of either the alcohol or the biological fluid (Figure 8b). The peaks of urine and ethanol were as distinct as in the Raman spectrum of the sample recorded without the transparent membrane despite the presence of an additional small peak at 1063 cm^{-1} corresponding to the membrane (Figure 8b). No interference from the small and broad peak of the transparent membrane at 883 cm^{-1} was detected in the Raman spectrum of the sample, which was covered with the membrane (Figure 8b). Therefore, the Vileda Freshmate® cling film was selected as the most appropriate transparent coating of the cavity.

2.4.4. Focus Optimization

Different laser focusing spots were investigated. The first was adjusted on the surface of the transparent cling film (A), the second in the sample just under the membrane (B), and the third one targeted the main volume of the liquid sample (C) (Figure 9a). The Raman spectrum recorded from the spot (A) revealed the significant effect of the membrane material on the spectrum. However, the peaks of ethanol at 880 cm^{-1} and urine at 1003 cm^{-1} were also detected (Figure 9b). Focusing under the membrane (B) resulted in a Raman spectrum in which the effect of the membrane on the spectrum was reduced significantly, while the signal from the peaks of both analytes was increased to a great extent (Figure 9b). When the main volume of the sample was focused (C), the peaks of the membrane were barely detectable; only the peak at 1063 cm^{-1} could be distinguished from noise level. On the contrary, the intensities of the peaks of ethanol at 880 cm^{-1} and urine at 1003 cm^{-1} reached maximum values compared to (A) and (B) (Figure 9b). Thus, the peaks of the Raman spectrum of Vileda Freshmate® cling film did not overlap the peaks of ethanol and urine and did not obstruct the detection of both constituents.

a.

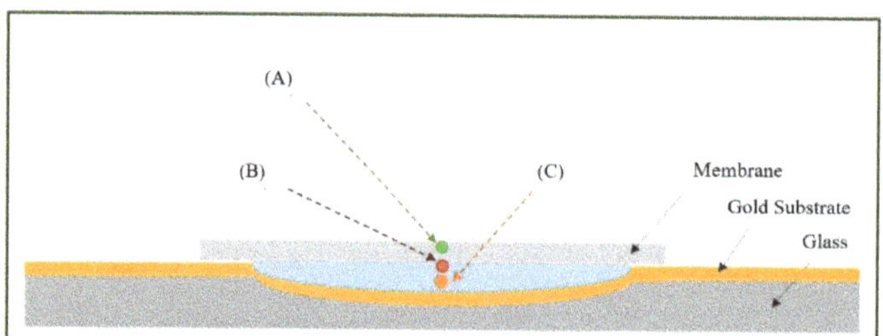

b.

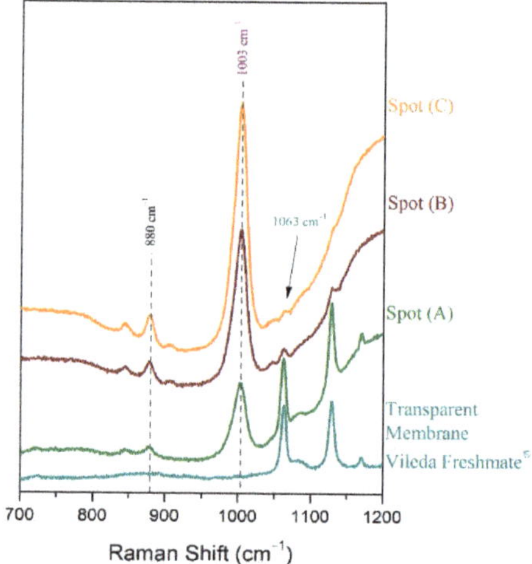

Figure 9. The method of placing the sample droplet in a cavity of a home-designed glass microscope slide coated with a gold highly reflective substrate and covering the sample with a piece of transparent cling film (Vileda Freshmate®) was applied for the determination of 5 µL/mL ethanol in human urine. (**a**) Schematic illustration of the focus on three different spots of the sample in the cavity; (**b**) the respective Raman spectra from the three different spots of the sample in the cavity were recorded.

2.4.5. LoD Determination

The LoD of ethanol in urine was also determined for the method of the gold-coated glass slide with cavity. For this purpose, human urine sample was spiked with 0.30 µL/mL, 0.50 µL/mL, 1.00 µL/mL, 2.00 µL/mL, 3.50 µL/mL, and 5.00 µL/mL ethanol. The Raman spectra of these samples were recorded after placing them in the cavity of the gold substrate carrier and after covering the fluids with the Vileda Freshmate® transparent membrane. The LoD of ethanol in human urine was found 1.00 µL/mL by visual evaluation (Figure 10), as this was the minimum concentration at which the analyte could be distinguished from noise level [40]. Moreover, the LoD of this method was found equal to 1.00 µL/mL, even when the method of signal-to-noise ratio was used, as this was the minimum concentration at which the signal-to-noise ratio was higher than 3:1 (Table 3) [51]. Thus, the determined LoD for the gold-coated glass slide with cavity method was found between the LoDs estimated for the droplet method and the cuvette method.

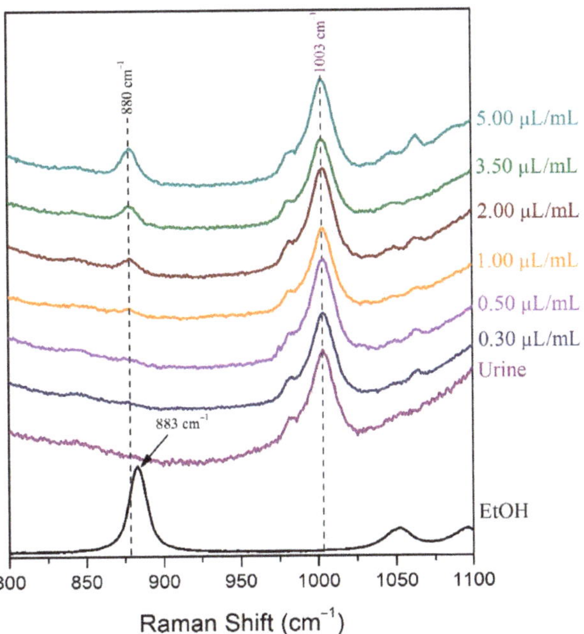

Figure 10. The LoD of ethanol in human urine using the method of gold-coated glass slide with cavity was determined by visual evaluation.

Table 3. The LoD of ethanol in human urine for the method of gold-coated glass slide with cavity was determined based on the signal-to-noise ratio.

Sample	Signal-to-Noise Ratio
Urine	1.00
0.30 µL/mL	1.96
0.50 µL/mL	1.96
1.00 µL/mL	3.17
2.00 µL/mL	6.86
3.50 µL/mL	9.44
5.00 µL/mL	16.79

2.4.6. Kinetic Study of Ethanol Evaporation

In order to study the kinetics of ethanol evaporation from the sample, the Raman laser was focused on spot (C), i.e., under the membrane and in the main volume of the sample in the cavity. Sequential Raman spectra were acquired. The duration of each scan was set at 5 min and 16 s. The evaporation rate of the alcohol from the 5 µL/mL ethanolic sample in human urine was found to be prevented significantly when the sample, placed in the cavity of the novel highly reflective carrier, was covered with the transparent cling film (Supplementary Materials, Figure S3) in comparison with the evaporation rate of the ethanol of an identical sample placed as a droplet on a gold substrate flat carrier. More specifically, the rate of change of the height of the ethanol peak at 880 cm^{-1}, with respect to the height of the urine peak at 1003 cm^{-1}, was calculated for four different time intervals (0 min and 0 s, 5 min and 16 s, 15 min and 48 s, and 31 min and 43 s) (Supplementary Materials, Table S3). These rates of ethanol evaporation in the specific methodology are slightly elevated compared to the reduction rates of ethanol in the method of the cuvette (Supplementary Materials, Table S2); however, they are dramatically lower than the respective rates in the method of the droplet (Supplementary Materials, Table S1).

It is, thus, proved that the proposed carrier, consisting of a gold-coated glass microscope slide with a cavity of 150 μL volume covered with a cling film of appropriate material, meets all the required conditions set. The sample volume necessary is minimized; evaporation of volatile constituents is well prevented; it is simple, not expensive, and easily constructed while offering an acceptable LoD of the analyte.

3. Discussion

Raman spectroscopy offers a novel perspective in the characterization of biological fluids. Multiple applications of Raman spectroscopy for the biochemical analysis and identification of compounds and substances in body fluids could be enumerated substituting the officially established methods [15,17,18,23–36,39,40,46]. However, most of the methods described in these studies involve the drying of the sample of biological fluid prior the acquisition of the Raman spectra, and, consequently, much time is consumed for the sample preparation. The dried sample also suffers from a lack of homogeneity, because the hydrophilic substances are separated from the hydrophobic constituents during the drying process. [25–30,33,34,40]. On the contrary, in our study all three proposed methodologies make use of the biological samples without any further preparation, reducing the required time of analysis and eliminating issues of sample homogeneity. Moreover, the proposed methods should offer simple, quick, less expensive solutions with LoDs lower than the maximum acceptable alcohol levels in drivers [1], requiring smaller amounts of biological sample for the detection of volatile compounds in body fluids than the official GC methods [10–12] while preventing the evaporation of the volatile constituents at the same time.

Each one of the three methods described in this study, the method of droplet, the method of cuvette, and the method of gold-coated glass slide with cavity, has its own advantages and drawbacks (Table 4). The method of droplet on a microscope slide coated with gold is the simplest method among those three, requiring only a few seconds for preparation of the sample. Then, the Raman spectrum of the sample is acquired from the droplet, which is placed on the highly reflective carrier, completing the analysis in a few minutes. In addition, no special equipment is required, and the production of gold-coated glass slides is quite economical, while only a few μL (<15 μL) of sample are required for the analysis. The LoD of the volatile alcohol was rather low as well. This method would be ideal for non-volatile samples of small available volume. Nevertheless, when the aim of analysis is the identification of volatile compounds, such as ethanol, in body fluids, the method of droplet is not suitable, because it suffers from significant errors and repeatability issues. These problems are due to the rapid evaporation of great amounts of the volatile compound from the droplet before the acquisition of the Raman spectrum is completed. When the method of droplet was applied on a 5 μL/mL ethanolic sample in urine, more than 15% and more than half of the ethanol was evaporated in approximately 2 and 10 min, respectively (Figure 11).

The method of cuvette offered a solution in reducing the evaporation of the alcohol from the sample significantly. More specifically, only 5.42% of ethanol was evaporated after 15 min and 52 s and 12.28% after 31 min and 42 s from the moment the 5 μL/mL ethanolic sample in urine was added in the commercially available Raman cuvette with a mirror on the back side (Figure 11). This very slow rate of evaporation of the volatile compounds suggests a useful method for the characterization of such volatile substances in biological fluids. However, the glass of the cuvette had some effect on the signal of the acquired Raman spectra, increasing significantly the LoD of the volatile alcohol in human urine from 0.50 μL/mL (method of droplet) to 2.00 μL/mL (method of cuvette). Moreover, the volume of the required sample to fill the cuvette (1.75 mL) is significantly enhanced compared to a few μL required in the method of droplet. Even though a smaller cuvette, which is commercially available, is selected, the required volume of the sample is no less than 0.35 mL. Such great amounts of sample are not usually available, as far as the body fluids are concerned. Partially filled cuvettes did not result in prevention of the evaporation of the volatile compound. Another important drawback is the cost of the cuvettes for Raman spectroscopy, as they are far more expensive than the microscope slides coated with the gold substrate. The cleaning of the cuvettes also may be challenging and time-consuming, especially when biological fluids are used, because they are usually viscous and stick on the cuvette's walls. Insufficient cleaning of the cuvette leads to impurities in the following samples, a decrease in the reflectiveness of the mirror adjusted to the back side of the cuvette, and, consequently, poor quality of the Raman spectrum.

Table 4. The advantages and disadvantages of each method.

Factor	Method of Droplet	Method of Cuvette	Method of Cavity
Simplicity	The simplest	Very simple	Very simple
Required Time of Sample Preparation	A few seconds	Some seconds	Some seconds
Required Time of Analysis	A few minutes	Some minutes	Some minutes
Required Volume	<15 μL	350–1750 μL	150 μL
Cost	Low	High	Low
Cleaning	Easy	Difficult	Easy
Interference of Carrier's Material on the Raman Spectrum	No effect	High interference of the glass	Very low interference of the transparent membrane
LoD	Very Low (0.50 μL/mL ethanol in urine)	High (2.00 μL/mL ethanol in urine)	Low (1.00 μL/mL ethanol in urine)
Evaporation Rate of Volatile Samples	Very fast	Very slow	Slow

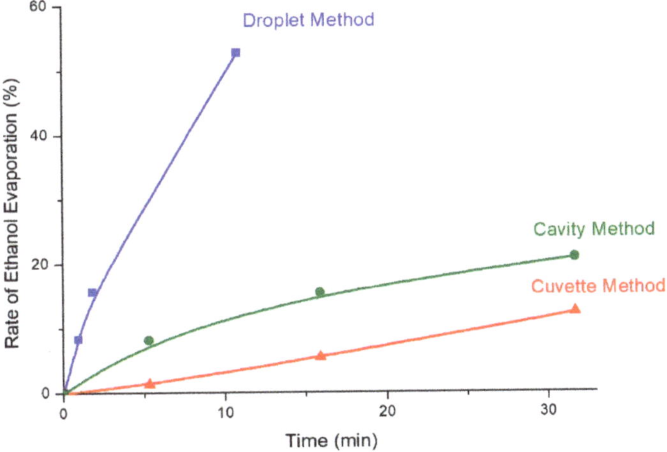

Figure 11. Rate of ethanol evaporation as a function of time for the method of droplet, the method of cuvette, and the method of the gold-coated glass slide with cavity.

The third method included the addition of the sample of biological fluid with the volatile compound in the cavity of a gold substrate carrier. The sample in the cavity was covered with a transparent cling film, which prevented the evaporation of the alcohol from a 5 μL/mL ethanolic sample in urine to a great extent. After 15 min and 48 s, only 15.44% of ethanol was evaporated, while after 31 min and 43 s approximately 21% of the total ethanol was evaporated (Figure 11). Even though the reduction rate of the volatile alcohol from the sample was higher than in the method of the cuvette sealed with the lid, it was still significantly lower than in the method of the uncovered droplet on the highly reflective carrier. In addition, the specific method was suitable for detecting the analytes below the transparent membrane, as its effect on the Raman spectrum was not as great as the effect of the glass in the case of the cuvette. However, in order to avoid the interference of the membrane in the Raman spectrum, a suitable membrane, whose peaks will not obstruct the detection of the analytes, should be selected, while at the same time the Raman laser should be focused on the main volume of the sample in the cavity and not just below the membrane. The highly reflective gold substrate of the slide assisted in the acquisition of high-quality Raman spectra. Furthermore, the LoD of ethanol in human urine when using the gold-coated glass slide with cavity method was estimated at 1.00 µL/mL. Although the LoD is higher than the one found in the method of droplet (0.50 µL/mL), it is half of the respective LoD determined for the method of cuvette (2.00 µL/mL). The estimated LoD of 1.00 µL/mL, however, is acceptable for identifying individuals intoxicated by ethanol, as light intoxication is observed for blood ethanol concentrations higher than 1.27 µL/mL, corresponding to a urine alcohol concentration of 1.65 µL/mL [52]. In addition, the determined LoD is lower than the highest acceptable limit of alcohol in urine when driving in most countries, which is 1.30 µL/mL [1]. Thus, it can be concluded that the proposed setting can be applied as a quick, easy, and non-destructive tool for on-site intoxication control before more quantitative official approaches are applied. Another advantage of this method was the relatively low volume of the required sample. Only 150 µL of the sample could completely fill the cavity. Although this volume was higher than the volume of the droplet, it was at least three-fold lower than the required volume in the method of the cuvette. In addition, compared to the GC methods that are officially employed for the determination of ethanol in biological samples, such as HS-GC, the volume of the sample required in the method of the gold-coated slide with cavity is equal to or less than the volume of biological sample required in HS-GC (100–500 µL) [10–12]. Moreover, the cleaning of the carrier was not as challenging as the cleaning of the cuvette. The sample was removed by washing the slide with ultra-pure water and ethanol after each usage. In order to dry the liquid drops after cleaning, warm air was preferred instead of soft tissue so that the gold substrate would not be destroyed. Finally, the cost of purchasing and coating with gold the microscope slide with the cavity was significantly lower than the cost of purchasing the commercially available cuvette for Raman spectroscopy.

At this point, it is important to mention that this is the first time in literature that such a carrier and sample preparation method for Raman spectroscopy was designed, proposed, and used for the identification of liquid samples, especially for volatile compounds in biological fluids. Furthermore, the coating of flat microscope slides with gold [26,53,54] and the usage of variable types of cuvettes for the Raman analysis of liquid samples [36,41–44] are referred to in previous studies; however, there is no reference in literature about using as substrate for Raman spectroscopy the cavity of a microscope slide which was coated with gold.

4. Materials and Methods

4.1. Samples

Absolute ethanol of analytical reagent grade (Fisher Scientific UK Ltd., Loughborough, UK) has been used for recording ethanol Raman spectrum and for preparing the mixtures with urine. Urine samples from three volunteers of the Laboratory of Instrumental Pharmaceutical Analysis, Department of Pharmacy, University of Patras, Greece were collected. Their age ranged from 20 to 55 years, and they were not suffering from any severe or infectious disease, such as cancer or AIDS. The urine samples were collected in a sterile urine collector. Permission for the implementation of this study was administered by the Bioethics Committee of University of Patras, Greece (Protocol Number: EB 45). No further processing of the urine samples was applied. All three volunteers have assured that they did not consume any alcoholic drinks for the previous 48 h prior the urine collection.

The 5 µL/mL ethanolic sample in human urine was prepared by adding the appropriate quantity of absolute ethanol in a volume of human urine. The final volume of each sample was 3 mL. For transferring the appropriate volume of absolute ethanol, a 10 µL automated pipette was used (Pipetman Neo®, Gilson Inc, Middleton, WI, USA), while a 1 mL or 5 mL automated pipette was employed for the human urine (BioPette® Autoclavable Pipettes, Labnet International Inc, Edison, NJ, USA). The samples were homogenized by shaking in a vortex (MS2 Minishaker, IKA®-Werke GmbH & CO., KG, Staufen im Breisgau, Germany). The samples were analyzed immediately after their production, and, subsequently, they were stored at $-20\,^{\circ}\mathrm{C}$ until the next measurement.

Xerostom® oral spray 15 mL (Pharmaserve-Lilly, Kifisia, Athens, Greece), which is used for the relief of dry mouth, was purchased from a local drugstore.

4.2. Raman Spectroscopy

The Raman spectra were acquired by using an InVia Raman spectrometer (Renishaw, Wotton-under-Edge, UK) coupled with optical microscope (DM Leica, Leica Microsystems, Wetzlar, Germany). The samples were excited with a 785 nm wavelength laser diode. The spectral resolution was 2 cm^{-1}, and a charge-coupled device (CCD) detector was used. The nominal value of laser power was 250 mW, while the laser power on the edge of the microscope lens was measured at 32 mW. A 20x, 0.4 NA objective lens (model 566026, Leica Microsystems, Wetzlar, Germany) was employed for the measurements. The Raman spectra of ethanol, human urine, and the materials used as substrates were recorded in the spectral region of 100–2000 cm^{-1}, while the Raman spectra of the ethanolic samples in human urine were acquired in the spectral region of 700–1200 cm^{-1}. The spectra were acquired through the Windows-based software WiRE© 2.0.

For the daily calibration of the spectrometer, the Raman spectrum of a silicon reference standard was obtained before the acquisition of the Raman spectra of the samples. When recording the silicon reference standard spectrum, the laser power was set at $5 \times 10^{-8}\%$ of the maximum power, and the time of scan was adjusted to 1 s. Each spectrum of the reference standard was a result of two accumulated scans. The Raman shift of silicon peak at 520 cm^{-1} and its intensity were used to validate the calibration of the spectrometer.

4.2.1. Raman Spectra Acquisition Using the Method of Droplet

The first method for the acquisition of Raman spectra involved the usage of a special highly reflective slide designed by coating a glass microscope slide with gold substrate (EMF Corporation, Ithaca, NY, USA). The coating of the slides was constituted by two layers; the first one was a layer of titanium (50 Å) binding the glass of the slide with the gold coating, while the external one was a bare gold layer (1000 Å). This gold layer offered a high reflection on the carrier. The dimensions of the slides were 2.6×7.6 cm, and their thickness was 1.0 mm. A droplet of each sample (approximately 5–10 μL) was placed on the carrier with the gold substrate using a 1 mL syringe. Movement of the point of focus to predetermined distances was achieved using the high precision levers of the microscope stage following focusing on the top of the droplet.

4.2.2. Raman Spectra Acquisition Using the Method of Cuvette

A commercially available macro-cuvette 100-QX with a polytetrafluoroethylene (PTFE) lid (Hellma, Müllheim, Germany) with a mirror on the back side which contributed to the reflection of the scattered radiation was the second carrier used for the acquisition of the Raman spectra. A volume of approximately 1750 μL was necessary for filling completely the cuvette. The cuvette was created from synthetic quartz glass (Suprasil® 300, Heraeus Quarzglas GmbH & Co., Hanau, Germany), and its path length was 5 mm. The height of the cuvette was 45 mm, the width 12.5 mm, and the depth 7.5 mm; its inside width was 9.5 mm, and its base thickness was 1.5 mm. The cuvette was filled with the sample completely, and the cap was placed carefully on top and covered with Parafilm M® (Bemis Company, Inc., Neenah, WI, USA) in order to seal it tightly so that no air bubbles would be trapped in the sample. The Raman laser was focused on the sample in the cuvette at an angle of 90° by using an angle mirror lens. Movement of the point of focus to predetermined distances was achieved using the high precision levers of the microscope stage following focusing on the outer wall of the cuvette.

4.2.3. Raman Spectra Acquisition Using the Method of Gold-Coated Glass Slide with Cavity

For the third proposed methodology, a microscope slide with a 1-well cavity was coated with gold (EMF Corporation, Ithaca, NY, USA) as the simple microscope slide in order to obtain a highly reflective carrier. Thus, a 50 Å titanium layer was deposited on the glass of the slide, and a 1000 Å bare gold layer was added as the external layer. The slides' dimensions were 2.6 × 7.6 cm with 1.25 mm thickness. The diameter of the cavity was 1.5 cm, while the well's depth was 0.6 mm. The sample was placed in the well, which required 150 μL to be filled. The cavity with the sample was covered using a piece of a 0.006 mm thick polyethylene low-density transparent cling film (Vileda Freshmate®, FHP Hellas, Kifisia, Athens, Greece) in order to prevent ethanol evaporation. A second 0.006 mm thick poly(vinyl chloride-vinyl acetate-vinyl alcohol) transparent membrane (Sanitas®, Sarantis S.A., Marousi, Greece) was also tested. Movement of the point of focus to predetermined distances was achieved using the high precision levers of the microscope stage following focusing on the cling film.

4.2.4. Repeatability of the Method of Gold-Coated Glass Slide with Cavity

The repeatability of the method was tested using a human urine sample spiked with 5.00 μL/mL ethanol. Three consecutive measurements were performed. A fresh quantity of sample was used each time and placed in the cavity of the gold-coated glass slide and covered with cling film. The laser was focused on spot C, i.e., in the main volume of the sample. The cavity slide was cleaned and dried between measurements. The ratio of the intensity of the ethanol peak at 880 cm^{-1} to the intensity of the urine peak at 1003 cm^{-1} was calculated for each measurement. For the evaluation of the repeatability of the method, the relative standard deviation was determined (Table 5).

Table 5. Repeatability of the method of gold-coated slide with cavity for a urine sample spiked with 5 μL/mL ethanol.

Measurement	I880/I1003
1st Measurement	0.317
2nd Measurement	0.313
3rd Measurement	0.303
Average	0.311
Standard Deviation	0.007
Relative Standard Deviation (%)	2.25

4.2.5. Determination of the LoD

The LoD of ethanol in human urine was estimated by visual evaluation and signal-to-noise ratio for each of the three proposed methods [51]. Regarding, the method of droplet and the method of cuvette, five samples of human urine spiked with ethanol (0.25 μL/mL, 0.50 μL/mL, 2.00 μL/mL, 3.50 μL/mL, and 5.00 μL/mL) were prepared. The samples were homogenized by shaking in a vortex (MS2 Minishaker, IKA®-Werke GmbH & CO. KG, Staufen im Breisgau, Germany), and, subsequently, their Raman spectra were acquired. The same procedure was also followed for the determination of the LoD in the method of gold-coated glass slide with cavity. In this case, six mixtures of ethanol with human urine (0.30 μL/mL, 0.50 μL/mL, 1.00 μL/mL, 2.00 μL/mL, 3.50 μL/mL, and 5.00 μL/mL) were prepared and homogenized, and their Raman spectra were recorded.

4.2.6. Study of the Kinetics of Ethanol Evaporation

For each of the three proposed methodologies, the rate of ethanol evaporation from a 5 μL/mL ethanolic sample in human urine was determined. For this purpose, the ratio of the intensity height of ethanol's peak at 880 cm^{-1} to the intensity height of the urine's peak at 1003 cm^{-1} was calculated for the initial measurement and for each time interval. For the determination of the evaporation rate of ethanol, the rate of change of the ratio at each time interval with respect to the initial ratio was determined according to the following equation:

$$\text{Rate of Change} = \frac{\frac{I_{EtOH,t}}{I_{Urine,t}} - \frac{I_{EtOH,t0}}{I_{Urine,t0}}}{\frac{I_{EtOH,t0}}{I_{Urine,t0}}} \times 100\%, \tag{1}$$

where $I_{EtOH,t}$ and $I_{Urine,t}$ are the intensities of ethanol's peak at 880 cm^{-1} and urine's peak at 1003 cm^{-1}, respectively, at a specific time interval, while $I_{EtOH,t0}$ and $I_{Urine,t0}$ are the intensities of ethanol's peak at 880 cm^{-1} and urine's peak at 1003 cm^{-1}, respectively, at the initial time point (t = 0 min and 0 s).

5. Conclusions

All three proposed methods in this study could be used for the acquisition of the Raman spectra of liquid samples. However, each one is limited by certain drawbacks and is suitable for certain types of sample. The usage of the gold-coated slide with the cavity, which is covered with a transparent cling film, demonstrated certain advantages over the other two methods for the identification of volatile liquid samples. Particularly, the use of the proposed methodology for the determination of other volatile compounds (ethanol, methanol, acetone, isopropyl alcohol, etc.) in variable body fluids (blood, serum, urine, saliva, semen, vaginal fluids, etc.) could be proved beneficial, as it is a simple, fast, inexpensive, user-friendly, and environmentally friendly method requiring only a few μL of sample, offering low LoDs and high protection from evaporation of the volatile constituents. Moreover, the proposed method could be used as a quick approach for detecting the alcohol levels in drivers who have consumed alcoholic beverages above the maximum acceptable limit. However, for the detection of small amounts of alcohol in human bodies, the conventional methods cannot be replaced.

Supplementary Materials: The following supporting information can be downloaded at: https://www.mdpi.com/article/10.3390/molecules27103279/s1, Figure S1: Raman spectra of the droplet of a 5 μL/mL ethanolic sample in human urine were acquired immediately (t = 0 min and 0 s) after droplet deposition on the gold-coated glass slide, after 0 min and 57 s, 1 min and 50 s, and 10 min and 49 s. The Raman spectra are normalized according to the urine peak at 1003 cm^{-1} and smoothed; Table S1: The rate of change of the ethanol peak to urine peak intensities ratio with respect to time. The sample used consisted of 5 μL/mL ethanol in human urine, and the method used was the deposition of a droplet on gold-coated glass slide; Figure S2: Raman spectra of a 5 μL/mL ethanolic sample in human urine were acquired immediately (t = 0 min and 0 s) after placing the sample in the commercially available cuvette for Raman spectroscopy with a mirror on the back side, after 5 min and 17 s, 15 min and 52 s, and 31 min and 42 s. The Raman spectra are normalized according to the urine peak at 1003 cm^{-1} and smoothed; Table S2: The rate of change of the ethanol peak to urine peak intensities ratio with respect to time. The sample used consisted of 5 μL/mL ethanol in human urine, and the method used was the addition of the sample in a commercially available cuvette; Figure S3: Raman spectra of 5 μL/mL ethanol in human urine were acquired immediately (t = 0 min and 0 s) after placing the sample in the cavity of the home-designed glass microscope slide coated with a gold highly reflective substrate and covering the sample with a piece of transparent cling film (Vileda Freshmate®), after 5 min and 16 s, 15 min and 48 s, and 31 min and 43 s. The Raman spectra are normalized according to the urine peak at 1003 cm^{-1} and smoothed; Table S3: The rate of change of the ethanol peak to urine peak intensities ratio with respect to time. The sample used consisted of 5 μL/mL ethanol in human urine, and the method used was the deposition of the sample on gold-coated glass slide with cavity, covered with transparent cling film.

Author Contributions: Conceptualization, M.O. and C.K.; methodology, M.O. and P.P.; validation, P.P. and M.O.; investigation, P.P., M.L. and M.O.; data curation, P.P. and M.O.; writing—original draft preparation, M.L.; writing—review and editing, M.O.; supervision, M.O.; project administration, M.O. All authors have read and agreed to the published version of the manuscript.

Funding: This research received no external funding.

Institutional Review Board Statement: The study was conducted according to the guidelines of the Declaration of Helsinki and approved by the Institutional Review Board (or Ethics Committee) of University of Patras (protocol code eb 45 and 10.09.2015).

Informed Consent Statement: Informed consent was obtained from all subjects involved in the study.

Conflicts of Interest: The authors declare no conflict of interest.

Sample Availability: Samples of the compounds are not available from authors.

References

1. Dasgupta, A. Alcohol a double-edged sword: Health benefits with moderate consumption but a health hazard with excess alcohol intake. In *Alcohol, Drugs, Genes and the Clinical Laboratory*; Dasgupta, A., Ed.; Academic Press: Cambridge, MA, USA, 2017; pp. 1–21.
2. Biasotti, A.A.; Valentine, T.E. Blood alcohol concentration determined from urine samples as a practical equivalent or alternative to blood and breath alcohol tests. *J. Forensic Sci.* **1985**, *30*, 194–207. [CrossRef] [PubMed]
3. Caplan, Y.H.; Levine, B. The Analysis of Ethanol in Serum, Blood, and Urine: A Comparison of the TDx REA Ethanol Assay with Gas Chromatography. *J. Anal. Toxicol.* **1986**, *10*, 49–52. [CrossRef] [PubMed]
4. Jones, A.W.; Holmgren, P. Urine/blood ratios of ethanol in deaths attributed to acute alcohol poisoning and chronic alcoholism. *Forensic Sci. Int.* **2003**, *135*, 206–212. [CrossRef]
5. Jones, A.W.; Kugelberg, F.C. Relationship between blood and urine alcohol concentrations in apprehended drivers who claimed consumption of alcohol after driving with and without supporting evidence. *Forensic Sci. Int.* **2010**, *194*, 97–102. [CrossRef]
6. Berger, A. How Does it Work? Alcohol breath testing. *BMJ* **2002**, *325*, 1403. [CrossRef]
7. Ashdown, H.F.; Fleming, S.; Spencer, E.A.; Thompson, M.J.; Stevens, R.J. Diagnostic accuracy study of three alcohol breathalysers marketed for sale to the public. *BMJ Open* **2014**, *4*, e005811. [CrossRef]
8. Labianca, D.A. Uncertainty in the Results of Breath-Alcohol Analyses. *J. Chem. Educ.* **1999**, *76*, 508–510. [CrossRef]
9. Mihretu, L.D.; Gebru, A.G.; Mekonnen, K.N.; Asgedom, A.G.; Desta, Y.H. Determination of ethanol in blood using headspace gas chromatography with flameionization detector (HS-GC-FID): Validation of a method. *Cogent Chem.* **2020**, *6*, 1760187. [CrossRef]
10. Macchia, T.; Mancinelli, R.; Gentili, S.; Lugaresi, E.C.; Raponi, A.; Taggi, F. Ethanol in biological fluids: Headspace GC measurement. *J. Anal. Toxicol.* **1995**, *19*, 241–246. [CrossRef]
11. Dorubeț, D.; Moldoveanu, S.; Mircea, C.; Butnaru, E.; Astărăstoae, V. Development and validation of a quantitative determination method of blood ethanol by gas chromatography with headspace (GC-HS). *Rom. J. Legal Med.* **2009**, *4*, 303–308. [CrossRef]
12. Sutter, K. Determination of Ethanol in Blood: Analytical Aspects, Quality Control, and Theoretical Calculations for Forensic Applications. *Chimia* **2002**, *56*, 59–62. [CrossRef]
13. Boswell, H.A.; Dorman, F.L. Uncertainty of Blood Alcohol Concentration (BAC) Results as Related to Instrumental Conditions: Optimization and Robustness of BAC Analysis Headspace Parameters. *Chromatography* **2015**, *2*, 691–708. [CrossRef]
14. Skoog, D.A.; Holler, F.J.; Crouch, S.R. *Principles of Instrumental Analysis*, 6th ed.; Thomson Brooks/Cole: Pacific Grove, CA, USA, 2007.
15. Khandasammy, S.R.; Fikiet, M.A.; Mistek, E.; Ahmed, Y.; Halámková, L.; Bueno, J.; Lednev, I.K. Bloodstains, paintings, and drugs: Raman spectroscopy applications in forensic science. *Forensic Chem.* **2018**, *8*, 111–133. [CrossRef]
16. Raman, C.V.; Krishnan, K.S. The Negative Absorption of Radiation. *Nature* **1928**, *122*, 12–13. [CrossRef]
17. Baker, M.J.; Hussain, S.R.; Lovergne, L.; Untereiner, V.; Hughes, C.; Lukaszewski, R.A.; Thiéfin, G.; Sockalingum, G.D. Developing and Understanding Biofluid Vibrational Spectroscopy: A Critical Review. *Chem. Soc. Rev.* **2016**, *45*, 1803–1818. [CrossRef]
18. Mitchell, A.L.; Gajjar, K.B.; Theophilou, G.; Martin, F.L.; Martin-Hirsch, P.L. Vibrational spectroscopy of biofluids for disease screening or diagnosis: Translation from the laboratory to a clinical setting. *J. Biophotonics* **2014**, *7*, 153–165. [CrossRef]
19. Buckley, K.; Matousek, P. Non-invasive analysis of turbid samples using deep Raman spectroscopy. *Analyst* **2011**, *136*, 3039–3050. [CrossRef]
20. Orkoula, M.; Kontoyannis, C.G.; Markopoulou, C.K.; Koundourellis, J.E. Quantitative analysis of liquid formulations using FT-Raman spectroscopy and HPLC The case of diphenhydramine hydrochloride in Benadryl®. *J. Pharm. Biomed. Anal.* **2006**, *41*, 1406–1411. [CrossRef]
21. Jentzsch, P.V.; Ciobotă, V. Raman spectroscopy as an analytical tool for analysis of vegetable and essential oils. *Flavour Fragr. J.* **2014**, *29*, 287–295. [CrossRef]
22. Boyaci, I.H.; Genis, H.E.; Guven, B.; Tamer, U.; Alper, N. A novel method for quantification of ethanol and methanol in distilled alcoholic beverages using Raman spectroscopy. *J. Raman Spectrosc.* **2012**, *43*, 1171–1176. [CrossRef]
23. Açikgöz, G.; Hamamci, B.; Yildiz, A. Determination of Ethanol in Blood Samples Using Partial Least Square Regression Applied to Surface Enhanced Raman Spectroscopy. *Toxicol. Res.* **2018**, *34*, 127–132. [CrossRef] [PubMed]

24. Mikoliunaite, L.; Rodriguez, R.D.; Sheremet, E.; Kolchuzhin, V.; Mehner, J.; Ramanavicius, A.; Zahn, D.R.T. The substrate matters in the Raman spectroscopy analysis of cells. *Sci. Rep.* **2015**, *5*, 13150. [CrossRef] [PubMed]
25. Sikirzhytskaya, A.; Sikirzhytski, V.; McLaughlin, G.; Lednev, I.K. Forensic Identification of Blood in the Presence of Contaminations Using Raman Microspectroscopy Coupled with Advanced Statistics: Effect of Sand, Dust, and Soil. *J. Forensic. Sci.* **2013**, *58*, 1141–1148. [CrossRef] [PubMed]
26. Muro, C.K.; Doty, K.C.; Fernandes, L.S.; Lednev, I.K. Forensic body fluid identification and differentiation by Raman spectroscopy. *Forensic Chem.* **2016**, *1*, 31–38. [CrossRef]
27. Sikirzhytski, V.; Sikirzhytskaya, A.; Lednev, I.K. Advanced statistical analysis of Raman spectroscopic data for the identification of body fluid traces: Semen and blood mixtures. *Forensic Sci. Int.* **2012**, *222*, 259–265. [CrossRef]
28. Sikirzhytskaya, A.; Sikirzhytski, V.; Lednev, I.K. Determining Gender by Raman Spectroscopy of a Bloodstain. *Anal. Chem.* **2017**, *89*, 1486–1492. [CrossRef]
29. Sikirzhytski, V.; Sikirzhytskaya, A.; Lednev, I.K. Multidimensional Raman spectroscopic signature of sweat and its potential application to forensic body fluid identification. *Anal. Chim. Acta* **2012**, *718*, 78–83. [CrossRef]
30. Sikirzhytski, V.; Sikirzhytskaya, A.; Lednev, I.K. Multidimensional Raman Spectroscopic Signatures as a Tool for Forensic Identification of Body Fluid Traces: A Review. *Appl. Spectrosc.* **2011**, *65*, 1223–1232. [CrossRef]
31. Boyd, S.; Bertino, M.F.; Seashols, S.J. Raman spectroscopy of blood samples for forensic applications. *Forensic Sci. Int.* **2011**, *208*, 124–128. [CrossRef]
32. Kammrath, B.W.; Glynn, C.L.; Schlagetter, T. The Use of Raman Spectroscopy for the Identification of Forensically Relevant Body Fluid Stains. *Spectroscopy* **2017**, *32*, 19–24.
33. Virkler, K.; Lednev, I.K. Forensic body fluid identification: The Raman spectroscopic signature of saliva. *Analyst* **2010**, *135*, 512–517. [CrossRef] [PubMed]
34. Sikirzhytski, V.; Virkler, K.; Lednev, I.K. Discriminant Analysis of Raman Spectra for Body Fluid Identification for Forensic Purposes. *Sensors* **2010**, *10*, 2869–2884. [CrossRef] [PubMed]
35. Bonnier, F.; Petitjean, F.; Baker, M.J.; Byrne, H.J. Improved protocols for vibrational spectroscopic analysis of body fluids. *J. Biophotonics* **2014**, *7*, 167–179. [CrossRef] [PubMed]
36. Parachalil, D.R.; McIntyre, J.; Byrne, H.J. Potential of Raman spectroscopy for the analysis of plasma/serum in the liquid state: Recent advances. *Anal. Bioanal. Chem.* **2020**, *412*, 1993–2007. [CrossRef] [PubMed]
37. Spizzirri, P.G.; Fang, J.H.; Rubanov, S.; Gauja, E. Nano-Raman spectroscopy of silicon surfaces. *arXiv* **2010**, arXiv:1002.2692.
38. Wellner, A.; Paillard, V.; Coffin, H.; Cherkashin, N.; Bonafos, C. Resonant Raman scattering of a single layer of Si nanocrystals on a silicon substrate. *J. Appl. Phys.* **2004**, *96*, 2403–2405. [CrossRef]
39. Zapata, F.; Gregório, I.; García-Ruiz, C. Body Fluids and Spectroscopic Techniques in Forensics: A Perfect Match? *J. Forensic Med.* **2015**, *1*, 1–7. [CrossRef]
40. McLaughlin, G.; Sikirzhytski, V.; Lednev, I.K. Circumventing substrate interference in the Raman spectroscopic identification of blood stains. *Forensic Sci. Int.* **2013**, *231*, 157–166. [CrossRef]
41. Gojani, A.B.; Palásti, D.J.; Paul, A.; Galbács, G.; Gornushkin, I.B. Analysis and Classification of Liquid Samples Using Spatial Heterodyne Raman Spectroscopy. *Appl. Spectrosc.* **2019**, *73*, 1409–1419. [CrossRef]
42. Kalen, F. Development of Raman Spectroscopy for Thermometry in Liquids. Master's Thesis, Lund University, Lund, Sweden, 2007.
43. Zhao, Y.; Ji, N.; Yin, L.; Wang, J. A Non-invasive Method for the Determination of Liquid Injectables by Raman Spectroscopy. *AAPS PharmSciTech* **2015**, *16*, 914–921. [CrossRef]
44. Sato-Berrú, R.Y.; Medina-Valtierra, J.; Medina-Gutiérrez, C.; Frausto-Reyes, C. Quantitative NIR Raman analysis in liquid mixtures. *Spectrochim. Acta Part A* **2004**, *60*, 2225–2229. [CrossRef] [PubMed]
45. Izake, E.L. Forensic and homeland security applications of modern portable Raman spectroscopy. *Forensic Sci. Int.* **2010**, *202*, 1–8. [CrossRef] [PubMed]
46. Premasiri, W.R.; Clarke, R.H.; Womble, M.E. Urine Analysis by Laser Raman Spectroscopy. *Lasers Surg. Med.* **2001**, *28*, 330–334. [CrossRef] [PubMed]
47. Keuleers, R.; Desseyn, H.O.; Rousseau, B.; Van Alsenoy, C. Vibrational Analysis of Urea. *J. Phys. Chem. A* **1999**, *103*, 4621–4630. [CrossRef]
48. Burikov, S.; Dolenko, T.; Patsaeva, S.; Starokurov, Y.; Yuzhakov, V. Raman and IR spectroscopy research on hydrogen bonding in water-ethanol systems. *Mol. Phys.* **2010**, *108*, 2427–2436. [CrossRef]
49. Parke, S.A.; Birch, G.G. Solution properties of ethanol in water. *Food Chem.* **1999**, *67*, 241–246. [CrossRef]
50. Bozorgmehr, B.; Murray, B.T. Numerical Simulation of Evaporation of Ethanol–Water Mixture Droplets on Isothermal and Heated Substrates. *ACS Omega* **2021**, *6*, 12577–12590. [CrossRef]
51. ICH Q2 (R1) Validation of Analytical Procedures: Text and Methodology. European Medicines Agency. 1995. Available online: https://www.ema.europa.eu/en/documents/scientific-guideline/ich-q-2-r1-validation-analytical-procedures-text-methodology-step-5_en.pdf (accessed on 10 March 2022).
52. Pohorecky, L.A.; Brick, J. Pharmacology of ethanol. *Pharmacol. Ther.* **1988**, *36*, 335–427. [CrossRef]

53. Algahtani, M.S.N.; Davies, M.C.; Alexander, M.R.; Burley, J.C. Establishing a new method to evaluate the recrystallization of nano-gram quantities of paracetamol printed as a microarray using ink-jet printing. *Cryst. Growth Des.* **2019**, *19*, 638–647. [CrossRef]
54. Tantra, R.; Brown, R.J.C.; Milton, M.J.T.; Gohil, D. A Practical Method to Fabricate Gold Substrates for Surface-Enhanced Raman Spectroscopy. *Appl. Spectrosc.* **2008**, *62*, 992–1000. [CrossRef]

MDPI

St. Alban-Anlage 66

4052 Basel

Switzerland

Tel. +41 61 683 77 34

Fax +41 61 302 89 18

www.mdpi.com

Molecules Editorial Office

E-mail: molecules@mdpi.com

www.mdpi.com/journal/molecules

9 783036 557229